Location	
RECKITT & COLMAN Pharmaceutical Division Dansom Lane, Hull	
SCIENTIFIC & TECHNICAL LIBRARY	Doctype
	Order

WITHDRAWN

25883

D1776148

Please return to:

PHARMA
R & D LIBRARY

Digestion and the Structure and Function of the Gut

Please return to:

PHARMA
R & D LIBRARY

Karger Continuing Education Series

Donal F. Magee
Arthur F. Dalley II
Omaha, Nebr.

Please return to:

PHARMA
R & D LIBRARY

Digestion and the Structure and Function of the Gut

124 figures and 8 tables, 1986

 KARGER

Donal F. Magee

MA, PhD, DM, Professor and Chairman, Department of Physiology,
Creighton University School of Medicine, Omaha, Nebr.

Arthur F. Dalley, II

BS (Zoology), PhD (Anatomy), Associate Professor, Department of Anatomy,
Creighton University School of Medicine, Omaha, Nebr.

Karger Continuing Education Series, Vol. 8

Topics covered in the Karger Continuing Education Series are selected to help improve clinical skills and introduce the reader to health-related areas undergoing exceptional growth. Produced as compact instructive texts, volumes set forth information which serves to heighten the general awareness and command of current medical procedures and practice. The concise textbook format enhances the value of these books as convenient teaching and training tools for medical scientists, medical clinicians, and health professionals.

National Library of Medicine, Cataloging in Publication
    Magee, Donal F. (Donal Francis), 1924 –
    Digestion and the structure and function of the gut/
    Donal F. Magee, Arthur F. Dalley II. –
    Basel (Switzerland): Karger, 1986. –
    (Karger continuing education series; vol. 8) Includes bibliographies and index.
    1. Digestion   2. Gastrointestinal System – anatomy & histology
    3. Gastrointestinal System – physiology
    I. Dalley, Arthur F. II. Title. III. Series
    W1 KA821P v.8
    WI 102 M191d
    ISBN 3–8055–4204–6

Drug Dosage
    The author and the publisher have exerted every effort to ensure that drug selection and dosage set forth in this text are in accord with current recommendations and practice at the time of publication. However, in view of ongoing research, changes in government regulations, and the constant flow of information relating to drug therapy and drug reactions, the reader is urged to check the package insert for each drug for any change in indications and dosage and for added warnings and precautions. This is particularly important when the recommended agent is a new and/or infrequently employed drug.

All rights reserved.
    No part of this publication may be translated into other languages, reproduced or utilized in any form or by any means, electronic or mechanical, including photocopying, recording, microcopying, or by any information storage and retrieval system, without permission in writing from the publisher.

©   Copyright 1986 by S. Karger AG, P.O. Box, CH–4009 Basel (Switzerland)
    Printed in Switzerland by Friedrich Reinhardt AG, Basel
    ISBN 3–8055–4204–6

Contents

1	**Introduction**	1
	Glandular Epithelial Tissue	3
	Muscle of the GI System	10
	General Structure of the Gastrointestinal Canal	17
	References	24
2	**Mouth and Saliva**	25
	Chewing	25
	Saliva	26
	Taste	46
	References	53
3	**Swallowing: Pharynx and Esophagus**	55
	The Pharynx	55
	The Esophagus	63
	References	74
4	**Stomach: Gastric Motility**	75
	The Muscular Coat of the Stomach	75
	Vagal Innervation of the Stomach	77
	Storage Function	79
	Digestive and Mixing Activity	80
	References	87
5	**Stomach: Gastric Secretion**	88
	The Gastric Mucosa	88
	Gastric Digestion and Secretion	105
	Regulation of Secretions	114
	Peptic Ulcer	125
	References	128

6	**The Biliary Tract**	131
	Gross Structure and General Function	131
	Microstructure	133
	Gallbladder Emptying	139
	Bile	142
	References	151
7	**The Pancreas**	153
	Structure of the Pancreas	153
	Pancreatic Secretion	157
	Intraluminal Digestion and Pancreatic Enzymes	170
	The Pancreas in Protein Deficiency Disease	177
	Digestion and Absorption in the Infant	179
	References	180
8	**Intestinal Motility**	183
	Intestinal Motility Patterns in Intact Animals	183
	Extrinsic Nerves	193
	Intestinal Paralysis	195
	Central Effects	196
	Peptidergic Transmitters	199
	References	203
9	**Interdigestive Activity**	205
	Periodic Motility of the Gut	205
	Periodic Gallbladder Contraction	206
	Periodic Secretion	206
	Pattern Alteration	206
	Hypothetical Mechanisms of Interdigestive Activity	208
	Interdigestive Activity and Secretory Studies	210
	References	210
10	**Absorption**	211
	The Small Intestine	211
	Intestinal Secretion	228
	Absorption	231
	References	252
11	**The Colon and Defecation**	256
	The Large Intestine	256
	The Anorectal Region	277
	References	298

12 The Gut and the Diffuse Endocrine System ... 303
Introduction and Historical Perspective ... 303
GEP Cells ... 306
GEP Cell Secretions: The Gut Peptides and Amines ... 311
References ... 317

13 Hunger and Appetite ... 320
Hunger vs. Appetite ... 320
Socioeconomic Factors ... 320
Hunger and Satiety ... 321
Diet and Body Weight ... 328
Thirst ... 333
References ... 334

Subject Index ... 336

1 Introduction

The prime concern of this text is the process of digestion in simple stomached animals and man. Our principal interest is, therefore, the secretion into and absorption from the gut and the disposition of dietary residues. This means that our physiological preoccupation will be with the whole animal. Since an understanding of function requires an appreciation of structure we emphasize the structure and organization of the organs we are about to consider. The modern trend in physiology is to consider function at cellular and subcellular levels, but we do not emphasize this. This modern approach is justified as one which excludes variables, but these are the very things in which we are interested. The digestive process which keeps us alive is a matter of variables; for example, change from the fasted to the fed state, the influence of one organ upon another and the effect on organs and systems of changes in nervous activity or hormone concentration. We favor measurements of actual secretions and their constituents and of motility over indirect and presumed measurements. If there are discrepancies between in vitro experiments and those in conscious animals or man our credence is given to the latter. We do not, for example, accept binding to tissue as evidence for the existence of physiologically important receptor unless the binding is proportional to a physiological response. We look askance at interpretations of activities elicited from isolated preparations in response only to astronomical concentrations of drugs or hormones.

We express opinions from time to time in the text which may be contrary to the current, the orthodox or the fashionable. These are clearly designated.

The listed references at the ends of chapters are not intended to be anymore than an introduction to more extensive reading. Many of them are to extensively referenced sources. Many are to very old publications. This is not because we are antiquated, but does indicate that much important work has simply become unfashionable and has subsequently been ignored or forgotten. Such work is being repeated today by people who do not know of its existence.

Physiology is, as are, I suppose, most fields of research, very much subject to the dictates of fashion. We have tried to reduce these to our idea of a proper perspective and thereby undoubtedly we run the risk of accusations of bias. Fads and fashions are often thought to constitute the direction of progress and this then justifies their emphasis in even elementary texts. In actual fact remarkable discoveries usually precede and thus initiate fashions. Progress usually comes from the solitary iconoclast rather than from the mob of followers.

Fashions originate with original discoveries or more often with new techniques, but they all carry the danger inherent in enthusiasm of obscuring the wood with the trees. The current obsession with transmucosal potentials has meant an immense number of publications on the absorption of things like sodium, chloride, amino acids, and glucose which generate and alter potentials. There is comparatively little interest in ions such as lead, magnesium, sulfate or phosphate. No one bothers much any longer about movement of the intestinal villi. The obsession with isolated cells and receptors is now so great that earlier work using whole animals is losing credibility among the cognoscenti. We all wish to appear intelligent. It seems that in physiology intelligence is inversely proportional to the size of the object of one's research. Obviously, therefore, the physiologist who claims to be isolating receptors from the cells is much more intelligent than he who collects pancreatic juice from conscious sheep. The biomathematician of course leads the pack since the object of his concern is dimensionless.

Physiologists pursuing a phenomenon naturally choose the object in which it is seen best and then extrapolate to things in general. One must be on the alert for this. The pancreata of the rat and guinea pig are favored for in vitro studies on enzyme secretion. Do these behave as they do in intact cats, dogs, and man? Rat parotid glands are favored for studies on stimulus secretion coupling following catecholamines, but in many other animals the secretory stimulus to the salivary glands is almost entirely cholinergic. Guinea pig colon has been popular for studies of smooth muscle electrophysiology, but there is evidence now that it is far from typical smooth muscle. We have attempted to keep the wood in the forefront in this text. It is after all the object of physiology to understand it even though at times enthusiasm may tempt investigators to think of isolated receptors for example as ends in themselves. Fashions in physiology are impermanent,

1 Introduction

as are those in clothes, music and art, and, quite often, they end without any resolution at all. These are usually in the nature of controversies. As time passes the opposing teams get larger: eventually they either retire, die, or get tired of the whole business and the issue remains unresolved. A few of these have been pointed out in the text. In affairs of this sort as a rule both sides are correct in their observations, but some fundamental facts have been overlooked. Exact duplication of experiments by the contending sides is rare; journals are loathe to publish straightforward confirmations.

Nowadays the word proof is seen and heard oftener in physiological circles, so much so that this also needs to be put in perspective. The word is usually used loosely. Things in biology are seldom proved. We amass evidence. Occasionally, the evidence is overwhelming. Oftentimes, however, erstwhile overwhelming evidence, in the light of new ideas and techniques, is seen to be no longer so overwhelming.

With these observations we are ready to start on our progress from mouth to anus with a few side excursions to the brain, liver and spinal cord.

Glandular Epithelial Tissue

Glandular epithelial tissues are those formed by cells specialized in producing a secretion which differs in composition from blood or intracellular fluid. The glands associated with the GI tract are classified as *exocrine* in type – at least so far as they relate to the function of the GI tract – in that their secretion passes through a duct system to the epithelial surface from which they were derived. At this point their products are discharged into the 'inward extension of the outside world' which is the lumen of the alimentary apparatus (fig. 1/1). More specifically, the GI glands are *merocrine* glands in that their secretory product is formed within and is discharged from the cell without loss or 'pinching off' of cytoplasm.

Some glands associated with the GI tract are unicellular (e.g. the 'goblet cell' of the intestine), but most are multicellular. Such 'compound' glands consist of two functional portions: (1) the *secretory* ('acinar') portion or 'endpiece', consisting of the cells which produce the secretory product, and (2) the *duct* portion, which provides transport of the secretory product out of the gland (fig. 1/1). Glands are classified according to the configuration or form of the two portions (fig. 1/2).

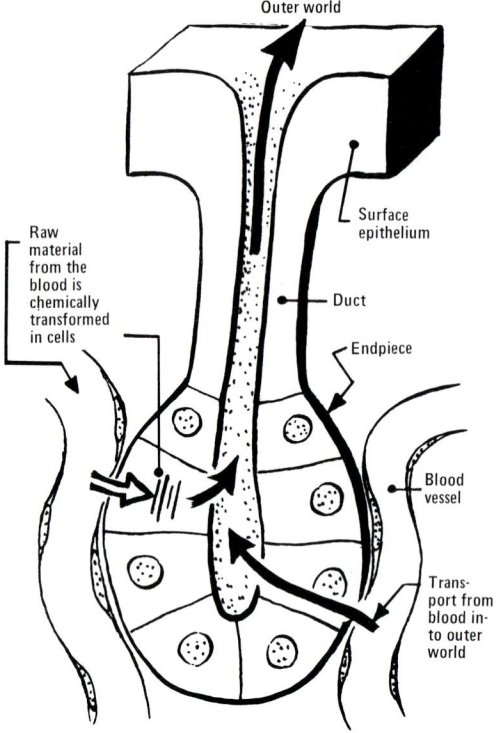

Fig. 1/1. Schematic diagram of the basic structure and function of a multicellular gland. [Reproduced, with permission, from Elias, H.; Pauly, J.; Burns, E.: Histology and human microanatomy; 4th ed. (Wiley, New York 1978).]

In addition to classification by histological organization, compound glands are often classified according to the consistency and appearance of the secretion they produce. *Mucous glands* secrete a viscous glycoprotein with a lubricating/protective function; *serous glands* have a watery secretion rich in proteins, such as digestive enzymes. While some cells are distinctly mucous or distinctly serous, there is in fact an almost continuous gradient from serous to mucous cells. Thus it is sometimes impossible, on the basis of either appearance or product, to neatly classify certain cells (and, hence, the glands they compose) as one or the other, and these are classified as *seromucous* cells or glands.

1 Introduction

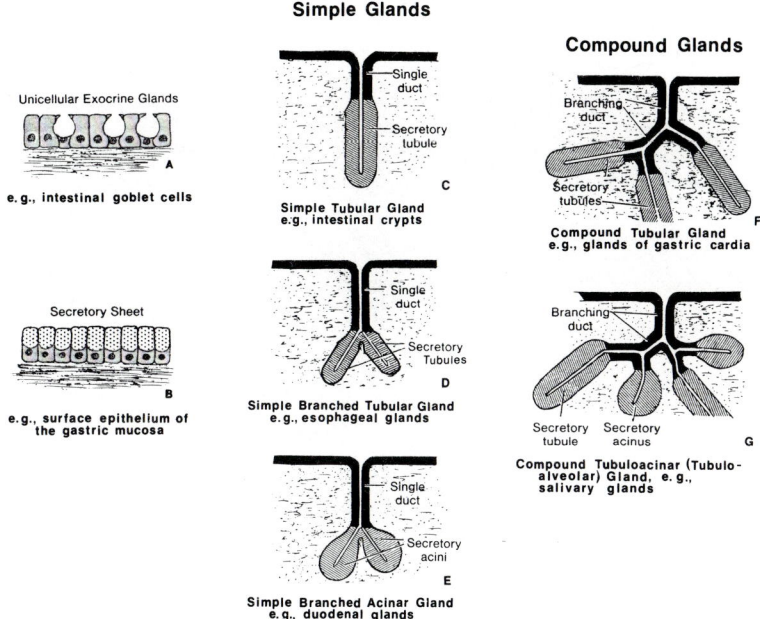

Fig. 1/2. Schematic diagram of the basic morphological types of glands associated with the GI tract. [Modified from Krause, W.; Cutts, J.: Concise text of histology (Williams & Wilkins, Baltimore 1981).]

The cells of the secretory portion of the glands are characterized by cellular components specialized for synthesizing, transporting, storing and discharging the secretory products.

The Structure of Protein-Secreting (Serous) Cells
Endoplasmic Reticulum. The protein synthesizing apparatus of the cell is the *rough* (or 'granular') *endoplasmic reticulum* (rER). The rER is a three-dimensional parallel array of interconnecting flat cisternae and channels bound by a unit membrane heavily studded on its outer (cytoplasmic) surface with ribosomes (fig. 1/3). Peptides assembled by the attached ribosomes are vectorially transferred across the rER membranes into the reticular cavities where a flocculent, electron-dense material is often observed. The intensely basophilic rER is a major characteristic of serous cells, in

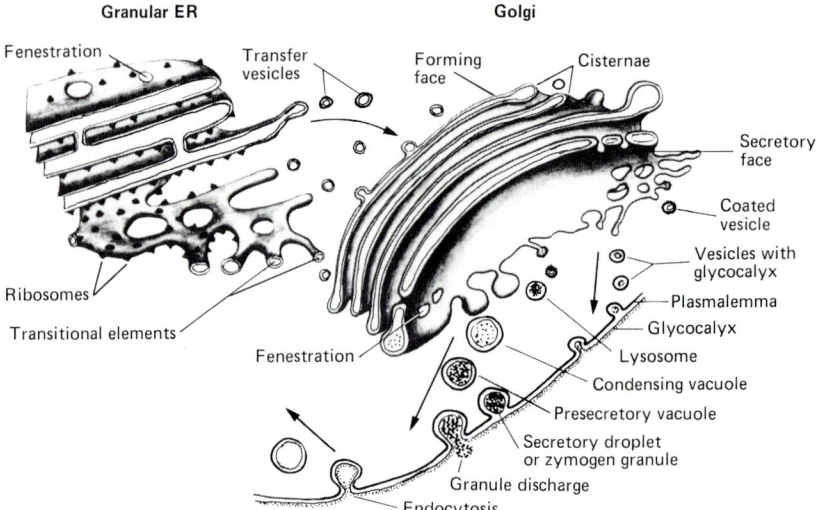

Fig. 1/3. Schematic diagram of the protein-synthesizing apparatus of a secretory cell. See text for details of the synthesizing sequence. Note the trilaminar structure of the cytoplasmic (unit) membrane. The dark line represents the 'cytoplasmic leaflet', the lamina always adjacent to the cytosol; thus it is the external lamina of the cytomembranes (the unit membrane of the organelles) and the internal lamina of the plasmalemma. The thickness of the cytomembranes is not uniform: the membranes of the ER and of the immature cisternae at the forming face of the Golgi apparatus are thinner than those of the mature cisternae of the secretory face and of the plasmalemma. [Reproduced, with permission, from Leeson, C.; Lesson, T.: Histology; 3rd ed. (Saunders, Philadelphia 1976).]

which it is located basally in the generally pyramid-shaped cells, surrounding the centrally-placed spherical nucleus and occupying approximately half of the cell volume.

More centrally-located portions of the rER – particularly those adjacent to the Golgi complex – often show evaginations which lack the attached ribosomes and protrude toward the Golgi complex (fig. 1/3). These portions are called the *transitional elements* of the rER. It is believed that rER-contained proteins diffuse toward the transitional elements which then 'bud' or 'pinch off' to form *transfer* (or 'transport') *vesicles*. These vesicles convey the contained rER product to the convex *forming face* of the Golgi apparatus where the secretory substance accumulates through vesicular fusion.

Golgi Apparatus. The *Golgi apparatus* or *complex* is most highly developed in protein-secreting cells in which it assumes a supranuclear position (i.e. between nucleus and secretory or 'apical' surface). In electron micrographs it appears as a stacked assembly of flattened saccules centrally with numerous vesicles located peripherally (fig. 1/3). These elements are both bound by smooth membranes. Together they form a shallow bowl with the concave *secretory face* directed toward the cell apex. Within the concavity of the bowl immature storage granules *('condensing vacuoles')* are located.

The secretory substances are altered as they pass through the Golgi complex. After the proteins synthesized in the rER have accumulated at the forming face, they are 'packaged' by the Golgi apparatus into the condensing vacuoles which appear at the secretory face. Within the condensing vacuoles, the secretory substance is concentrated into a smaller space, probably by the elimination of water. In cells the product of which consists both of protein and carbohydrate, much of the carbohydrate component of their sections is added within the Golgi apparatus, to the membranes of which the enzymes of carbohydrate synthesis and transfer are attached. As specific examples, in pancreatic and parotid exocrine gland, the Golgi apparatus is most likely involved in the addition of polysaccharides to the glycoprotein secretory substances (e.g. RNase B and DNase in the pancreas and amylase in the parotid).

Secretory Granules. The apical cytoplasm of merocrine cells not actively secreting (i.e. 'resting cells', as during periods of fasting) is largely occupied by numerous *mature storage* or *secretory ('zymogen') granules*. These consist of densely packed secretory products surrounded by a single smooth unit membrane. During secretion, the secretory granules migrate to the secretory surface, fuse with the plasma membrane and then discharge their contents into the acinar lumen by a sort of reversal of phagocytosis (exocytosis).

In following the progress of the secretory products through the cell, we see that from the time the ribosome-assembled peptides are sequestered into the cisternae of the rER to the ultimate discharge of the completed product from the cell, the secretory substances are contained within membrane-bound compartments, isolated from the cell sap. Thus, not only is their transport through the cell – in a generally baso-apical direction – controlled,

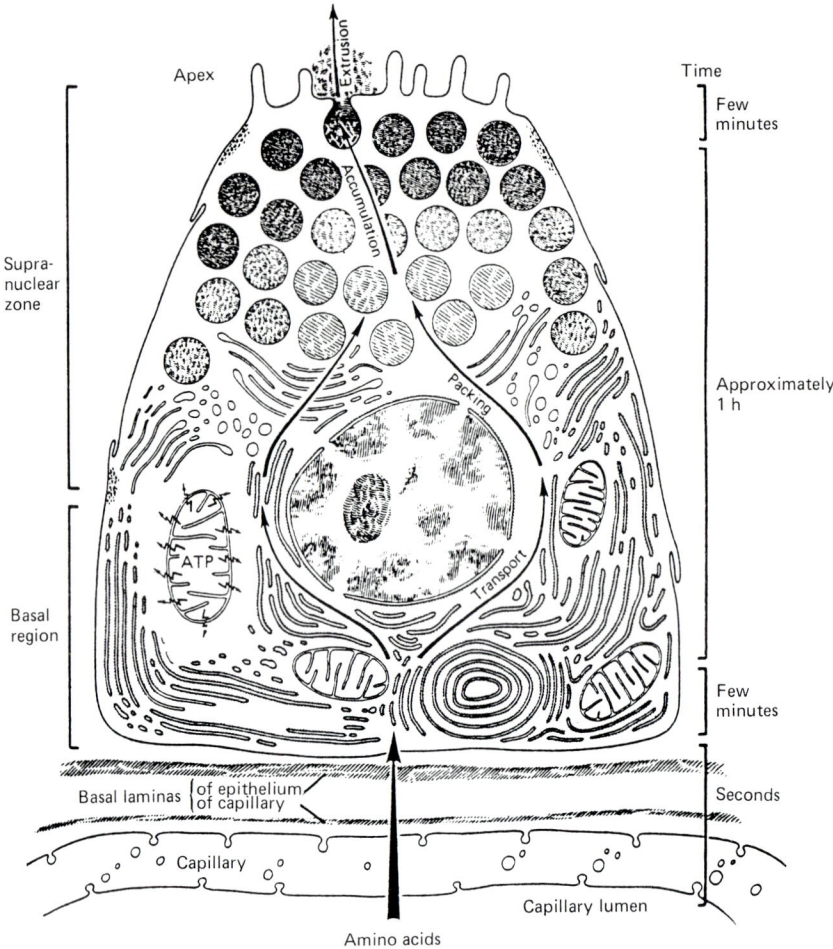

Fig. 1/4. Schematic diagram of a protein-secreting (serous) cell. This diagram emphasizes the polarity of the cell, the cell organelles being arranged in a sequential 'assembly line' for the manufacture of the secondary product. On the right side a time scale indicates the approximate duration of each step in the process. [Reproduced, with permission, from Junqueira, L.; Carneiro, J.; Contopoulos, A.: Basic histology; 4th ed. (Lange Medical Publications, Los Altos 1983).]

but the elements of the cell sap are protected from the potentially destructive hydrolases.

In summary, then, the protein-secreting cells of the GI tract are characterized by (1) intensely basophilic, densely-packed rough endoplasmic reticulum – located basally and occupying approximately half of the cell volume; (2) regularly spherical or ovoid nuclei, located centrally (or toward, but not at, the base); (3) highly-developed Golgi apparatus, supranuclear in position, and (4) storage and secretory granules, of varying density, abundant in the apical cytoplasm of the resting cell (see fig. 1/4).

Among cells sharing these morphological features, there are variants that can be distinguished only by the presence or absence of histochemically demonstrable carbohydrate polymer in the cytoplasmic granules. According to this classification [5] the pancreatic acinar cells and the gastric chief cells are termed as true *serous cells* on the basis of the relative absence of such carbohydrate; in the salivary glands, where the granules contain sialomucin and sulfomucin, the cells are termed *seromucous cells* [3, 4], although distinctly more 'serous' than 'mucous' in appearance.

The Structure of Mucous-Secreting Cells

Mucous acinar cells vary in appearance according to the phase of the secretory cycle. Following secretion, when the cells has discharged its mucous and is beginning a new cycle of mucous production, the most conspicuous feature of the cytoplasm is the extensive rER. Its cisternae show more dilatation than that of the serous cells, and it, too may contain a moderately electron-dense material. In this phase, the Golgi apparatus is relatively inconspicuous. The nucleus is displaced far to the base of the cell, and frequently demonstrates intranuclear inclusions during this phase of the secretory cycle.

As the cycle continues, the rER involutes and the supranuclear Golgi apparatus becomes more prominent. Its stacked cisternae are filled with a substance which increases in electron density as the secretory face of the organelle is approached. The process of concentrating the secretory product apparently occurs within the Golgi complex itself rather than in condensing vacuoles, which are absent in mucous cells. The apical half of the cell becomes increasingly filled with poorly staining mucous droplets surrounded by single membranes to the point that the cell develops a 'foamy' or 'moth-eaten' appearance (fig. 1/5). As the secretory material accumulates, the intranuclear inclusions disappear and the nucleus becomes flattened against the basal plasma membrane. Close electron microscopic examination shows fusion occurring between adjacent mucous droplets. It is thought that such fusion, rather than being artifact, may be an integral part of the extrusion sequence. Details of the discharge mechanism remain controversial. In general, mucous-secreting cells are somewhat larger than their serous counterparts.

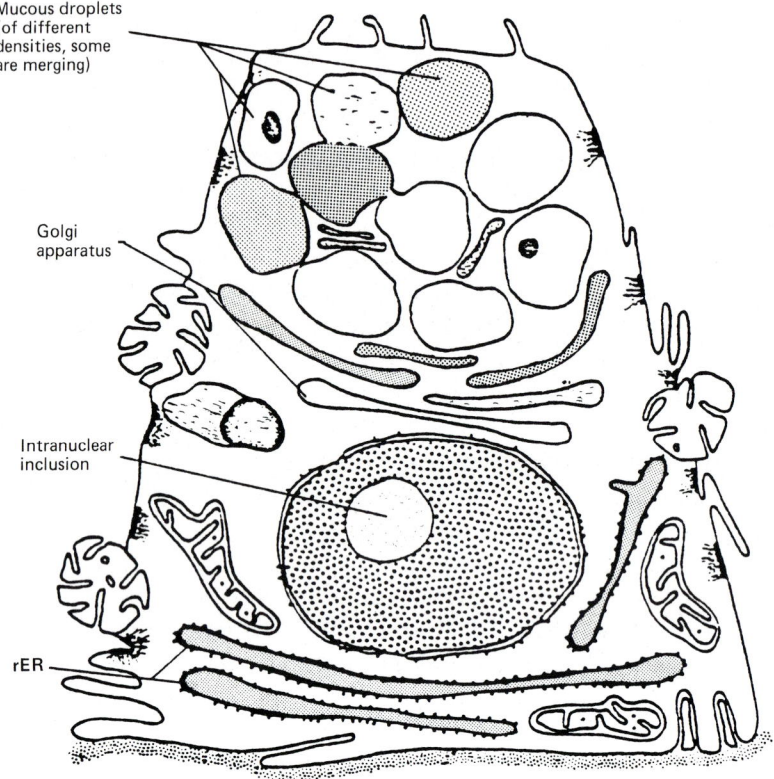

Fig. 1/5. Schematic diagram of a mucous-secreting cell. The accumulating mucous droplets give the apical half of the cell a 'foamy' appearance. Prior to discharge, the intranuclear inclusions disappear and the abundance of mucous droplets causes the nucleus to become flattened against the basal plasma membrane. [Reproduced, with permission, from Tandler, B.: Microstructure of salivary glands; in Rowe: Proceedings of symposiums on salivary glands and their secretion, pp. 8–21 (University of Michigan, Ann Arbor 1972).]

Muscle of the GI System

Most of the muscle associated with the GI tract and its glands is smooth (nonstriated or unstriped) muscle, so-called because it lacks the cross-striations or striped appearance characteristic of skeletal muscle at both the light and electron microscopy levels. It has also been designated as 'involuntary muscle' since we *normally* do not willfully control it (except for the ciliary muscle of the eye). Through biofeedback techniques and/or yoga, a few individuals have learned an ability to control this 'involuntary' muscle to some degree.

Distribution

Smooth muscle is mainly visceral in distribution, forming the contractile portion of the wall of the GI tract from the midpoint of the esophagus to the internal sphincter of the anus. It provides for the churning of ingested food, mixing it with digestive juices, as well as for its propulsion through the absorptive and excretory portions of the tract. It is also found in the walls of the ducts of the glands associated with the system.

Smooth muscle composes most of the walls of the vessels forming the circulatory system from the great vessels down to but not including the capillary level. It is also found in the integumentary, respiratory, urinary and genital systems and in the iris and ciliary body of the eye.

Smooth Muscle Cells (Myocytes)

When organized into sheets or bundles, the boundaries of individual cells are difficult to distinguish through the optical microscope. However, shaking a bit of muscle in a weak acid solution dissolves the intercellular substance and allows isolation and study of individual cells. The myocytes are seen to be mononucleate, elongated fusiform or spindle-shaped cells, thicker centrally where the nucleus is found, and then thinning into finer, tapering ends (fig. 1/6). Smooth muscle cells may range in length from 15 μm in small arterioles to 500 μm (0.5 mm) in gravid myometrium, but generally throughout the G.I. tract they are approximately 200 μm long and about 6 μm in diameter at the midpoint of the cell (a red blood cell is 7 μm in diameter).

The cytoplasm of muscle cells (of any type) is given the special designation *sarcoplasm* and the cell membrane is called *sarcolemma*. The sarcoplasm of smooth muscle cells in stained preparations (optical microscopy) is acidophilic and homogenous, seeming to be nearly devoid of structure. The centrally-placed solitary nucleus conforms to cell shape, being ovoid in a 'resting' cell, elongated along the long axis in a stretched myocyte and wrinkled or pleated ('crenated') in outline in a contracted fiber. It contains one or more nucleoli and, having a fine chromatin network, does not stain darkly. Electron microscopy (EM) reveals that the few cytoplasmic organelles present (elements of agranular and granular endoplasmic reticulum (ER), a small Golgi apparatus, lysosomes and glycogen granules and mitochondria) are confined primarily to the juxtanuclear zone, capping the two

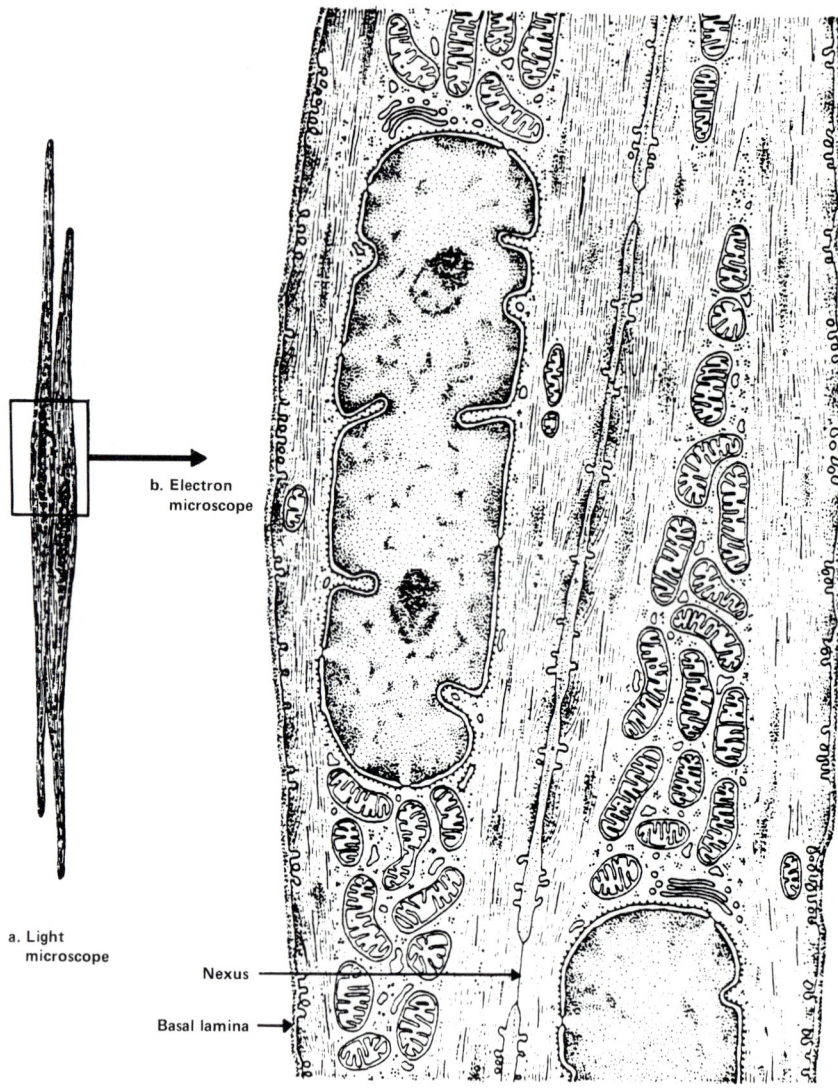

Fig. 1/6. Morphology of smooth muscle cells. **a** Two isolated myocytes as they would appear under light microscopy. **b** Detail of **a** showing the juxtanuclear zones of the myocytes as they would appear under electron microscopy. [Reproduced, with permission, from Poirier, J.; Dumas, J.: Review of medical histology (Saunders, Philadelphia 1977).]

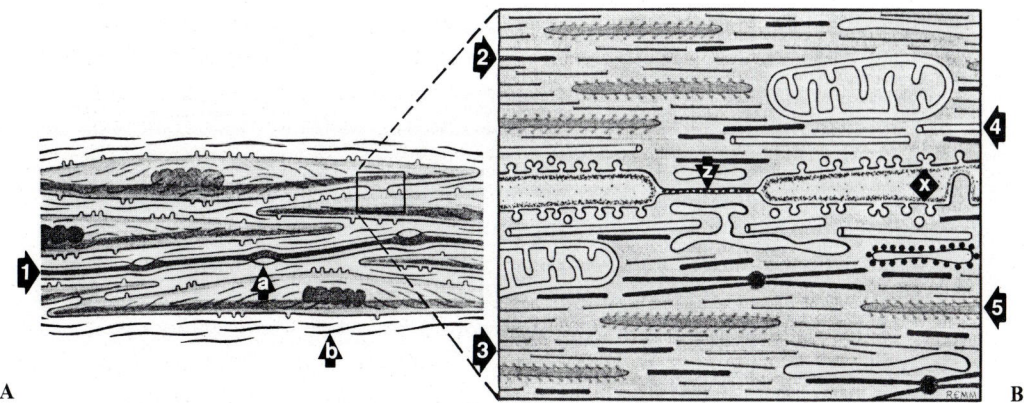

Fig. 1/7. Structural details of smooth muscle cells. **A** A group of myocytes. The nuclei of the fusiform cells demonstrate the crenated outline seen during contraction. A visceral efferent (i.e. autonomic) nerve fiber (1) passes through the tissue block demonstrating the 'en passant' mode of innervation. The Schwann cell surrounding the nerve fiber is retracted at each varicosity of the fiber (a) where transmission occurs; this point is not closely applied to the surface of the myocyte as nerve endings are in skeletal muscle. **B** Detail of (**A**) in which components of the cytoplasm and details of the cell membranes including a 'gap junction' or 'nexus' (z) – are demonstrated. 2 = 'Intermediate' filament; 3 = fine actin filament; 4 = microtubule; 5 = thick myosin filament. Endocytotic vesicles are seen along the cell membrane on each side of the nexus. The extracellular space is occupied by a dense network of delicate reticular and elastic fibers ('b' in **A**) embedded in an amorphous glycoprotein matrix ('x' in **B**). [Reproduced, with permission, from Williams, P.; Warwick, R.: Gray's anatomy; 36th British ed. (Saunders, Philadelphia 1980).]

nuclear poles (fig. 1/6). The remainder of the sacroplasm consists mainly of a closely-packed, filamentous, contractile, proteinaceous material (the *myofilaments*) grouped into irregular bundles (*myofibrils*) and disposed parallel to the long axis of the cell. Between the myofibrils, a few microtubules and mitochondria are found. Small dense areas along the cell membrane (*attachment plaques*) and within the sarcoplasm (*dense bodies*) appear to be sites of insertion for the myofilaments (fig. 1/7).

Contraction of Myocytes

The exact mode of contraction of smooth muscle cells is not clearly established. Both thick and thin filaments have been demonstrated in smooth muscle cells by electron microscopy, and *myosin* and *actin* have been identifical biochemically (fig. 1/7B). However, myosin filaments have been difficult to demonstrate in standard EM preparations. It has been suggested that in smooth muscle myosin is labile, being normally in a depolymerized state and polymerizing into aggregates or thick filaments only just prior to contraction. Myosin and actin of smooth muscle are capable of interaction as in skeletal muscle, and thus it is believed by most that contraction is based on a mechanism wherein myosin forms sliding cross-linkages between adjacent actin filaments. The whole cell may contract throughout its length at one time, or a wave of contraction may pass over it, one part of the myocyte being contracted while the remainder is at rest.

Smooth muscle tissue as a whole is characterized by a slow and sustained mode of contraction with a relatively small expenditure of energy, The cells maintain some degree of tension even during the 'resting phase' between distinct contractions; this tone is maintained throughout life and even after section of its nerve supply. After death, intestinal smooth muscle greatly elongates.

Organization of Smooth Muscle Cells in Tissue

Smooth muscle cells are always disposed with their long axes parallel to the direction of contraction. They may be present as isolated units or arranged in small fasciculi with loose connective tissue, as in the lamina propria of the intestinal villi (where their contraction shortens the villi and thus helps to expel lymph – or chyle – from the lacteals). Smooth muscle cells may be organized into a single sheet with the myocytes oriented circumferentially about the walls of vessels or hollow viscera so that contraction results in the constriction of the lumen to either restrict flow (vasoconstriction) or to increase pressure on the contents, usually leading to its expulsion from the lumen (expression of a glandular secretion from a duct). In the wall of the GI tract (the tunica muscularis), smooth muscle is arranged in two sheets with the fiber axes of the sheets lying at the right angles to each other. Described as an outer 'longitudinal' lamina and an inner 'circular' lamina, some investigators [1] have claimed that these orientations are actually spiral in type, the 'circular' being a close helix and the 'longitudinal' being an open helix. More recent observations [2] disagree. At any rate, the arrangement of the musculature allows complex movements such as the churning motion or the regular peristaltic waves associated with digestion. Contraction of a portion of the circular muscle lamina forms a narrow ring of constriction, decreasing the diameter of the lumen at that point. Simulta-

1   Introduction

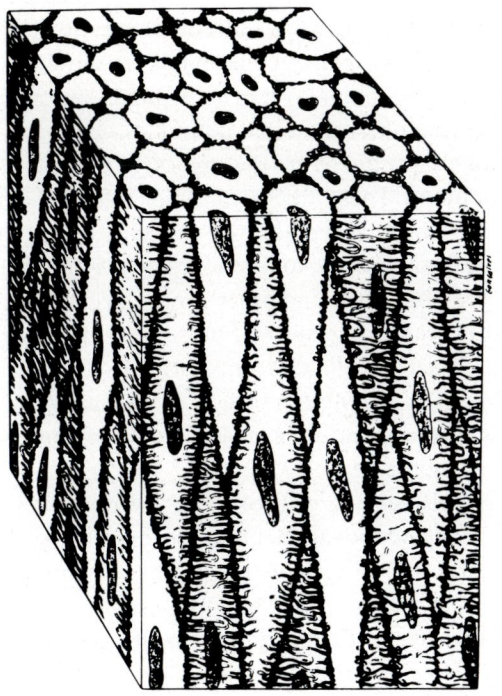

Fig. 1/8. Diagram of a tissue block cut from smooth muscle. Uncut cell surfaces are covered with a network of elastic and reticular fibers. Where the cells are cut transversely (cross-sectioned) the cells exhibit varying diameters and only a few profiles include nuclei. [Reproduced, with permission, from Junqueira, L.; Carneiro, J.; Contopoulos, A.: Basic histology; 4th ed. (Lange Medical Publications, Los Altos 1983).]

neous contraction of the longitudinal lamina results in a localized shortening of the tube surrounding the constriction together with a luminal dilatation just downstream from the constriction. Whether they are organized as a sheet or a fasciculus, when observed in longitudinally-sectioned tissue, the spindle-shaped cells are seen to be packed together in a parallel but offset arrangement so that the thick middle portion of one cell lies adjacent to the tapering ends of neighboring fibers (fig. 1/8). Thus, in transversely-sectioned tissue the smooth muscle cells appear as discs or polygons of cytoplasm with diameters varying according to where along its length each cell was sectioned. The plane of section will pass through the thick nuclear portions of only a relatively few of the myocytes, and so nuclei will be correspondingly few and present only in the largest cellular profiles.

In the narrow (400–800 Å) space between individual muscle cells is a delicate but dense network of fibers, partly collagenous or reticular and partly elastic, embedded in an intercellular layer of amorphous glycoprotein (*glycose aminoglycan*) which resembles the basal lamina of epithelia. The extracellular fibers are probably produced by the myocytes themselves at some stage in their development. These fibers continue into the denser fibroconnective tissue septae which ensheathe the fascicles or sheets of muscle cells. Included among the abundant collagenous and elastic fibers of these *interfascicular septae* are fibroblasts and other connective tissue cells, as well as networks of blood vessels and nerve fibers. This arrangement allows the traction of each contracting cell to be transmitted first to the finer reticuloelastic network around the cells and in turn to the stronger connective tissue of the muscle bundle, permitting a steady, uniform contractile force to be generated by the smooth muscle sheet.

In certain limited areas between the cells of visceral smooth muscle, intercellular substance is lacking and the sarcolemmas of adjacent myocytes come into close approximation, the intercellular space narrowing to only 20 Å. These regions – called *nexuses* or *gap junctions* – are believed to be sites of low resistance pathways permitting the rapid passage and spread of electrical excitation over the cellular sheet (fig. 1/6, 1/7).

Innervation and Initiation of Contraction of Smooth Muscle

Nonmyelinated plexuses, formed of sensory fibers and postsynaptic motor fibers of both the sympathetic and parasympathetic divisions of the autonomic nervous system, send branches to smooth muscle. On the basis of their motor innervation and related behavior, two functional types of smooth muscle are described: *multi-unit* and *unitary*. In the multi-unit type, rich nerve plexuses supply an abundance of motor terminals so that most or all muscle cells are directly innervated. Contraction is thus *neurogenic*, being initiated by direct nervous stimulation. There is a little evidence of intercellular conduction, and small groups of muscle cells may act independently. The smooth muscle of the iris, larger arteries and ductus deferens are of this type.

Smooth muscle of the unitary type is supplied by sparser plexuses which provide much fewer nerve terminals to the tissue. Only a few of the cells appear to be in contact with a nerve ending. The relatively slower contractions of unitary muscular tissue are *myogenic*, being initiated *by the myocytes* either spontaneously or in response to stretch. The smooth muscle of the GI tract, ureter, uterus and some smaller blood vessels is of the unitary type. Spontaneous impulses are generated which spread through the muscle tissue – presumably by way of the intercellular contacts which occur at the nexuses – accompanied by a wave of contraction. The electrical coupling of the myocytes via the nexuses allows a single nerve terminal to influence many myocytes without direct innervation excitation, as well as ensuring that the myocytes act in concert. The multi-unit and unitary types of smooth muscle described probably represent the extremes of a continuum since a wide variety of intermediate types and behaviors have been described.

General Structure of the Gastrointestinal Canal

The following account provides an overview of the general structure of the alimentary tract, viewing it as a typically tubular structure running from the mouth (rima oris) to the anus. Specific details concerning the specialized structure of particular regions of the tract are provided in table 1/I, with elaboration in the appropriate chapters.

Component Layers (Tunicae) of the Canal Wall
Throughout its length, the alimentary tract presents common structural features in the form of four coats (tunics) or layers (laminae): (1) the *mucous membrane* or *mucosa;* (2) the *submucosa;* (3) the *muscular layer,* and (4) the *serosa* or *adventia* (fig. 1/9A). These can thus be considered as basic components which, though fundamentally similar throughout the tract, vary somewhat in nature and thickness from region to region in relationship to the functional requirements.

The Mucous Membrane (Tunica mucosae). The mucous membrane is the innermost of the four tunics. It consists of a lining epithelium and an underlying sublayer of fine, interlacing connective tissue fibers, the lamina propria. A third component, the muscularis mucosae, is found only in the tubular portion of the digestive tract (see fig. 1/9B).

The *epithelium* is the primary site of physical and functional interaction between the body and ingested materials. Thus, its main functions are (1) to facilitate the passage of the bolus while protecting underlying tissues, and (2) to provide a selectively permeable barrier capable of both secretion and absorption as required for digestion. In areas such as the pharynx-esophagus and lower anal canal where the former functions are primary, it is composed of stratified, squamous cells; in regions where the latter functions predom-

Table 1/I. Histology of the wall of the digestive tube [from ref. 6]

Digestive tube (viscus)	Mucosa		Muscularis mucosae	Submucosa	Muscularis	Peripheral connective layer
	epithelium	lamina propria				
Esophagus	nonkeratinized stratified squamous	cardiac glands (mucous) at the upper and lower extremities	absent in the upper third	esophageal glands (chiefly mucous, but also serous)	two layers (inner circular and outer longitudinal); striated muscle in the upper third	adventitia
Stomach	simple columnar with closed mucous pore	gastric glands: fundic glands in the body and fundus pyloric glands in the antrum	two layers with radial processes entering the mucosa	no distinguishing features	three layers (inner oblique middle circular, and outer longitudinal)	peritoneal serosa (except the posterior aspect of viscus where secondarily retroperitoneal – duodenum)
Duodenum	simple columnar with enterocytes having a striated border and goblet cells	crypts of Lieberkühn; forms the axis of the villi; lymphoid follicles; Peyer's patches (at the end of the ileum)	elevated by the plicae circularis	Brunner's glands (acinous mucous)	two layers (inner circular and outer longitudinal)	
Jejunum-ileum				no distinguishing features		

1 Introduction

				lymphoid nodules		peritoneal serosa (except the posterior aspect of viscus where secondarily retroperitoneal – colon, rectum)
Appendix	simple columnar with goblet cells and enterocytes having a striated border	few crypts of Lieberkühn: lymphoid nodules especially in the appendix	discontinuous	no distinguishing features	two layers (inner circular and outer interrupted longitudinal (colic bands))	
Colon and rectum			no distinguishing features			
Anal canal						
Anorectal area	nonkeratinized stratified squamous	large venous plexus				
Anocutaneous area	keratinized stratified squamous	no distinguishing features			inner smooth muscle sphincter (thickening of the inner circular layer)	adventitia
Cutaneous area		pilosebaceous follicles and apocrine sweat glands			outer striated muscle sphincter	

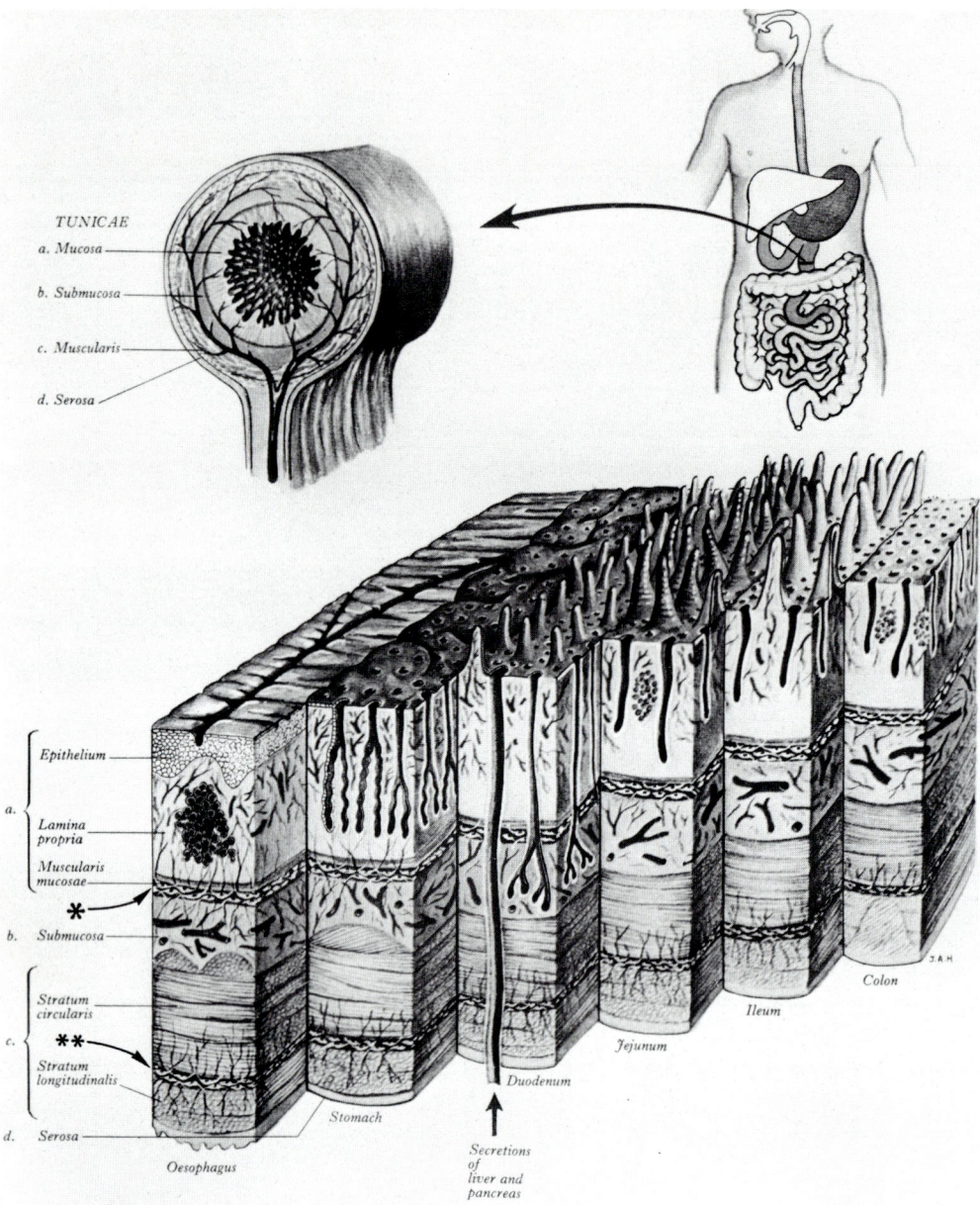

Fig. 1/9. Diagrammatic representation of the organization of the gut wall, showing (top) the four basic tunics and (bottom) variations on the common theme at each level of the GI tract (see also table 1/I). Deep to the tunica mucosa (a), the submucosal (Meissner's)

inate, it is but a single cell in thickness, being composed of both secretory cells and columnar absorptive cells with modified luminal surfaces. It is wet throughout the length of the GI tract, lubricated by the mucous it secretes.

The epithelium rests upon a basal lamina which in turn is supported by the lamina propria. The *lamina propria* is a compact stroma, composed mainly of reticular fibers but often rich in elastin fibers as well. It may contain glands, scattered fibers of smooth muscle and lymph nodules ('solitary follicles'). The latter are often sufficiently large to extend into the submucosa. Extensive beds of blood and lymph capillaries are present, and it is into these that the absorbed food matter passes in the absorptive sections of the tract.

Close study reveals a small but nearly continuous layer of macrophages and lymphoid cells in this sublayer just below the epithelium. In fact, because of their abundance, the entire lamina propria of much of the gut might be considered as modified lymphatic tissues. Some of the lymphoid cells have been shown (by immunofluorescence techniques) to actively produce antibodies (gammaglobulins). The lymphoid material found here presumably enables the lamina propria to function as a defense barrier, attempting to prevent bacterial invasions of the deeper tissues and hence the body in general.

Where present, the *muscularis mucosa* is a thin but continuous bilaminar plexus of circular (inner) and longitudinal (outer) smooth muscle fibers which clearly demarcates the mucosa and submucosa. The muscularis mucosa enables alteration of the local conformation of the mucosa independently of other movements of the digestive tract, increasing its contact with the food. Where the muscularis mucosa is absent, there is a gradual transition from lamina propria to the submucosa.

The Submucosa (Tunica submucosa). The second coat is comprised of a matrix of coarse, loosely interwoven collagenous fibers and scattered elastic fibers in which larger vessels, lymphatics and a nerve plexus ramify. It may

---

nerve plexus (*) is located. The two layers of the tunica muscularis (c) are separated by a layer of connective tissue which includes the myenteric (Auerbach's) plexus of nerves (**) as well as vascular and lymphatic plexuses. [From Williams, P.; Warwick, R.: Gray's anatomy; 36th British ed. (Saunders, Philadelphia 1980).]

also contain (portions of) glands and lymphoid tissue. The *submucosal (Meissner's) nerve plexus,* composed of visceral afferent and efferent fibers, includes enteric parasympathetic ganglia (cell bodies) (fig. 1/9). It is not inaccurate to consider the submucosa as a sort of 'neurovascular service area' or conduit conveying the larger blood vessels and nerves that send their finer vessels into the adjacent layers (mucosa and muscularis) that embody the specific organ functions.

The submucosa varies greatly in thickness. Where it is thick, it provides the mucosa with considerable mobility, allowing the independent movement of the mucosa produced by the muscularis mucosa. Where it is thin or absent, the mucosa is immobile, being firmly adherent to underlying structures (for example, in the gums). Where its thickness is very irregular, it creates the macroscopic folds and ridges of the internal aspect of the gut ('plicae semi-circulares'), increasing the surface area of the mucosa (and hence the amount of absorption and/or secretion possible).

The Muscularis externa (Tunica muscularis). The muscularis externa forms an outer supporting wall for the two inner tunics in a tubular portion of the tract. It consists of two sublayers or sheets of muscle, an inner circularly-oriented layer and an outer longitudinally-oriented layer. For most of the tube, the two sublayers are composed of involuntary (smooth) muscle, but at both ends (upper esophagus and lower rectum) the muscle is of the voluntary (striated) type. The two sublayers are separated by a thin layer of connective tissue which includes the *myenteric (Auerbach's) plexus of nerves* as well as vascular and lymphatic plexuses (fig. 1/9).

The activity of the muscle coat is largely responsible for the propulsion of the luminal contents ('peristalsis' – the mechanism of which is explained under 'Organization of Smooth Muscle Cells in Tissue', this chapter, p. 14) as well as for local churning of the food mass, mixing it with the digestive enzymes.

Relative to the other tunics, the muscular tunic is a thick layer, but its actual thickness varies along the length of the tube to meet specific functional demands. To enable the more complex movements required of the stomach – where churning and mixing of the food mass takes precedent over its transport – a third sublayer of obliquely-oriented muscle is added between the two usual sublayers. At particular points along the tract the circular layer is markedly thickened to form sphincters – 'gates' which have been presumed to halt or permit progression of the food mass into, along or out of the tract.

Serosa or Adventitia (Tunica serosa or Tunica adventitia). The fourth or outermost layer of the tract is composed of a relatively dense areolar tissue which frequently includes adipose (fat) tissue.

In the extraperitoneal portions of the tract (oral cavity, pharynx, esophagus and rectum) this outermost layer blends with the surrounding fascia or connective tissue of adjacent structures. Here it is termed an *adventitia*.

In the portions of the tract which have evaginated the peritoneal (abdominopelvic) cavity to become suspended within it by mesenteries (i.e. stomach and intestines), the areolar tissue of the fourth layer is covered on its outer surface by a single layer of squamous mesothelial cells. This moist serous membrane is the visceral portion of the peritoneum. It permits the organs thus enclosed to slide freely over one another during the movements of peristalsis. The outermost tunic of these portions of the tract – including the layer of visceral peritoneum – is termed a *serosa* rather than an adventitia.

Again, blood and lymphatic vessels and nerves are present; indeed, the neurovascular structures found in the other layers are usually ramifications or tributaries of the distributing and collecting vessels and nerves running in this outermost layer.

Vasculature of the Gastrointestinal Tract

The wall of the digestive tube is highly vascular. A rich arterial supply provides it with oxygen and the metabolites necessary to sustain its muscular and secretory activities. In the intraperitoneal portion of the tract, this abundance is especially apparent as the mesenteries convey the vasculature to and from the tube. Here tributaries of the portal vein and a profusion of lymphatics carry away the absorbed products of digestion.

Nerves of the Gastrointestinal Tract

Intricate ganglionated intrinsic nerve plexuses are found in the wall of the alimentary tract from esophagus to rectum. The subdivisions, mentioned above in relationship to the specific layers in which they are found, are rather artificially divided since all parts are interconnected. Extrinsic nerves reach the abdominal gut by coursing with the arteries which serve it. Postsynaptic sympathetic, presynaptic parasympathetic and visceral afferent nerve fibers are thus conveyed to the gut and are for the most part distributed initially to the myenteric (Auerbach's) plexus. Both the myenteric and submucosal (Meissner's) plexuses contain scattered ganglionic cells (cell bodies of postsynaptic parasympathetic neurons). They are most numerous in regions where motility is greatest. The autonomic fibers in these plexuses permit extrinsic regulation of the gut (parasympathetics promote peristalsis and secretion; sympathetics inhibit both while restricting blood flow, diverting it for use by skeletal muscle) but rhythmic contractions of the gut still occur even when it is cut off from the extrinsic nerve supply. In other words, the gut is largely autonomous, with stimulation and coordination of peristalsis managed by the intrinsic nerve plexuses.

Gastrointestinal complaints and disorders are a frequent and often major component of psychosomatic syndromes. An anatomical basis for the link between emotional and gastrointestinal disturbances is found in the abundant innervation of the digestive tract by the autonomic nervous system.

Most of the gut is insensitive to ordinary tactile, painful and thermal stimuli, although it is responsive to tension (stretch), anoxia and chemical stimuli. The structural counterparts of the physiologically-identified receptors have for the most part evaded observations thus far, although there has been considerable postulation.

References

1. Carey, E.: Studies on the structure and function of the small intestine. Anat. Rec. *21:* 189–216 (1921).
2. Elsen, J.; Arey, L.: On spirality in the intestinal wall. Am. J. Anat. *118:* 11–20 (1966).
3. Leblond, C.: Distribution of periodic acid-reactive carbohydrates in the adult rat. Am. J. Anat. *86:* 1–50 (1950).
4. Leblond, C.; Glegg, R.; Eidinger, D.: Presence of carbohydrates with free 1,2 glycol groups in sites stained by the periodic acid-Schiff technique. J. Histochem. Cytochem. *5:* 445–458 (1957).
5. Munger, B.: Histochemical studies on seromucous and mucous-secreting cells of human salivary glands. Am. J. Anat. *115:* 411–429 (1964).
6. Poirier, J.; Ribadeau Dumas, J.-L.: Review of medical history, pp. 144–145 (Saunders, Philadelphia 1977).

2 Mouth and Saliva

Chewing

People find it difficult to swallow food if they cannot chew. This explains why the edentulous may suffer malnutrition. Chewing is a matter of training and habit rather than physiological regulation. Habit requires that pills be chewed before swallowing, hence, the difficulty; this disappears with practice and pills and large stomach tubes can then be swallowed. The closing force exerted by the muscles of mastication is very high, but the opening force is relatively weak in most animals. By taking advantage of this fact, there are people who earn a living wrestling with and immobilizing crocodiles and alligators.

The maximal force measured between the incisor teeth in man has been recorded as ranging between 11 and 25 kg and between 27 and 90 kg for the molars of the two sides. These are biting forces and are much larger than those actually needed during ordinary eating. The contact surfaces between the molar teeth and the movements between the two surfaces seem to be more critical than the total force between them.

Chewing animals have large flat molars and often, cattle for example, have only lower incisors. Carnivores are bolters rather than chewers and their molars are usually narrow and pointed. The shearing movement in chewing, which is so important in reducing particle size and in breaking up cellulose, is very evident in cattle, but although not as obvious to the casual observer, is present in man also.

In biting easily the closing force between the incisor teeth is important. This explains why it is that people with properly fitting false teeth who can chew easily and well often have difficulty biting apples. Perhaps, this is why it is considered refined to cut one's apple rather than to bite into it.

No one will argue that it is not nutritionally beneficial to chew one's food, but like many things taken for granted, the evidence is slight. One cannot digest grain, for example, unless the outer cellulose hull has been split. (There are no cellulases in the vertebrate digestive secretions.) Chewing will do this, at least partially. It is a common experience that after eating cooked maize (US corn) to find whole kernels in the stool and if one puts manure from horses that have been fed whole oats on one's garden, a growth of oats can be expected in the spring.

The question remains, however, is it nutritionally better to swallow finely macerated food than large lumps? Is chewing simply to reduce food to a size that will fit into the esophagus having first lubricated it well with saliva? On this point, there is no evidence. Many years ago, it was determined that dogs gained weight better on lumps of meat, which being carnivores they do not chew at all, than when fed the same amount as homogenates in water. As mentioned above, people are apprehensive of swallowing large lumps and will often suffer malnutrition rather than run the risk of choking. Excessive chewing recommended by some food faddist is, indeed, just a fad without any physiological evidence to sustain it. Some animals, like ruminants, chew a great deal because they must and others, carnivores, chew hardly at all because they do not need to; neither is intrinsically healthier than the other.

Saliva

Microstructure of Salivary Glands

Broadly speaking, any cell or organ discharging a secretion into the oral cavity is a salivary gland. By convention, however, the *major salivary* glands – lying at a distance from the oral mucosa but communicating with it via one or more extraglandular ducts – are distinguished from the *minor salivary glands,* which lie within or just deep to the oral mucosa and open on to its epithelial surface via many short ducts.

Both major and minor salivary glands develop early in prenatal life (6–7th week) as ingrowths of – or sproutings from – thickenings of the oral epithelium which are thought to be accumulations of presumptive glandular epithelium. The sprouts, retaining their continuity with the epithelium, develop into branching cords and thus the racemose duct system is present by birth. The acini, however, do not fully differentiate until shortly after birth.

Unlike most other organs, the salivary glands display extreme species variation at the histological, histochemical and ultrastructural levels. In a number of species, the salivary glands even display sexual dimorphism [31].

We shall concern ourselves here specifically with adult human glands, unless otherwise stated. The major salivary glands, of which there are three pairs, shall be considered first.

Major Salivary Glands

Secretory Units. In terms of the configuration of their secretory portions and ducts, the major salivary glands are all compound tubulo-acinar (tubulo-alveolar) glands (fig. 1/2, 2/1). Reduced to the lowest common denominator, each gland is composed of an aggregate of *secretory units* (morphological and functional units which have been referred to as 'salivons' or 'adenomeres'). In simplest terms, these consist of a *secretory end-piece* (an 'acinus' or 'terminal tubule'), an *intercalated* (intercalary) *duct,* and a *striated duct.* Actually, the secretory units (in aggregate) constitute the parenchyma or functional tissue of the glands, which is surrounded and subdivided (into lobes and lobules) by a supporting interstitial connective tissue framework, the *stroma.* Within the stromal capsule and septae, fibrocytes, macrophages, lymphocytes and fat cells are found, as are the lymphatics, nerves and blood vessels associated with the gland. The latter structures are organized in definite pattern relative to the duct system. An abundance of fat cells associated with the stroma is particularly characteristic of the parotid.

Following the direction of salivary flow, several acini (or terminal tubules) converge on each intercalated duct. Many intercalated ducts drain into each striated duct. The intercalated and striated ducts lie within the parenchymal lobules and thus constitute the *intralobular duct system.* Numerous intralobular systems empty into fewer *interlobular* ('excretory' or 'collecting') ducts running between the lobules in the stromal septae. These, in turn, all converge at the hilus of the gland where, generally, a single *major excretory duct* leaves the gland, extending to the oral mucosa.

By contrast, the vessels and nerves enter the gland mainly at the hilus and from there follow the ramifications of the duct system retrograde, branching gradually to the lobules where rich capillary networks surround the components of the secretory units. Even for the major excretory duct, flow of blood occurs in a direction opposite that of the saliva. This countercurrent arrangement of arterial flow to the direction of salivary secretion will be seen to have important functional significance.

Cells of the Secretory End Pieces. The acini account for the greatest part of the mass of the salivary glands, and the glandular cells of the acinus are the

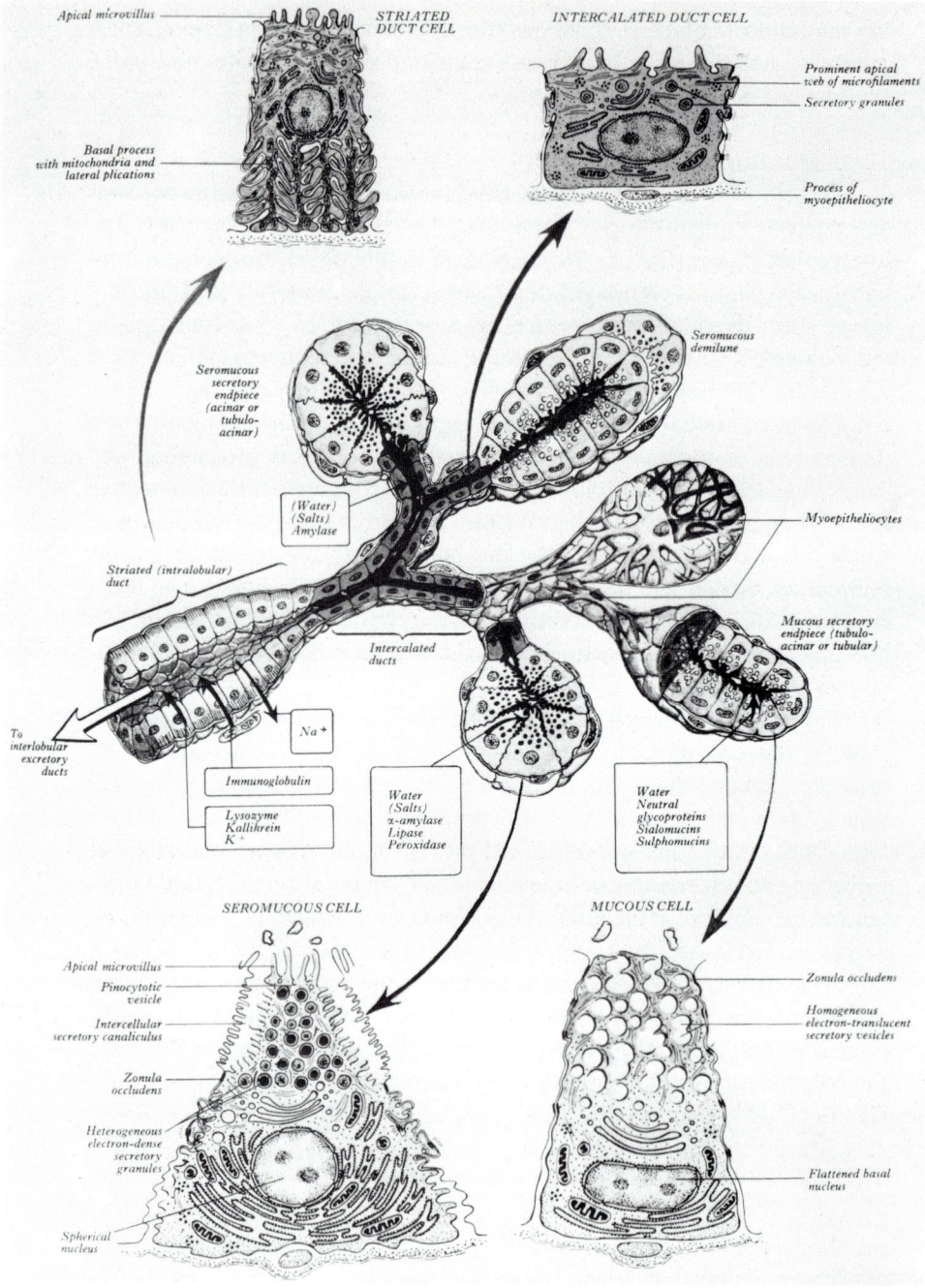

largest cells within the gland. These cells are the primary producers of salivary protein and glycoprotein and are customarily described as serous, seromucous or mucous according to the nature of the secretion and/or the appearance of the glandular cells themselves (see chapter 1). Unfortunately, there is no unanimity in the use of these terms. A gland – or an acinus – containing only one type of glandular cell may be referred to as *homocrine,* while one containing more than one type of acinar cell may be termed *heterocrine* or 'mixed'.

Technically speaking, all the major human salivary glands are heterocrine. The parotid, with its secretory end-pieces almost exclusively in the form of seromucous acini in adults (frequently referred to as 'serous' only – see 'Glandular Epithelial Tissue', chapter 1), comes nearest to being homocrine. In newborns, the glandular cells often stain positively for mucous, but small nests of cells demonstrating such staining are widely scattered and few in number in the adult parotid. Thus the parotids in man secrete a thin, watery ('serous') fluid containing amylase but almost no mucin.

The submandibular (formerly 'submaxillary') and sublingual glands are distinctly heterocrine in humans. The acini of the submandibular glands are either composed entirely of seromucous cells, or of mucous cells with seromucous cells present in the form of *demilunes* – crescentic seromucous cells or groups of such cells at the base of a mucous acinus (see fig. 2/1). Demilunes are really nothing more than seromucous cells which have been squeezed out by the large mucous acinar cells. The demilune cells lie between the mucous cells and the basal lamina, appearing to be separated entirely from the acinar lumen by the mucous cells but actually communicating with it by means of fine canaliculi which conduct the secretion through clefts between the mucous cells. The seromucous cells are undoubtedly responsible for the (weak) amylolytic activity present in the gland and its saliva. Demilunes are less common in the submandibular glands of man than in other species. In humans the submandibular gland consists of approximately 80% seromucous and 5% mucous cells; the remainder consists of striated duct cells (5%), vessels, nerves and other ducts [15].

Fig. 2/1. Schematic representation of the structure of a generalized salivary gland with ultrastructural details of cellular components. Outlined arrow indicates direction of salivary flow; small solid arrows indicate direction of transmembrane or transcellular transport; large solid areas indicate enlargements demonstrating cellular morphology. Reproduced, with permission, from *Williams and Warwick* [33].

Mucous cells predominate in the sublingual glands. There are few acini formed exclusively of seromucous cells; rather, those present usually take the form of thick demilunes in the otherwise mucous acini. The sublinguals consist of approximately 60% mucous, 30% seromucous and less than 3% intralobular duct cells [15]. Sublingual saliva shows little amylolytic activity.

Intercalated and Striated Ducts. The intralobular intercalated and striated ducts consist of a single layer of cells – i.e. a simple epithelium (fig. 2/1). The cells of the intercalated ducts are cuboidal, often somewhat flattened. Their cytoplasma presents an unspecialized ultrastructure which suggests they play little part in protein synthesis. They also lack the membrane specializations associated with electrolyte transport and are much less well supplied with blood vessels than are the more distal portions of the duct system [29]. On the basis of their morphology, it may be concluded that the intercalated ducts serve primarily as conduits, the saliva flowing through without sustaining major changes in composition. While the evidence to date supports this conclusion, the possibility that these cells may play a role in adding water to the saliva has not definitely been precluded. Although it is usually assumed that the acinar cells have this function, the problem of which cell type of the secretory unit is the main secretor of water is as yet unresolved. This will be discussed further under 'Secretion of water and electrolytes' which follows. Some investigators [30] have suggested that these ducts may also be reservoirs of pleuripotent cells capable of differentiating into acinar cells if demand so dictates.

The intercalated ducts are particularly prominent in the parotid where they are long and abundant, accounting for approximately 5% of the cell volume. In the sublingual glands, on the other hand, the intercalated ducts are extremely short and poorly developed (or, according to some investigators, entirely lacking).

In mammals other than rodents, the transition from intercalated to striated duct is abrupt, with no intervening cells of intermediate structure. Under the light microscope, striated duct cells are seen as tall, columnar cells, not clearly demarcated from one another and showing characteristic basal striations (fig. 2/1). Under transmission electron microscopy, the striations are found to consist of abundant columns of packed, vertically-oriented, rod-shaped mitochondria alternating with elaborate infoldings of the plasma membrane. Cytoplasmic processes or pockets have thus been formed which are occupied by mitochondrial columns. Three-dimensional analysis has shown the individual striated duct cells to be incredibly complex in form, their arrays of cytoplasmic processes being extensively plicated in turn so that they interdigitate with neighboring processes.

Such infolding of the plasmalemma and localized abundance of mitochondria is characteristic of epithelial cells engaged in electrolyte transport (as, for example, the gastric parietal cells). As will be seen further on, micropuncture studies have shown that the cells of the striated duct are no exception to this rule, being responsible for the changes in electrolyte concentration which occur as the secretory rate changes. The presence of at least half a dozen capillaries paralleling each striated duct – conveying blood in a direction retrograde to the salivary flow – facilitates the rapid removal of resorbed electrolytes. *Suddick and Dowd* [29] determined that, at the capillary/cellular level, the striated ducts were much more vascularized than were the intercalated ducts and acinar units.

Striated ducts are most conspicuous in the submandibular gland, where they are branched and longer than in the parotid. They are extremely short (only a few cells long) and thus not commonly seen in the sublingual glands.

Yet another cell type found in association with both the secretory end-pieces and the intercalated ducts is the myoepithelial ('basket') cell (fig. 2/1). Because they are usually alkaline-phosphatase- and ATPase-positive [28], these contractile cells – and their three-dimensional form – are recognized with little difficulty in light microscopic preparations stained for these enzymes. Their shape varies with their location. Myoepitheliocytes associated with the acini are stellate, with four to eight long, branching cytoplasmic processes extending from a central perikaryon. These tentacle-like processes, together with those of one or two other such cells, encircle the acinus forming a basket-like network. Myoepitheliocytes associated with the intercalated ducts are spindle-shaped, their processes seldom branching, and extend along the ducts longitudinally. Both types of myoepithelial cells and their processes lie between the acinar or duct cells and the adjacent basal lamina.

Ultrastructurally, the cytoplasm of the myoepithelial cells shows an almost point-for-point correspondance with smooth muscle cells (see 'Muscle of the GI System', chapter 1). Furthermore, both cell types respond to the same pharmacologic stimuli [30] and share common antigenic determinants [1].

The effects of the contraction of the myoepitheliocytes has been summed up as follows [12]: it speeds up the outflow of saliva, reduces the luminal volume of the intercalated ducts and secretory end-pieces, contributes to the secretory pressure, supports the underlying parenchyma, helps salivary flow to overcome increases in peripheral resistance, and may, in certain circumstances, help to expel the contents of the associated secretory cells. In short, the myoepithelial cells are capable of ultimately causing a gush of saliva into the mouth.

The Excretory Ducts. The interlobular excretory ('collecting') ducts have a pseudostratified columnar epithelium containing occasional goblet cells. The main excretory ducts, extending from the hilus of the gland to the oral cavity, have a stratified columnar epithelium which changes to the stratified squamous epithelium of the oral mucosa just prior to their orifices. The excretory ducts were long considered only as conduits to convey the saliva to the mouth. Recent studies [35] have shown, however, that in the rat, at least, they too take an active part in transporting salivary electrolytes. Epithelial modifications (perpendicular basal striations containing packed, elongated mitochondria), the high vascularity (blood capillaries here are more closely opposed than in any other part of the duct system) and the retrograde blood supply of these ducts lends morphological support to this functional possibility.

Minor Salivary Glands

Besides the major salivary glands, numerous others empty into the oral cavity. Scattered deep to the mucous membrane of the upper and lower lips and of the cheeks are the small, pea-sized *labial* and *buccal glands,* respectively. Although technically these are mixed glands, the seromucous cells are confined to their blind ends; the mucous cells lying along and secreting directly into their short, unbranched striated ducts provide most of their secretion.

There are several groups of glands associated with the tongue. The *glands of von Ebner* open into the groove surrounding the circumvallate papillae. These are compound tubular glands (see fig. 1/2) consisting of serous cells only. The duct system is poorly developed. Their thin, watery secretion probably serves to both distribute tasteful substances over the taste buds associated with the circumvallate groove and to wash them away. The *anterior lingual glands* (on the ventral surface of the tongue to either side of the frenulum near the apex) are mixed seromucous/mucous glands, while the *posterior lingual glands* (near the root of the tongue) are of the purely mucous variety.

2 Mouth and Saliva

The *palatine glands* (of the soft and posterior hard palate) and the *glossopalatine glands* (located just posterior to the sublingual glands) are also purely mucous glands.

The minor salivary glands have no amylolytic activity. They serve to moisten and lubricate the mucous membranes of the oral cavity; however, even though they are believed to secrete continuously, they do not secrete enough saliva to keep the mouth comfortably moist. A man having no functioning major salivary glands must frequently rinse his mouth with small drinks of water.

Tables 2/I and 2/II summarize much of the preceeding material concerning the structural features of the various salivary glands.

Functions of Saliva

Swallowing is difficult and may be impossible if the mouth is dry. Saliva is the lubricant which makes swallowing easy. It has this property in part simply because it is fluid, but also because it contains mucin which is the universal lubricant throughout the respiratory and digestive tracts.

Mucins are glycoproteins; some of these are phosphorylated and others sulfated. It is these glycoproteins which make saliva slimy. These, in the cells which produce them, following the sequence of events outlined in chapter 1. Glycosylation takes place in the Golgi region where droplets, which are stored in the apical region of the cell, are formed. Discharge of these seems to result from rupture of the cell membrane (apocrine secretion). Amylase granules by contrast are secreted as a result of a merger between their membranes and the cell membrane. This is merocrine secretion.

A second function of saliva is taste. All substances tasted must be in solution to reach or activate the taste buds. Those not already in solution are solubilized in saliva. For the survival of species, taste is a sensation of immense biological importance since it is by virtue of this sensation that species are able to avoid potentially poisonous food. Nearly all naturally occurring pharmacologically active alkaloids are bitter and are avoided by foraging animals.

Saliva and Digestion. In man and in animals, other than carnivores, the secretion from the major salivary glands (parotids and submandibular) contains the amylolytic digestive enzyme amylase. Amylase zymogen granules are a prominent feature of the resting parotid acinar cells and of the demilune cells of the submandibular glands. Following vigorous secretion of amylase rich saliva these stored granules disappear. The only other digestive

Table 2/I. Comparison of the major salivary glands [modified from *Finerty and Cowdry,* 11]

	Parotid	Submandibular	Sublingual
Size and shape	largest; main and accessory parts both encapsulated; compound, branched tubuloalveolar	intermediate; well-limited and encapsulated; compound, branched, tubuloalveolar	smallest; major gland and several minor ones; no capsule; compound, branched, tubuloalveolar
Position	filling retromandibular fossa between mandibular ramus and sternomastoid/ext. aud. meatus	deep (medial) to mandible extending around posterior border of m. mylohyoideus	below mucosa of the floor of the mouth
Secretory epithelium	seromucous alveoli; mucous alveoli rare (except in the neonate)	seromucous alveoli predominate; some mucous alveoli have seromucous demilunes	major gland: mucous alveoli predominate; many seromucous demilunes and some alveoli minor glands: nearly all mucous
Secretory granules	PAS-positive (indicates carbohydrate-protein polymer)	seromucous cells: PAS-positive rich in sialo-glycoproteins although some cells contain sulfated polysaccharides mucous cells: either sialomucin or sulfomucin or a mixture of both	seromucous cells: sulfated glycoproteins mucous cells: sulfated polysaccharides
Intercalated ducts	narrow, branching channels made of a single layer of cuboidal cells; long and abundant here	similar structure but much shorter and therefore less conspicuous	absent
Striated ducts	single layer of tall, basally-striated columnar cells; less elaborate than intercalated ducts here	same structure but somewhat longer, branched and may contain yellow pigment; most highly developed here	rare or absent
Main excretory ducts	parotid (Stenson's) duct – opens opposite second upper molar; stratified columnar cells on marked basement membrane (changes to stratified squamous just before orifice)	submandibular (Wharton's) duct – opens on caruncula on either side of frenulum of tongue; structure same	major sublingual (Bartholin's) duct opens near submandibular, sometimes by common aperture; also several minor sublingual (Rivinian) ducts; structure same.

Table 2/I (continued)

	Parotid	Submandibular	Sublingual
Interstitial tissue	fat cells abundant in stroma		connective tissue septa most abundant and conspicuous here
Nerve supply	sensory: trigeminal (V) n. via auriculotemporal + great auricular n. (to capsule) autonomic: (1) sympathetic: sup. cervical ganglion (vasoconstriction); (2) parasympathetic: glossopharyngeal (IX) n., otic ganglion	sensory: trigeminal (V) n. autonomic: (1) sympathetic: same; (2) parasympathetic: facial (VII) n. via chorda tympani, sub-mandibular ganglion	sensory: trigeminal (V) n. autonomic: same as for submandibular gland

Table 2/II. Minor salivary glands [reproduced, with permission, from ref. 16]

Name	Location	Type of gland
Labial glands	upper and lower lips	mixed
Buccal glands	cheeks	mixed
Anterior lingual glands	tip of tongue	mixed
Glands of von Ebner	near circumvallate papillae	serous
Posterior lingual glands	root of tongue	mucous
Palatine glands	palate	mucous

enzyme occurring in saliva is a trace of lipase secreted by the small von Ebner's glands at the base of the tongue.

Because the sojourn of food within the mouth is brief, oral digestion is unimportant, but both salivary amylase and lipase do exert their digestive action within the stomach where meals remain for a much longer time. This will be dealt with in chapter 4. The conditions within the mouth are right for amylase activity, should food be held there: the pH is over 4.5 and there are halogen ions (Cl⁻) present. Indeed, if saliva is added to a suspension of starch in a beaker in a very short time the blue starch iodine reaction will no longer be obtainable. The amylolytic action of saliva has been and still is used in some primitive communities to obtain a fermentable substrate for yeast in the manufacture of beer from grain. The biochemical characteristics of salivary amylase will be considered with pancreatic amylase in chapter 6 as the two are identical in action.

It might be suspected that a gland which secretes amylase would increase its output on high carbohydrate diets. Some glands, as will be seen later, adapt in this manner but not the salivary glands either in response to single meals of varying composition or to dietary regimes lasting months. The only changes seen in salivary composition are the secretory rate dependent changes in electrolyte concentration. Presumably, therefore, a sour orange will cause secretion of a less hypotonic saliva than a mouthful of oatmeal porridge because the former evokes a larger volume.

A curious property of salivary glands, especially the human parotid, is their ability to trap iodine and to establish much higher tissue than plasma concentrations. The glands also are capable of secreting a saliva containing iodine in higher concentrations than in plasma. Stomatitis is a common finding in poisoning with inorganic iodine compounds. Stomatitis and discoloration of the teeth is a common and often diagnostic consequence of heavy metal poisoning, e.g. lead, mercury, silver, bismuth cadmium. This leads to the suspicion that these also are secreted in significant quantities in saliva. Evidence for this, however, is scanty.

Salivary Secretion

Stimuli for Secretion. Since saliva has functions concerned with eating and others related to oral hygiene and swallowing, one would expect continuous secretion at rest and surges when food is eaten. The mouth is always wet; thus, it is clear that saliva is secreted continuously even during sleep, but measurable secretion during sleep has been obtained only from the submandibular gland in man and even this was tiny (0.02 ml/min). Investigations of this sort have been very rare, but secretion is well known to cease under deep anesthesia or coma. Measurements of 'resting' secretion have been much commoner; these have been made in people protected from outside stimuli but, of course, not from intrinsic stimuli. Substantial 'resting' secretion has been obtained (0.5 ml/min, which is 720 ml/day) but because of the impossibility of eliminating internal stimuli (even movement of the tongue will stimulate salivation) these figures are largely meaningless and certainly do not tell us if the major salivary glands have a basal, nonstimulated secretion. The work quoted above on secretion during sleep provides evidence by elimination that the mouth is moistened during sleep largely by the minor glands.

The normal stimuli for salivary secretion by the major glands are both intraoral and extraoral. In the latter category are the sight, smell or thought of appetizing food or anything connected with it. In the former category are taste (e.g. sour substances), mechanical stimulation of the oral mucosa and chewing movements. A tasteless substance like paraffin wax, for example, will cause some secretion. The intraoral stimuli obviously do not need to be appetizing. The impedimenta of the dentist are far from appetizing, yet they cause salivation. To evoke an extraoral or conditioned response an appetizing foodstuff must be involved. Nausea produces profuse salivation.

In man a little over a liter of saliva is secreted per day, most of it with meals. Two-thirds of it originates from the submandibular glands. Between 5 and 10% is contributed by the sublinguals and by the minor salivary glands. The rest is produced by the parotids. Carnivores have relatively smaller daily secretory volumes while ruminants secrete vast amounts of saliva estimated as up to from 50 to 100 liters daily in cattle. In ruminants, saliva is of vital importance nutritionally since it forms the largest part of the fermentation liquor in which ruminal digestion takes place.

Regulation of Secretion [9]. This has been investigated extensively only in the major glands, each of which has both sympathetic and parasympathetic nerve supplies, the latter in the form of discrete nerves or rami to each gland. The glandular rami to the parotid gland come from the auriculotemporal branch of the mandibular division of the trigeminal nerve, while those to the submandibular and sublingual glands come from the lingual branch of the same division (via the chorda tympani). These rami are virtually the entire parasympathetic nerve supply to these glands, but it would be foolish to imagine that there are no others. Indeed, there is substantial evidence that complete section of the auriculotemporal nerves leaves some parasympathetic innervation to the parotid glands. The cholinergic nerves synapse in discrete ganglia outside the substance of the gland so that direct electrical stimulation of the postganglionic fibres to the glands and surgical extirpation of the synapse is possible. Elsewhere within the gastrointestinal tract these synapses occur in diffuse ganglia arranged within the tissue innervated (enteric ganglia).

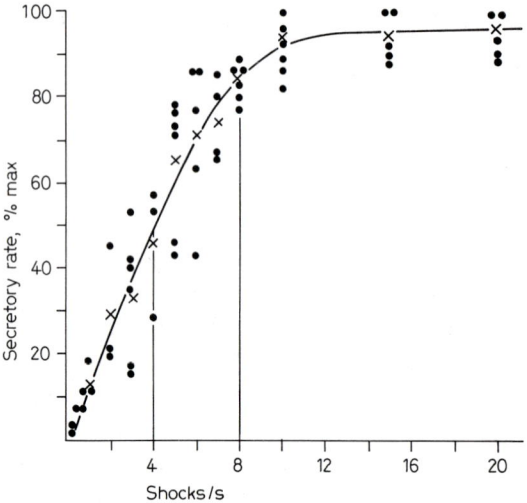

Fig. 2/2. The mean response of the submandibular glands of seven anesthetized dogs to electrical stimulation of the chorda-lingual nerve at various frequencies expressed as per cent of the maximal secretory rate. Reproduced, with permission, from *Emmelin and Holmberg* [9].

The sympathetic nerve supply to the glands arises from the upper thoracic cord, synapsing in the superior cervical ganglion; fibers then follow the arterial supply to the respective glands.

Denervation of the glands ends their secretion. Adequate doses of ganglionic blocking agents or anticholinergic drugs will also abolish secretion. Dryness of the mouth following therapy with antihistamines or antispasmodic agents, many of which are anticholinergic, is a common experience. Stimulation of the parasympathetic rami evokes the highest flow rates obtainable from the glands and section of these nerves reduces reflex stimulation of the glands (e.g. citric acid placed in the mouth) almost to zero followed in the course of time by atrophy. Atropine, the prototypical anti-cholinergic drug, does likewise with the difference that the reflex stimulation is *completely* abolished indicating that the innervation missed when the parasympathetic rami are sectioned is cholinergic also. In some glands in some animals, a single shock to the parasympathetic rami will elicit secretion, but slow repetitive stimulative (6–8 Hz) is usually necessary (fig. 2/2). Effective stimulation has been shown to be accompanied by an increase in acetylcholine in the venous effluent and agents which prevent the synthesis of acetylcholine (hemicholiniums) or its release abolish the secretory response to parasympathetic nerve stimulation.

The usual method by which autonomic nerves contact their effectors is known as *en passant* (in passing). The nerves ramify through the tissue surrounded by Schwann cells, the continuity of which is frequently interrupted by varicosities of the nerve axon which are packed with secretory vesicles. These are the nerve side of the en passant junction. The distance between varicosity and cell in the salivary glands of cats has been reported as 1,000 Å. In some instances, closer contact (200 Å) called *interacinar* or *direct contact* has been claimed. The most acceptable view from electron microscopic studies is that each secreting cell receives from 2 to 5 en passant nerve junctions. Calculations from electrical studies claim 5–10 contacts per cell.

Stimulation of the cholinergic trunks to the submaxillary glands of cats always results in a change in the transmembrane potential of the secretory cells. This is usually an increase in potential difference. Increasing the strength of stimulus from the lowest to the maximal usually results in a stepwise increase in cellular transmembrane potential. *Lundberg* [17] has claimed that there are usually 5–10 such steps. This, if correct, is an example of convergence. It is different from the convergence seen in the spinal cord since subliminal stimuli administered to two or more divisions of the nerve will not evoke secretion from the quiescent gland, but if one nerve is causing secretion the subliminal stimulus in the other will greatly augment secretion. This means that there is no overlap between the territories of individual preganglionic fibers; that one fiber innervates a string of cells (as one would expect of the en passant arrangement); and that transmitter (acetylcholine) can leak from cell to cell.

It looks, therefore, very much as if the stimulation of the salivary glands is a wholly cholinergic affair. There is no doubt that cholinergic agents and stimulation of cholinergic nerves do increase salivary secretion, and blockade does the opposite. The role of the sympathetics is much less clear. In some glands in some animals, sympathetic stimulation will cause a protein-rich secretion. In other glands, often in the same animals, no secretion at all is produced. In man, those circumstances which are associated with epinephrine release (fear, anger, and also medication with catecholamines) universally produces dryness of the mouth. A sympathetic nerve supply to the acini of certain glands can be detected, but this histological finding bears no relationship to a given gland's responsiveness to sympathetic stimulation. In glands which do respond to sympathetic stimulation, it is likely that each acinar cell receives both sympathetic and parasympathetic nerve supplies. Secretory potentials during intracellular recording from single cells of responsive glands can be evoked by stimulation of either system and sympathetic stimulation during maximal parasympathetic stimulated secretion will not increase secretion further. Reference has already been made to the myoepithelial cells which surround the acini. These appear to be contracted by sympathetic stimulation and thus secretion might be expressed from the acinar lumina. There is evidence for this in the submandibular gland of the dog where mediation seems to be α-adrenergic. Electron microscopy suggests that both adrenergic and cholinergic fibers supply secretory elements of the major salivary glands. Despite the apparently trivial role played by the sympathetics, they are the subjects of a great deal of research in rats. In this species, sympathetic stimulation and B-adrenergic agents increase protein secretion regularly with concomitant increases in cylic AMP. The role of this material in the subsequent intermediary steps in the manufacture, packaging and secretion of proteins is a very popular research exercise at the moment (see chapter 1). There have been numerous attempts made to show the existence of hormonal regulation of salivary secretion. These have all failed.

Secretion of Water and Electrolytes [26]. Some animals (ruminants for example) produce saliva which is isotonic with plasma, but many others including man produce a hypotonic secretion from the parotids and submandibulars at least. In most animals studied the sublinguals and the minor glands secrete an isotonic saliva. We will concentrate our attention on hypotonic salivas in the following because this is the product of the major glands, which produce most of the saliva in man and in the common experimental animals. Hypotonic secretions, moreover, are unique. The only other such secretion is sweat and the sweat glands have duct cells which resemble those of striated ducts very closely.

The most widely accepted theory for the secretion of water and electrolytes, in man and animals that produce a hypotonic saliva, is that of *Thaysen* et al. (fig. 2/3). This postulates the elaboration of an isotonic primary acinar fluid followed by the addition of potassium and bicarbonate and the reabsorption of sodium, chloride and bicarbonate, to a lesser extent, as it passes down the ducts, which are considered to be virtually impermeable to water. Since the reabsorption of sodium is far in excess of the addition of potassium, this results in a hypotonic secretion. At rapid rates of

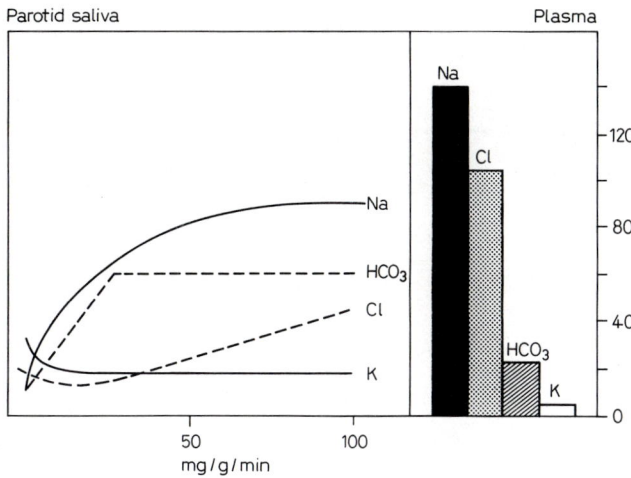

Fig. 2/3. Electrolyte composition of saliva at various flow rates (expressed as milligrams of saliva per gram of salivary gland per minute) compared with plasma. Reproduced, with permission, from *Bro-Rasmussen* et al. [6].

secretion, the primary secretion passes through the ducts too quickly to allow full sodium reabsorption, hence the saliva is less hypotonic at high secretory rates than it is at low ones. The reabsorptive capacity of the striated and collecting ducts reaches its maximum at approximately 50% of the maximal secretory rate; beyond this rate the osmotic pressure increasingly approaches that of plasma. It never reaches that of plasma, of course, because even at maximal secretory rates the ducts are still actively and maximally reabsorbing sodium ions. This is reflected clearly in the energy expenditure or oxygen consumption of the glands. It rises rapidly to about 50% of the maximal rate and thereafter much more slowly. This indicates that the tubular reabsorption requires much more energy than does the elaboration of the primary secretion.

Micropuncture techniques [27] have shown that the electrolyte concentration of saliva, in species with hypotonic saliva, decreases from isotonicity as it flows through the duct system towards the mouth. Microperfusion of ducts has shown that they have, indeed, a low permeability to water (fig. 2/4). A deficiency in the micropuncture studies is that the site of the puncture is always uncertain and that so far no one seems to have succeeded in placing a micropipette into an acinar lumen. The ducts can be poisoned

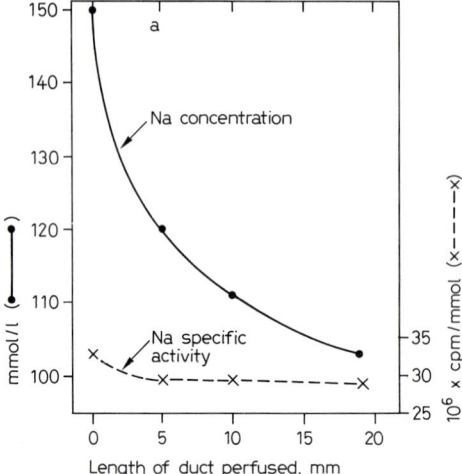

Fig. 2/4. The relationship between the length of salivary duct perfused and the Na concentration of the perfusate. Isotonic NaCl containing Na was used at a perfusion rate of 940 ml/min. Reproduced, with permission, from *Schneyer* [27].

by retrograde perfusion with mercuric salts. This ends their ability to produce a hypotonic saliva but does not impair the volume secreted. On the other hand, injury to the acini, be it produced by duct ligation, mercuric salts or X-rays, does diminish flow rates, but does not interfere with the ability of the gland to reabsorb sodium as long as the ducts remain undamaged.

This proves to us that the acini are the source of salivary water. The composition of the fluid collected after poisoning of the ducts (which is the unmodified primary secretion of *Thaysen*) makes it clear that this is not an ultrafiltrate of plasma even though isotonic with it and free of plasma protein. An ultrafiltrate would contain glucose and potassium in the same concentrations as in plasma; however, the former is virtually absent from saliva and the latter is much higher than in plasma. By 1851 *Ludwig* had already offered substantial evidence that ultrafiltration was unlikely when he showed that the submaxillary gland could secrete against a pressure higher than the arterial. It is clear, therefore, that the elaboration of the primary secretion requires energy and is an active process.

The potassium concentration of freely flowing saliva is approximately double that of plasma and it remains constant regardless of flow rate once flow is well established (fig. 2/3). At very low flows and in the first few ml, when flow is just starting, potassium concentrations are three or four times those of plasma. This passing high potassium phase is called the *potassium transient*. The initial burst of potassium, which is mainly from acinar cells, is derived from an intracellular pool which discharges both into the secretion and the blood, where it can easily be detected. The transient usually lasts about three minutes. The sum of the potassium found in blood and saliva closely matches the decrease seen in the glandular tissue during this period. After the initial burst, acinar potassium rises somewhat and salivary and venous potassium return to steady state levels which, in the former as stated earlier, is about double the plasma potassium.

The movements of ions in and out of cells are, of course, reflected in transmembrane potentials, but unlike excitable tissue (muscle and nerve) secreting cells of the gastrointestinal tract, in which membrane potentials have been measured, do not give a clear picture. In excitable tissue, the potential is the equilibrium potential for the K^+ ion. In resting acinar cells the potential is much too small to be an equilibrium potential for K^+ (20–35 mV inside negative) as the concentrations of K^+, Na^+ and Cl^- within the acinar cells are not different from other cells. The concentration of K^+ within the cell is thus maintained there against a large electrical gradient of 65 mV since the K^+ equilibrium potential for these cells otherwise would be −90 mV as in muscle and nerve. The hyperpolarization following stimulation is simply the tendency of the transmembrane potential to approach its resting level effected by the outward movement of K^+. Hence, the K^+ transient. Once the new equilibrium level is reached, secretion continues. When it stops, the acinar cells fill up again with K^+ ions and the transmembrane potential difference falls. In some glands, stimulation to secrete unexpectedly produces hypopolarization of varying magnitude from −35 to −6 mV, the size varying directly with the strength of the stimulus.

The origin of the K^+ in the steady state secretion is different from that in the transient. Its constancy suggests that it is of ductular origin. Micropipette and microperfusion studies are consistent, in the main, with this idea [34]. The concentration of K^+, for example in rat submandibular gland, has been shown to increase as fluid passes from the intercalated ducts to the opening of the main duct into the mouth. When the main excretory duct is perfused, Na and Cl leave the perfusate and K^+ is added to it [27]. This must be an active process since the K^+ concentration is higher than in the plasma. There is evidence that at least some of the Na^+ absorbed is in exchange for K^+ entering the ducts.

Bicarbonate ions behave unlike K^+ and Na^+ in that the concentration rises with flow rate, but reaches its plateau, which often exceeds plasma levels, before Na^+ (fig. 2/3). This again must be an active process. The Cl^- ion moves passively with Na^+. Some HCO_3^- may be absorbed passively also and indeed there is evidence for an exchange of HCO_3^- for Cl^-, but active secretion must be invoked to explain the early rise in HCO_3^- with increasing secretion and the plateau ultimately reached.

Salivary Blood Flow

In every gland when secretion starts blood flow increases. This is very evident in the major salivary glands. Stimulation of the auriculotemporal nerve to the parotid gland will increase the venous outflow 5-fold (to the extent indeed that it becomes arterial in color). Claude Bernard noticed this and ascribed it to specific vasodilator nerve fibres. If, however, there are specific vasodilator fibres their mediator is a puzzle as atropine will prevent nerve stimulated secretion but not nerve stimulated vasodilatation (fig. 2/5). Acetylcholine on the other hand will stimulate neither secretion nor vasodilation after atropine, and botulinun toxin, which blocks acetylcholine release, prevents the stimulating action of nerve stimulation on both.

Forty-eight years ago *Ungar and Parrot* [32] studied the fairly well known and then topical vasodilator activity of saliva itself. They attributed this to *kallikrein*. Others failed to find this factor in the venous blood from the gland. The scheme ultimately drawn up by *Hilton and Lewis* [13] is that when the gland is stimulated via its nerve an enzyme (kallekrein) escapes from gland cells into the interstitial fluid where it liberates a vasodilator peptide from plasma protein. The peptide is called *kallidin* or *bradykinin*. Such substances do exist in many tissues, including glands. Their structures are known and they are liberated as *Hilton and Lewis* have proposed. This makes a nice scheme to describe the vasodilatation in active glands in general, but it has been hotly disputed. Even in its original form there are difficulties such as the failure of acetylcholine to produce vasodilatation after atropine. Modern assay methods [10] have now resolved a major drawback which was that bradykinin could not be detected in the venous blood. Now it can be. It does rise following secretory nerve stimulation, but not proportionately with the vasodilatation. Some indeed claim that its concentration is depressed after atropine.

So whether or not there are specific vasodilator fibres to the salivary glands is still a moot point. The old evidence for these is the most convinc-

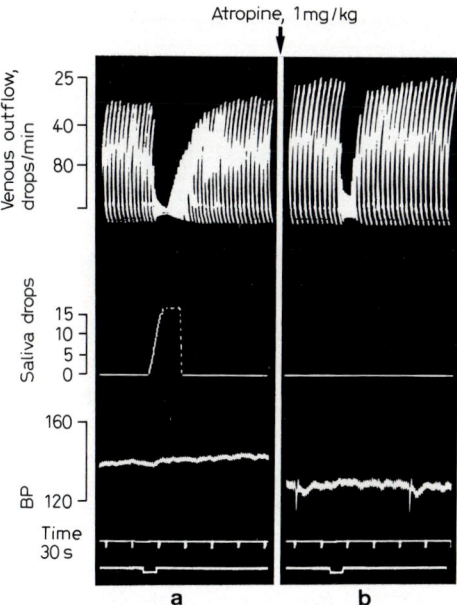

Fig. 2/5. From above down; venous outflow, salivary secretion and systemic blood pressure before (a) and with (b) atropine. The venous outflow device measures the time taken to fill a recording receptacle, therefore, the flow rate is inversely proportional to the length of the line. Reproduced, with permission, from *Hilton and Lewis* [13].

ing: weak nerve stimulation of secretory nerves can produce vasodilatation without secretion and peripheral stimulation of the nerves 3–5 days after section produces secretion without vasodilatation (as was claimed in 1929). The great *Bayliss* [3] could find no correlation between oxygen consumption and blood flow. One would expect a correlation if raised blood flow resulted indirectly from some product of the metabolism of the active gland.

The role of the sympathetics is, in contrast to the above, straightforward. Stimulation depresses blood flow and stimulation during copious secretion will depress that also; whether directly or consequent to vasoconstriction is unknown.

Saliva and Oral Hygiene

Structures within the mouth are not as susceptible to infection as are other tissues. Bacteria do not multiply easily in the mouth, unless salivary secretion stops [20]. When this happens the mouth becomes fetid and teeth will decay. Sterility of the mouth is impossible to achieve and, fortunately, not necessary for dental procedures. This is because saliva contains substances which suppress bacterial growth. The thiocyanate ion is one of these. Saliva contains direct antibacterial activity both bacteriostatic and bacteriocidal. It also has indirect activity in that it prevents binding of bacteria to both teeth and mucous membrane. The mechanisms by which these are effected are by the presence of immunoglobulins in saliva, by large glycoproteins called lectins capable of aggregating bacteria (the minor glands seem to have the most important role in this mechanism), by enzymes (lysozyme and lactoperoxidase) and by lactoferrin. Salivary bacteriocidal activity has often in the past been linked to the thiocyanate ion ($OSCN^-$) which it contains. This ion is an essential part of the lactoperoxidase system.

Taste [22, 23]

Taste and flavor are often confused, but they are quite different. Flavor is appreciated by the olfactory mucosa and is, therefore, really smell. Indeed, in some languages the word for flavor and aroma is exactly the same. For olfaction to occur, it is immaterial whether substances are in solution or not; for taste solution is absolutely essential. Sugar and salt would be absolutely tasteless unless in solution. If there is no other source of water, the salivary glands provide it.

Taste Buds

In the adult, the peripheral organs of taste (the *'taste buds'* or 'gustatory caliculi') are found for the most part in relation to the papillae on the dorsum of the tongue. They are also found on the oral surface of the soft palate, the epiglottis, the fauces and the posterior wall of the oropharynx of infants and children, but these undergo atrophy at a rate which increases as one ages. Those in the region of the root of the tongue and epiglottis are lost early in life. By age 45, both the number and sensitivity of the taste buds have been greatly diminished.

Most of the surface of the tongue is covered with papillae, but only the fungiform papillae at the tip and sides and the foliate and circumvallate papillae at the back of the tongue bear taste buds. They are usually multiple on those papillae which bear them, dotting the tops of the fungiform and the sides of the foliate and circumvallae papillae. Since the fungiform are the

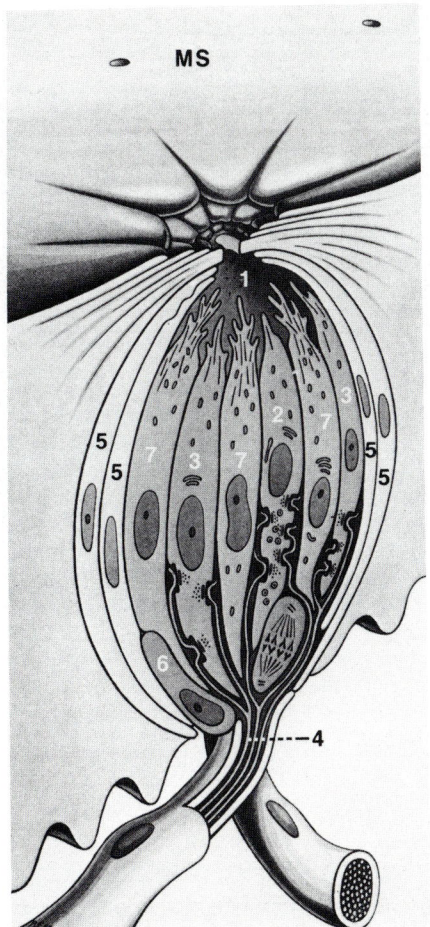

Fig. 2/6. The taste bud, schematically reconstructed. The mucosa has been sectioned perpendicular to its surface (MS), the cut passing through the taste bud and its overlying gustatory pore and apical cavity (1). Two types of sensory cells (2 and 3), one with dense-cored vesicles thought to contain catecholamine (2), are depicted. Nerve fibers (4) make multiple synaptic contacts with both types. Peripheral cells (5), basal cells (6) (one undergoing mitosis is not numbered) and supportive ('sustentacular') cells (7) are also demonstrated. Reproduced, with permission, from *Williams and Warwick* [33].

most accessible, most physiological studies have been carried out on them.

Taste buds are spherical or barrel-shaped bodies composed of modified epithelial cells (fig. 2/6). Under light microscopy, these generally appear as

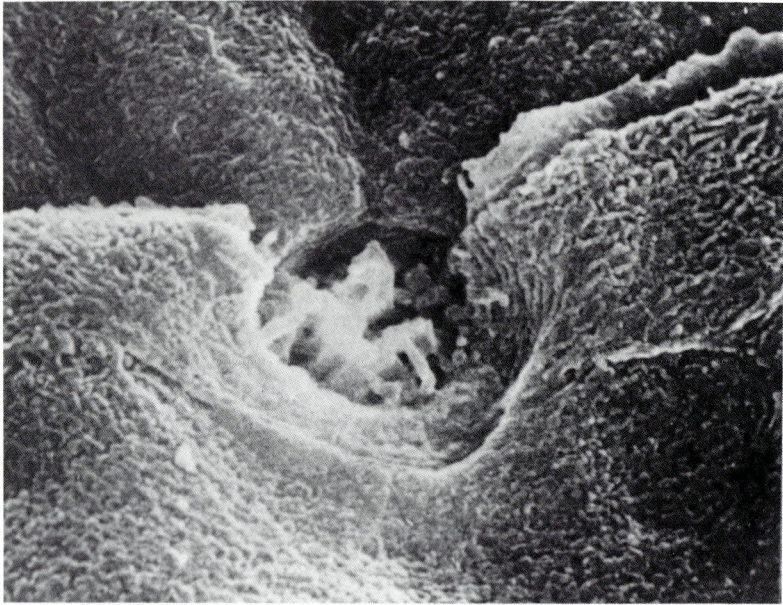

Fig. 2/7. Looking through the gustatory pore unto the apical cavity of a taste bud. Note the microvilli ('gustatory hairs') extending into the apical cavity. Reproduced, with permission, from *Bloom and Fawcett* [4].

ovoid collections of pale-staining cells within the darkly-staining oral epithelium. The elongated cells of the body of the taste bud appear crescentic when sectioned longitudinally. Their pointed apices converge on a small cavity which is filled with an amorphous fluid and which communicates with the cavity of the mouth through a *gustatory pore* (fig. 2/6, 2/7). The apices of some of these cells bear fine microvilli ('gustatory hairs') which protrude into the polysaccharide-rich fluid of the small cavity. Only substances dissolved in this fluid can reach the surfaces of the sensory cells and thus be tasted.

Unlike the sensory end-organs of the skin, which are modifications of the afferent nerve axons, the gustatory sensory cells are separate, self-contained, modified epithelial cells ('neuroepithelial cells'). They establish synapse-like contacts with the terminal axons of the afferent gustatory nerve fibers. They do degenerate, however, if the afferent nerve fibers are severed and they reappear when the nerve fibers regenerate. Each sensory cell bears about 50 such synapses, and a single axon may ramify to supply many sensory cells in the same and in other buds which may be quite distantly located.

Although most of the foregoing has been long established, there has been – and continues to be – considerable debate as to the identification of the gustatory sensory cells amongst the other types of cell in the taste bud. Three dimensional reconstructions from electronmicrographs of serially-sectioned taste buds [19] indicate five distinct cell types, two of which appear to be receptor cells characterized by the presence of synaptic-like contacts with afferent nerve terminals at various points on their surfaces. One of the two apparent receptor cells also contains vesicles of a size and character similar to those identified in the nervous system as catecholamine-containing vesicles.

The other cell types appear to be nonsensory supportive cells, ensheathing or generative in function: one type, the *peripheral cell,* appears to form a surrounding capsule for the taste bud; another type, the *basal cell,* underlies the basal aspects of the other cells of the bud and is probably blastemal in function, giving rise to new cells of the other types through mitosis (the life span of the receptor cell is only 1–13 days in rats – thus constant replacement is required); the third type, the true *supportive* ('sustentacular') *cells,* insulate each receptor cell from its neighbors and, at their basal end, form sheaths for the afferent nerve fibers which lose their Schwann's cell sheath upon entering the bud. The apical region of these latter cells also includes membrane-bound vesicles containing amorphous material similar in density and texture to the fluid of the taste bud cavity, and thus it is presumed to be a secretory product of these cells. It is surprising to find that the 'gustatory hairs' of light microscopy are found under electron microscopy to be microvilli of the supporting cells rather than of the apparent receptor cells, which bear few microvilli on their apices [18].

The sensory nerve for the anterior ('oral' or 'presulcal') part of the tongue, excluding the circumvallate papillae, is the lingual nerve. The lingual nerve contains fibers of general (exteroreceptive – pain, temperature and touch) sensibility from the mandibular division of the trigeminal (V) nerve and fibers of gustatory sensibility from the facial (VII) nerve (by way of the chorda tympani). The sensory nerve for the circumvallate papillae and the posterior ('pharyngeal' or 'postsulcal') part of the tongue is the glossopharyngel (IX) nerve. Taste buds at the root of the tongue and epiglottis are innervated by the vagus (X) nerve.

Taste Physiology

There are only four true tastes: salt, sweet, sour, and bitter. To be sour a solution must be acidic, that is, it must contain a dissociable H^+ but not all acids are sour. To be salty a substance must be a salt, but again not all salts are salty. Lead acetate is known as sugar of lead, because to anyone foolish enough to taste it, it is sweet. It is possible to distinguish between even those salts which are salty. If there is only one salt receptor, why the difference between NH_4Cl, KCl, $LiCl$ and $NaCl$? This has practical importance in that there is no completely satisfactory substitute for the taste of NaCl. The salty taste is influenced by both anions and cations. The potency of both anions and cations as mimics of NaCl forms a series similar to that discussed under absorption. This series *(Hofmeister)* was drawn up to represent the ability of various ions to precipitate proteins from solution or to adsorb to surfaces. These same considerations apply to acids, since the H^+ ion is responsible for the sour taste, but the tastes of acetic, citric, lactic and hydrochloric acids are unmistakably different from each other. In this case it is argued, in rebuttal, that if smell is eliminated these acids are indistinguishable. Intensity of sourness should be proportional to the dissociation constant, but weak organic acids are much sourer than they should be on this basis.

No such simple structural relationship holds for sweet and bitter tastes, but most toxic plant alkaloids are bitter. This taste has undoubtedly played an important part in the selection of food and therefore in the survival of species.

The human tongue, and that of most other animals tested, is more sensitive to the bitter taste than to any other. The threshold given for quinine in man is 8 μmol/l and for cane sugar is 0.01 mol/l. The thresholds for acetic acid and salt are similar to sugar. Salt presents a special problem in that saliva itself contains NaCl and saliva of course has no taste. Thresholds given for NaCl are usually determined after frequent rinsings with distilled water. Taste buds adapt very rapidly, hence, under normal circumstances, an NaCl solution will seem to be tasteless unless it is more concentrated than is the salt of saliva [21]. The taste buds adapt rapidly to each of the four basic tastes although it does appear at times as if bitter is very persistent. Cross-adaption occurs for sour, sweet and bitter substances, but not at all for the salt taste. Tastes do influence one another. An everyday example of this is the use of sugar to counteract sourness. Bitter substances enhance both salt and sour tastes.

The most heated controversy in taste physiology concerns regional specificity. Certain areas of the tongue are more sensitive to one taste than to others. The back of the tongue is the most sensitive to bitter. If the glossopharyngeal nerve is cut, appreciation of bitter tastes will be greatly impaired. It is generally agreed that the front of the tongue is most sensitive to sweet tastes and the sides to sour. Salty tastes are thought to be without any regional specificity. Thirty or forty years ago it was found that single fibers in the chorda tympani did not respond to single tastes, but usually to all four. This finding gave rise to the pattern theory of taste, i.e. all taste buds respond to a single taste with varying degrees of intensity and this pattern of discharge is interpreted by the brain as a specific taste. A variant of this is the labelled line theory which maintains that the pattern of discharge can be contained in a single fibre (fig. 2/8). The multiple innervation of taste cells by single branching fibres makes this plausible. Workers studying intact taste buds still maintain that the buds themselves do the coding and that there are taste buds sensitive to single basic tastes. Electrical stimulation of buds on the fungiform papillae is held to produce a single taste sensation. A recent study [2] has found that 60% of fungiform papillae tested produced none of the four basic tastes – 7% responded to only one test substance, 16% to two, 9% to three, and 14% to four. It was clear from this study that the most obvious regional specificity was for bitter, the front and sides of the tongue being rather insensitive to this taste.

Taste is most acute at body temperature. The buds become less and less sensitive as their temperature falls. This fact makes the current preference for iced drinks and iced water with potentially tasty meals rather illogical physiologically speaking.

A number of substances have been discovered with which interesting tricks can be played with the sensation of taste. An old one, gynemic acid, suppresses the bitter and sweet tastes in man and also the neural responses to sweet substances in animals. Miraculin is a newly reported taste altering protein. It makes sour things taste sweet and has been used to support the single fibre pattern theory of taste, since fibres designated as sweet, by those who adhere to the theory, fire in response to this protein. Two other proteins of this type are thaumatin and manellin, which like miraculin have been isolated from fruits. The latter are the sweetest tasting substances known to date.

The taste buds in man are sensitive to some blood-borne substances. A time-honoured method of measuring circulation time is to inject sodium dehydrocholate into the median cubital vein. When it reaches the tongue it can be tasted and the elapsed time between arm and tongue measured. Another influence of blood composition on taste is a reputed decrease in the

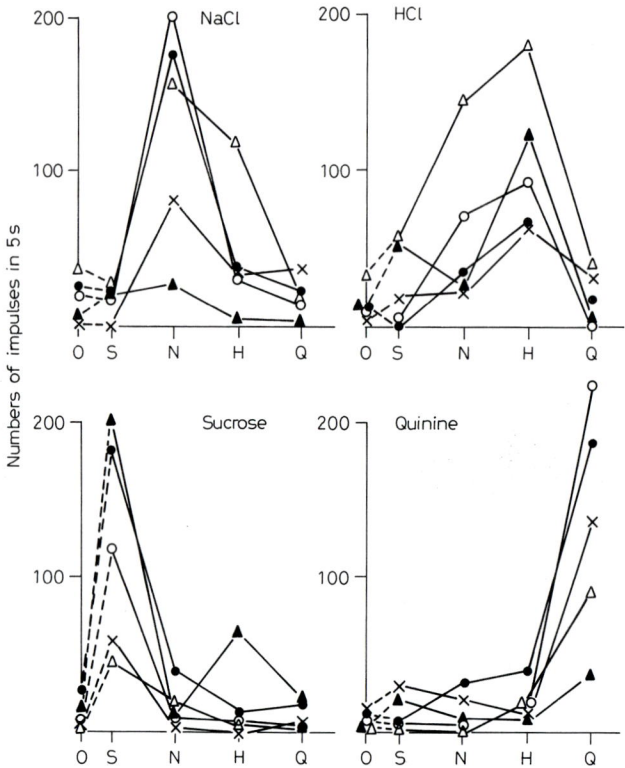

Fig. 2/8. Profiles of response in a macque monkey of fibers SNH and Q to the substances indicated on each graph. The fibers had earlier been most responsive as follows: S to sucrose, N to NaCl, H to HCl and Q to quinine. There were five fibers in each group. The graph shows that only in the case of HCl is there much spread. Reproduced, with permission, from *Sato* et al. [24].

threshold for salt in salt-depleted animals and man. This is used to explain the salt hunger characteristic, at times, of grazing animals. The probable explanation is increased aldosterone followed by decreased [NaCl] in saliva and elevated [K$^+$]. That this occurs in salt-depleted ruminants is well established. The lower [NaCl] of saliva means that the buds are adapted to the reduced salivary salt and consequently their threshold is reduced also.

References

1. Archer, F.; Kao, U.C.Y.: Immunochemical identification of actomyosin in myoepithelial cells of human tissues. Lab. Invest. *18:* 669–674 (1968).
2. Arvidson, K.: Morphological and psychophysical studies on taste receptors of the fungiform papillae in man and monkey (Academisk Avanhandling, Stockholm 1981).
3. Bayliss quoted by Anrep G.; Evans, G.: In the mode of action of vasodilation nerves. J. Physiol. *54:* 10–11 (1920).
4. Bloom, W.; Fawcett, D.: A textbook of histology; 10th ed. (Saunders, Philadelphia 1975).
5. Brooks, F.; Shelly, W.: In Botelho, The exocrine glands (University of Pennsylvania Press, Philadelphia 1969).
6. Bro-Rasmussen, F.; Killmann, S.; Thaysen, J.: The composition of pancreatic juice as compared to sweat, parotid saliva and tears. Acta physiol. scand. *37:* 97 (1956).
7. Burgen, A.; Emmelin, N.: Physiology of the salivary glands (Arnold, London 1961).
8. Burgen, A.: Secretory processes in salivary glands; in Code, Handbook of physiology, sect. 6, vol. 2. Alimentary canal: secretion (American Physiological Society, Washington 1967).
9. Emmelin, N.: Nervous control of salivary glands; in Code, Handbook of physiology, section 6, vol. II: Alimentary canal: secretion, pp. 595–632 (American Physiological Society, Washington 1967).
10. Ferriera, S.; Smaje, L.: Bradykinin and functional vasodilatation in the salivary gland. Br. J. Pharmacol. *58:* 201–209 (1976).
11. Finerty, J.; Cowdry, E.V.: A textbook of histology; 5th ed. (Lea & Febiger, Philadelphia 1960).
12. Garrett, J.; Emmelin, N.: Activities of salivary myoepithelial cells – a review. Med. Biol. *57:* 1–28 [1979].
13. Hilton, S.; Lewis, G.: The cause of the vasodilatation accompanying activity in the submandibular salivary gland. J. Physiol. *128:* 235–248 (1950).
14. Jenkins, G.N.: The physiology of the mouth (Davis, Philadelphia 1966).
15. Junqueira, L.; Carneiro, J.; Contopoulos, A.: Basic histology, 4th ed. (Lange Medical Publications, Los Altos 1983).
16. Krause, W.; Cutts, J.H.: Concise text of histology, p. 269 (Williams & Wilkins, Baltimore 1981).
17. Lundberg, A.: Electrophysiology of the salivary glands. Physiol. Rev. *38:* 21–40 (1958).
18. Miller, R.; Chaudhry, A.: Comparative ultrastructure of vallate, foliate and fungiform taste buds of golden Syrian hamster. Acta anat. *95:* 75–92 (1976).
19. Murray, R.; Murray, A.: The anatomy and ultrastructure of taste endings; in Wolstenholme, Knight, Taste and smell in vertebrates, CIBA Foundation symp., pp. 3–24 (Churchill, London 1970).
20. Nebrun, E.: What is the relationship between saliva and dental caries activity? Proc. Symp. Salivary Glands and Their Secretion, pp. 22–37 ([University of Michigan, Ann Arbor 1972).

21 O'Mahoney, M.: Taste adaptation: The case of the wandering zero. J. Food Technol. 9: 1–12 (1974).
22 Pfaffmann, C.; Frank, J.; Norgren, R.: Neural mechanisms and the behavioral aspects of taste. A. Rev. Psychol. 30: 283–325 (1979).
23 Pfaffmann, C.: The sense of taste; in Magoun, Handbook of physiology, vol. 1, sect. 1: Neurophysiology, pp. 502–533 (American Physiological Society, Washington 1959).
24 Sato, M.; Ozawa, H.; Yamashita, E.: Response properties of macaque monkey chorda tympani fibers. J. gen. Physiol. 66: 781–810 (1975).
25 Schneyer, L.; Schneyer, C.: Inorganic composition of saliva; in Code, Handbook of physiology, sect. 6: Alimentary canal, vol. 22, pp. 497–530 (American Physiological society, Washington, 1967).
26 Schneyer, L.; Emmelin, N.: Salivary secretion; in Jacobson, Shanbour, MTP International review of science, physiology, ser. 1, vol. 4, p. 183 (University Parks Press, London 1974).
27 Schneyer, L. H.: Secretion of Potassium by the perfused excretory duct of the rat submaxillary gland. Am. J. Physiol. 217: 1324–1329 (1969).
28 Shear, M.: The structure and function of myoepithelial cells in salivary glands. Archs. oral Biol. 11: 769–780 (1966).
29 Suddick, R. P., and Dowd, F. J.: The microvascular architecture of the rat submaxillary gland: possible relationship to secretory mechanisms. Archs oral Biol. 14: 567–576 (1969).
30 Tandler, B.: Ultrastructure of the human submaxillary gland. III. Myoepithelium. Z. Zellforsch. 68: 852–863 (1965).
31 Tandler, B.: Microstructure of salivary glands; in Rowe, Proc. Symp. Salivary Glands and their Secretions, pp. 8–21 (University of Michigan, Ann Arbor 1972).
32 Ungar, G.; Parrot, J.: Sur la présence de la callicréine dans la salive et la possibilité de son intervention dans la transmission chimique de l'influx nerveux. C. r. Séanc. Soc. Biol. 122: 1052–1055 (1936).
33 Williams, P. L.; Warwick, R.: Gray's anatomy; 36th British ed. [Saunders, Philadelphia 1980).
34 Young, J.; Fromter E.; Schlogel E.; Hamann, E.: A microperfusion investigation of sodium reabsorption and potassium secretion by the main excretory duct of the rat submaxillary gland. Archs Ges. Physiol. 295: 157–172 (1967).
35 Young, J. A.; Lennep, E. E. van: The morphology of salivary glands (Academic Press, London 1978).

3 Swallowing: Pharynx and Esophagus

The Pharynx

Food, having been mixed with saliva, lubricated and judged by its taste to be worth consuming, is now ready to be swallowed. This entails getting the mouthful of food into the esophagus and then from the esophagus to the stomach. The latter, the *esophageal phase of swallowing,* does not appear to be of impressive difficulty in upright animals like ourselves that have the assistance of gravity, but the former, *the pharyngeal phase,* might appear to present great difficulty. The pharynx is the posterior continuation of the nasal cavity above and the oral cavity below (fig. 3/1). Extending from the base of the skull to the level of the cricoid cartilage, it is continuous inferiorly with both the trachea (via the larynx) and the esophagus. The respiratory and digestive systems thus merge temporarily – and, in fact, cross – in the pharynx. An inhaled mouthful of porridge could easily be fatal; in actual fact inhalation of this sort (aspiration) does occasionally occur but not during swallowing.

Structural Features

The pharynx is subdivided into nasal (nasopharynx), oral (oropharynx) and laryngeal (laryngopharynx) portions (fig. 3/1). The nasopharynx is exclusively respiratory both structurally and functionally. It is the oro- and laryngo-pharynx which functionally and structurally correspond most closely to the general plan of the digestive tube (see 'General Structure of the Gastrointestinal Canal', chapter 1).

In the gastrointestinal tract, events usually occur slowly and often merge one into the other. The exception is the pharyngeal stage of swallowing. We concern ourselves here with events measured in fractions of a second. The structure of the oral and laryngeal pharynx thus reflects the functional demands of such rapid movement and passage. It is hardly surprising, therefore, that the muscular coat (muscularis externa) is com-

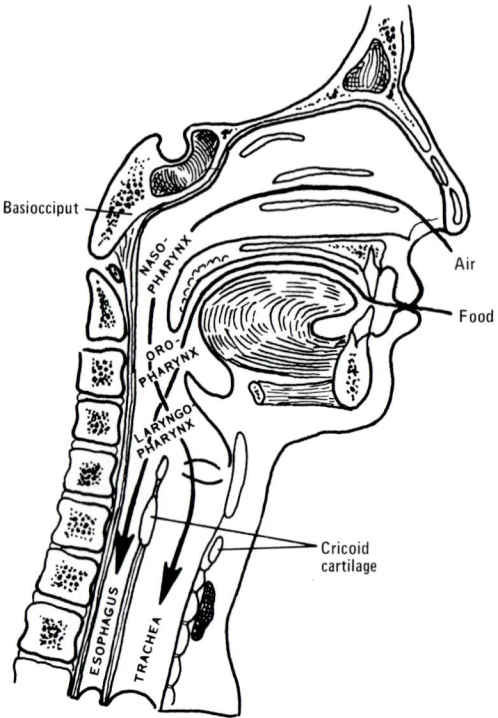

Fig. 3/1. Diagrammatic representation of the medial aspect of the bissected visceral portions of the head and neck, showing passageways for air and food, and the three divisions of the pharynx. After *Brantigan* [5].

Fig. 3/2. The muscular coat (muscularis externa) of the pharynx, as seen from posterior (**A, B**) and lateral (**C**) views. **A** The external, circular layer of musculature formed by the pharyngeal constrictors. The dotted line indicates the midline incision along the fibrous dorsal pharyngeal raphe necessary to display the interior of the pharynx, as seen in **B**. **B** The internal, longitudinal layer of musculature formed by the stylo- and palato-pharyngeus muscles. The muscular coat of the pharynx differs from the general structure of the gastrointestinal canal (see corresponding paragraph, chapter 1) in that (1) it is composed entirely of striated (voluntary) muscle, and (2) the circular and longitudinal layers are reversed in position. **C** In this lateral view, the anterior attachments (origins) of the pharyngeal constrictors are demonstrated. Gaps between the elements of the circular layer are filled with fascial membranes and some longitudinal muscle, but herniations may occur at these sites. **A** is modified from *Becker* et al. [3], **B** is adapted from *Basmajian* [2], and **C** is adapted from *Brantigan* [5].

3 Swallowing: Pharynx and Esophagus

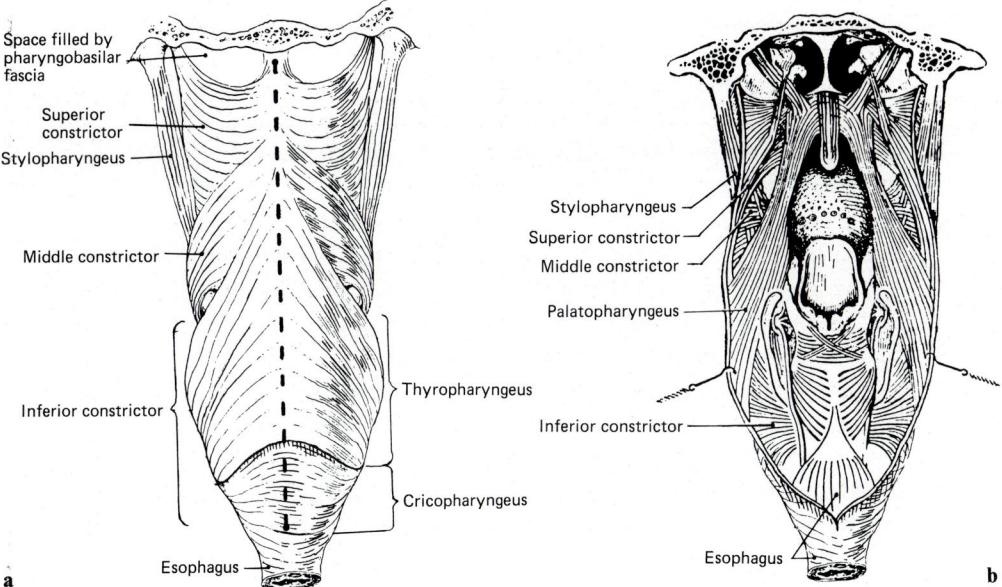

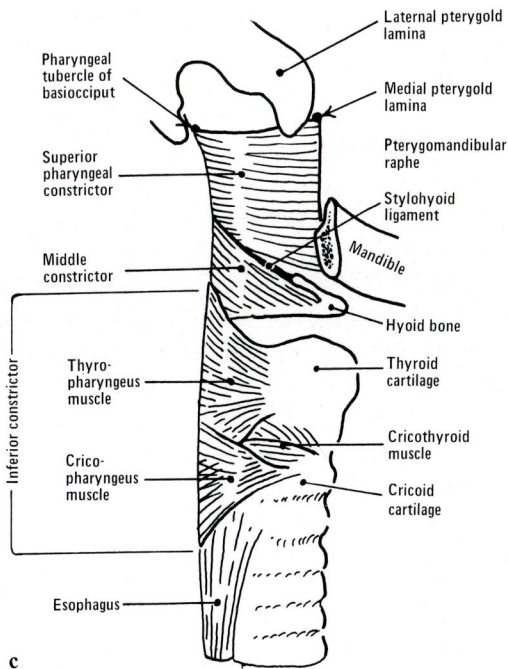

posed of striated (voluntary) muscle: an inner, longitudinal layer formed in part by the stylo- and palato-pharyngeus muscles, and an outer, oblique or circular layer composed of the three pharyngeal constrictors (fig. 3/2). Note that the circular and longitudinal layers of the muscularis externa are reversed in position from that found elsewhere in the GI tract. In order that the effect of the muscular contractions be maximized, the mucous membrane is directly applied to the muscular coat with little if any intervening submucosa, except below the level of the laryngeal aperture where the laryngopharynx is continuous with the esophagus. Since passage of the bolus and protection of the underlying tissues is of primary importance in this region of the digestive tube, the mucous membrane (tunica mucosa) consists of a nonkeratinizing stratified squamous epithelium and a dense, fibroelastic lamina propria. Immediately beneath the lamina propria is a well-developed layer of elastic fibers which is continuous with the muscularis mucosae of the esophagus (fig. 3/3). The abundance of elastic tissue supplies the lining of this part of the GI tract with the pliability and resiliance necessary to accommodate to the rapid movements of the underlying musculature. Glands of the pure mucous type lubricate and moisten the epithelial surface. With the nearly continuous passage of air over much of this surface, the latter function is especially important here. Nearly constant swallowing of saliva also aids in this function.

The Pharyngeal Phase of Swallowing [2, 9, 16]

The bolus of food to be swallowed is squeezed backwards towards the oropharynx between the dorsum of the tongue and the hard palate (fig. 3/4a). This, of course, is intended; in other words, it is voluntary. The tongue does not contract as a unit but rather like a worm with a contraction running from tip to base. At the entrance of the oropharynx, the pressure of whatever is to be swallowed against the palate and the pillars of the fauces sets off the swallowing reflex. From this point on swallowing is involuntary. This is clear in many domestic animals. To get the family dog to swallow a capsule, one only has to place it far enough back in the oral cavity. This initiates the swallowing reflex leaving the animal no option but to swallow its medicine. It is much harder in man since the gag reflex seems to be much more prominent and is elicited from the same region.

The afferent fibers subserving both reflexes travel in the vagal and glossopharyngeal nerves. Proprioceptive nerve endings have not been convincingly demonstrated in the pharyngeal muscles themselves, except for the tensor of the palate (tensor veli palati). Thus we have no knowledge of

3 Swallowing: Pharynx and Esophagus

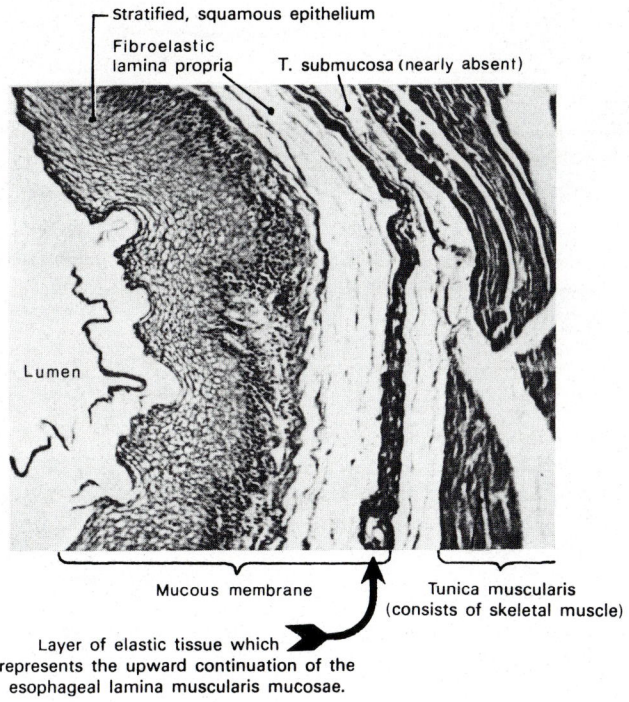

Fig. 3/3. The posterior wall of the oropharynx as seen on sagittal section. The tissue has been stained to demonstrate elastic tissue. Reproduced, with permission, from *Elias* et al. [8].

the position or state of contraction of these muscles except for the tensor palati [1].

The initiation of the reflex results in the contraction of the digastric and geniohyoid muscles which, with other small muscles, elevate and pull the larynx forwards so that its opening lies under the base of the tongue (fig. 3/4b). The rima glottis is closed by apposition of the vocal cords and, in addition, the inspiratory center in the CNS is inhibited. This is seen in decreased firing rates in the phrenic nerve during swallowing. The muscles which elevate and tighten the palate then contract closing off the nasal cavity (fig. 3/4b). This is so effective that there are usually no pressure changes in the nasal cavity during swallowing. The pressures in the oropharynx are substantial (fig. 3/5). The first wave is caused by the tongue. The next is caused by contraction of the muscles of the oropharynx itself. The pharynx

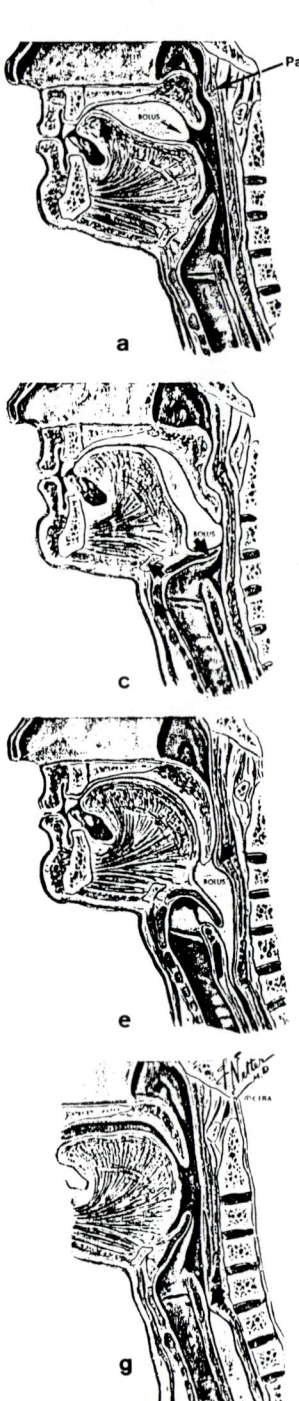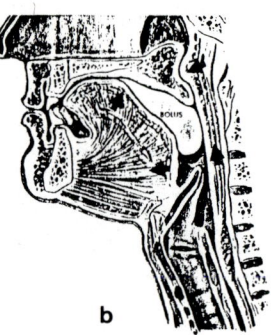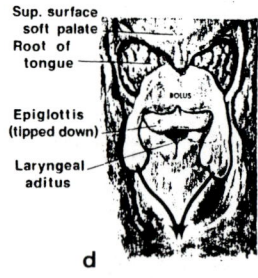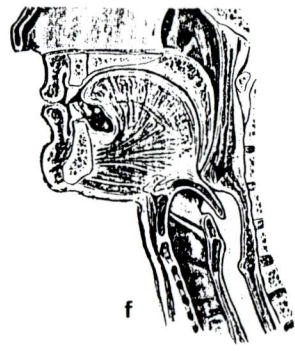

Deglutition

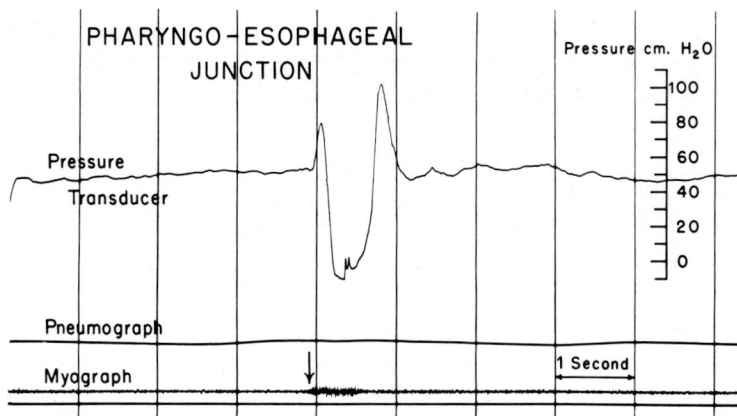

Fig. 3/5. Pressure changes at the pharyngo-esophageal junction in man. The pressure transducer is within the high pressure region of the upper esophageal sphincter. Reproduced, with permission, from *Code* et al. [7].

Fig. 3/4. The oral/pharyngeal stage of swallowing. **a** Pressure between tongue and hard palate drives bolus posteriorly. A constriction (Passavant's ridge) is forming on the post. Pharyngeal wall, marking the post. Boundary between naso- and oro-pharynx. **b** As pressure against hard palate continues, the root of the tongue and larynx are pulled anteriorly; a receptive space for the bolus forms in the oropharynx. Nasopharynx is sealed off. **c** Longitudinal muscles of pharynx (and floor of mouth) elevate larynx and draw pharyngeal wall upward to engulf the bolus. Bolus (and aryepiglottic mm) tip epiglottis downward. Peristaltic (ring) contractions of pharyngeal muscles have begun to migrate downward. **d** Posterior view of anterior pharyngeal wall; bolus has passed through isthmus of the fauces, tipping epiglottis down over laryngeal aditis, but does NOT seal it off. The bolus is thus deflected to either (or both) side(s). Aspiration is prevented by (1) closure of rima glottis; (2) sealed off nasopharynx, and (3) temporary inhibition of the respiratory center of the CNS. **e** As the stripping or peristaltic wave proceeds downward, the longitudinal mm (which elevated the pharynx in **c**) combine with it to pull the soft palate down against the tongue. This action continues in **f**. **f** Relaxation of the cricopharyngeus muscle allows bolus to enter esophagus. Esophageal phase of the swallowing process begins. **g** Soft palate and epiglottis ascend to initial positions. Modified from *Netter* [15].

is now a closed cavity. The nose is sealed off by the soft palate, the larynx is up under the base of the tongue and the opening of the esophagus is still in its resting (closed) position (fig. 3/4c). This is still only about 0.3 s since the beginning of the reflex. Now the muscles of the pharynx begin their peristaltic contraction (fig. 3/4) and the upper esophageal sphincter relaxes (fig. 3/4f).

The important muscles of the pharynx are the superior, middle and inferior pharyngeal constrictors (fig. 3/2a–c). These muscles originate from structures in front of them: the pterygomandibular raphe, the mandible, the hyoid bone, and the cricoid and thyroid cartilages (fig. 3/2c). The two components of the inferior constrictor, arising respectively from the thyroid cartilage and cricoid cartilages, are given separate names (thyropharyngeus and cricopharyngeus (fig. 3/2a). These two do not overlap as completely as do the other components of the pharyngeal muscles. From time to time, a pharyngeal pouch is pushed as a hernia ('pulsion diverticulum') through the thin spot ('Killian's dehiscence') between the two muscles.

Upper Esophageal Sphincter. The pharyngeal constrictors, except for the lower part of the cricopharyngeus, join left side with right as a fibrous raphe down the back of the pharynx (fig. 3/2a). The muscle fibers of the lower cricopharyngeus, which are horizontal, pass as an uninterrupted muscular band completely around the laryngopharynx. This is the most important component of the upper esophageal sphincter; it is not the only component since the adjacent muscles above it also play a part. Thus, by upper esophageal sphincter, we mean all the muscle responsible for keeping the upper esophagus closed between swallows ('pharyngoesophageal sphincter') [18]. (At rest the upper entrance to the esophagus appears as a tightly closed transverse slit). This muscle is striated and has the myoneural junction common to striated muscle. These fire continuously at rest. When the reflex phase of the swallow is 0.2–0.3 s old, impulses cease reaching the sphincter along its efferent nerves as a result of depression of the activity of its vagal cells of origin in the nucleus ambiguous. The sphincter thus relaxes and the pressure between the two sides of the sphincter falls (fig. 3/5). There is no evidence for the existence of inhibitory nerves to any of the striated swallowing muscles. Thus, relaxation of the sphincter is not active; the lumen of the esophagus is actually dragged open by the forward movement of the larynx and trachea, the posterior walls of which form and attach to the anterior walls of the pharynx and esophagus, respectively.

The peristaltic contractions or ring contractions of the pharyngeal muscles start above moving downwards from muscle to muscle at about 15 cm/s (fig. 3/4b–g). It reaches the sphincter muscles in a little more than 0.5 s from the start of the swallow (fig. 3/4g). These sphincter muscles, relaxed to admit the bolus, contract in their turn to push it on into the esophagus, generating a pressure of between 100 and 300 mm Hg before relaxing once more to the resting (which means a closing) pressure of about 50 mm Hg (fig. 3/5). This ends the pharyngeal phase of swallowing.

The Esophagus

The esophagus (= oesophagus) is a tubular food conveyor 20–35 cm (8–10 inches) in length and is the most muscular segment of the alimentary tract (see fig. 3/9). It also is, with the exception of the vermiform appendix, the narrowest region of the alimentary canal. Structurally, it conforms quite closely to the pattern described under 'General Structure of the Gastrointestinal Canal' (chapter 1).

Structural Features
Mucous Membrane and Submucosa. The esophageal epithelium is, like that of the pharynx with which it is continuous, a stratified squamous epithelium – a type which resists abrasion from swallowed material (fig. 3/6). However, the epithelial cells are in fact constantly abraded away and replaced, the continuous renewal evident by the frequent mitotic figures found in the basal layer of the epithelium. In man the epithelium is nonkeratinized, but in species accustomed to a coarse diet (e.g. ruminants) it undergoes extensive cornification. Characteristically, the esophageal epithelium is indented by peg-like protrusions (called 'papillae' by some authors) of the underlying connective tissue of the lamina propria (fig. 3/6). Since the epithelium is thick, there is relatively little need for lymphatic tissue to guard against invasion by pathogenic organisms. Thus, the lamina propria is less cellular than in other parts of the digestive tube, with what lymphatic tissue there is found in relation to penetration of the lamina propria by glandular ducts. Inferiorly, at the gastroesophageal junction, there is an abrupt transition from the stratified squamous epithelium of the esophagus to the simple columnar epithelium of the stomach (fig. 3/7). On macroscopic examination, the boundary separating the smooth, white mucous membrane of the esophagus from the pink, rugous surface of the

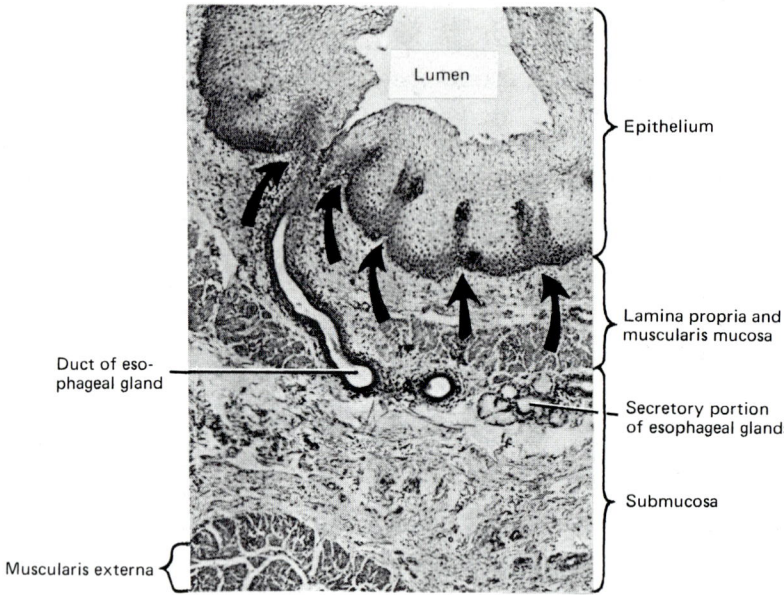

Fig. 3/6. Transverse section of lower one-third of the esophagus. Arrows indicate peg-like protrusions (papillae) of the lamina propria into the thick, stratified squamous epithelium. × 35. Reproduced, with permission, from *Leeson* et al. [13].

gastric mucosa appears as an irregular dentate or zigzag line – the Z line – which is an important clinical landmark in endoscopy.

At the level of the cricoid cartilage, the dense elastic layer of the pharynx (fig. 3/3) is succeeded by the muscularis mucosa. In the upper esophagus, these longitudinal bundles of smooth muscle are loosely arranged in a pattern resembling the irregularly-placed vertical slats of a crude picket fence. Further down the esophagus, the bundles become consolidated into a complete layer (fig. 3/8). Near the stomach, the esophageal muscularis mucosa attains a thickness which is greater than that of any other part of the GI tract (200–400 μm). The esophageal muscularis mucosae is also unique in that the fibers run only in one (longitudinal) direction.

Because of the tonus of the circular layer of the muscularis externa, the muscularis mucosae and the underlying submucosa (tunica submucosa) are thrown into extensive longitudinal folds which result in the characteristically pleated outline of the esophageal lumen in cross-section (see fig. 3/8).

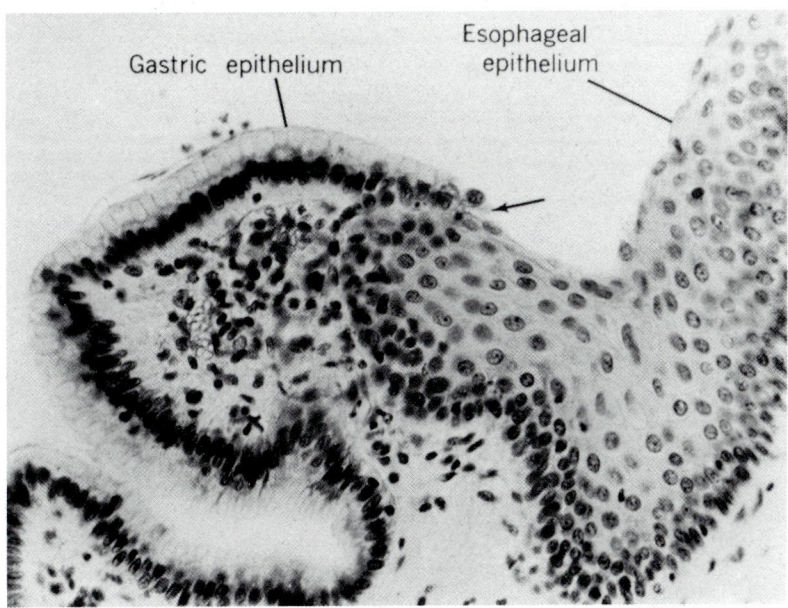

Fig. 3/7. The gastroesophageal junction. The abrupt transition from the stratified squamous epithelium of the esophagus to the simple columnar epithelium of the stomach (and typical of most of the GI tract) is indicated (arrow). HE. × 275. Reproduced, with permission, from *Bloom and Fawcett* [4].

During swallowing, the thick, loose and elastic connective tissue forming the submucosa allows the bolus of food to 'smooth out' these folds, temporarily dilating the lumen to a size sufficient to accommodate the swallowed mass.

Since the bolus has previously been mixed with saliva, little additional lubrication is required. There are, however, two types of glands present in the esophagus. The *esophageal glands* (proper) are small, compound tubuloalveolar mucous glands, unevenly distributed along the length of the esophagus. Their richly branched secretory portions lie within the submucosa. Glands closely resembling the glands of the cardiac portion of the stomach – and therefore called *esophageal cardiac glands* – are clustered at two sites: one group lies superiorly, between the level of the cricoid cartilage and that of the fifth tracheal ring; the other group lies inferiorly, near the gastroesophageal junction. These latter glands, like the former, secrete mucin. Unlike the proper esophageal glands, however, their secretory por-

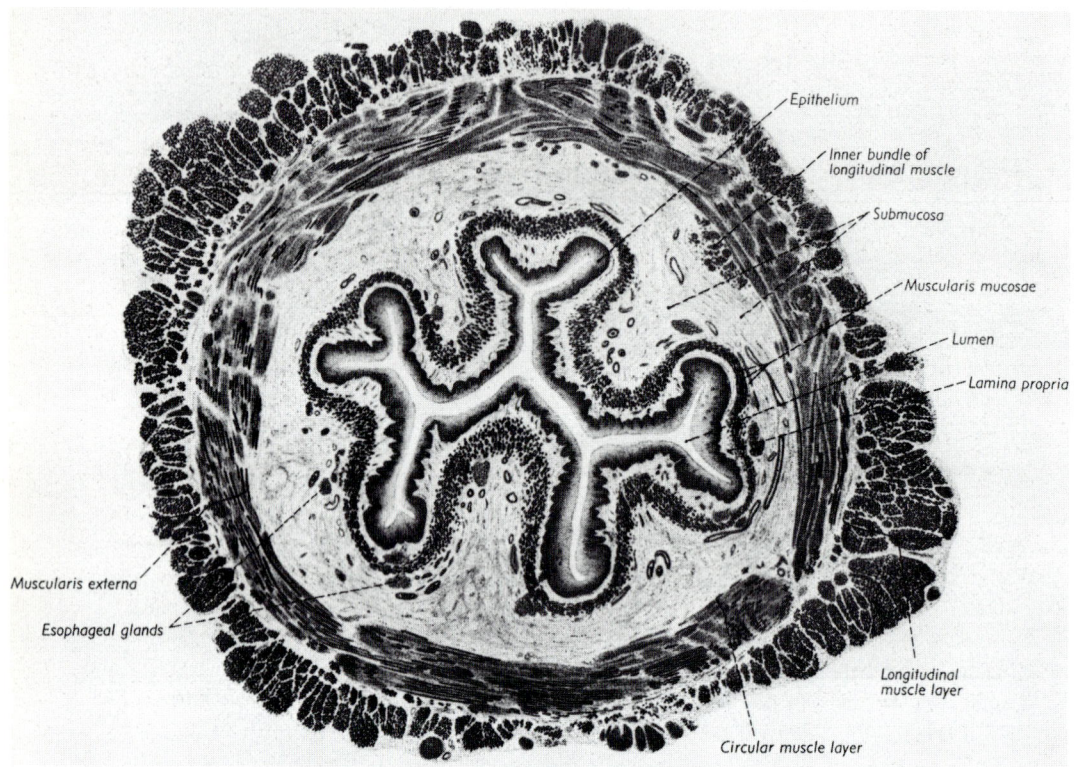

Fig. 3/8. Transverse section through the middle one-third of the esophagus (camera lucida drawing). Reproduced, with permission, from *Hammersen* [10].

tions are confined to the lamina propria. The number, distribution and development of glands in the esophagus is highly variable in man; in some species (e.g. the dog), they are very numerous and may be mixed rather than purely mucous; in other species (e.g. monkeys) they are very few in number; and in still other species (e.g. horses, cats and rodents) they are absent altogether.

Muscularis externa. The muscular coat (muscularis externa) is, functionally speaking, the most important tunic of the esophagus. It is particularly thick here (0.5–2.2 mm), especially its outer longitudinal layer. This thick muscularis, particularly its upper portions of striated muscle, provides the motive power for the rapid propulsion of food from the pharynx to the stomach. Indeed, swallowing is possible even when the muscularis must oppose the force of gravity, as when standing on one's head. The arrangement of the two layers of muscle is less regular than that of other parts of the GI tract. Two layers – a generally longitudinal outer layer and a generally circular inner layer are present as elsewhere, but both may have the fibers placed obliquely or otherwise irregularly in parts or most of the esophagus. In the dog, such irregularities may make it difficult to discern a predominant direction in either layer.

The upper third of the human esophagus is, as are all of the swallowing muscles to this point, striated muscle in both layers. The myoneural junction is typical of that in voluntary muscle, although in this case the innervation is from the vagus (X) nerve. The lower third to half is composed entirely of smooth muscle in both layers. Between these two regions is a transitional zone where there is a mixture of striated and smooth muscle. Often the striated muscle extends farther inferiorly in the longitudinal layer than in the circular layer. The innervation of the smooth muscle also is vagal and is again typical. That is, the cell bodies of the neurons actually terminating on the smooth muscle cells lying not in the medulla as do those of somatic motor neurons but in Auerbach's plexus (between the circular and longitudinal muscle layers) where pre- and postsynaptic parasympathetic fibers synapse as in the remainder of the gut. The esophagus from mouth to stomach serves no digestive function, other than as a simple conduit to the digestive organs.

The Esophageal Phase of Swallowing

The esophagus below the upper esophageal sphincter as far as the lower esophageal sphincter, which will be dealt with later, is flaccid between swallows. Its muscle is electrically silent. When normal, its diameter increases and decreases passively with inspiration and expiration, respectively. Pressures recorded from midesophagus follow intrapleural pressures very closely and even though they are a little higher (fig. 3/9), they are usually taken as the intrapleural pressure because they are so easy and convenient to obtain. Following a barium swallow, an X-ray of the normal esophagus demonstrates indentations where compressed by the adjacent aortic arch and primary bronchus (fig. 3/10).

The muscular composition of the esophagus varies greatly from species to species. It is entirely smooth in birds and entirely striated in dogs and ruminants. The dog, as has often been the case in GI physiology, has been used as the human model, but now the favored model is the oppossum which

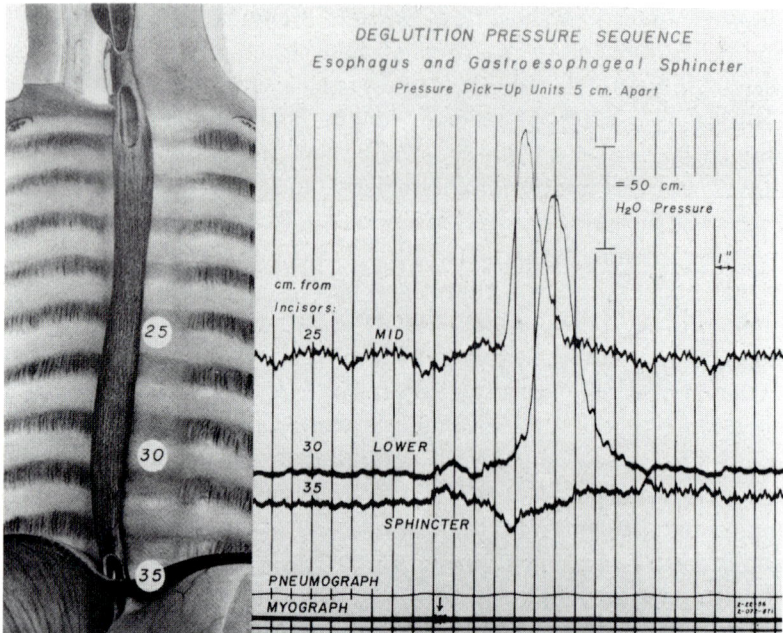

Fig. 3/9. Deglutition pressure sequence with the pressure sensors at the sites shown. Reproduced, with permission, from *Code* et al. [7].

has a muscular arrangement like that in man. Cats and primates also have the human arrangement. The upshot of the displacement of the dog has been that a nice simple story erroneously told of the human esophagus has become complicated.

The simple, descriptive part of the story is that with relaxation of the upper sphincter, the lower one also relaxes (the muscle between is, of course, relaxed at rest). The strong contraction of the upper sphincter which follows entry of the bolus into the esophagus then travels the length of the organ at a velocity of between three and four cms per sec to and involving the lower esophageal sphincter. This is called a *primary wave* since it follows a swallow. It is faster in the midesophagus than in either the upper or the lower parts and it takes 2-4 s to pass. The pressure generated by this wave is between 40 and 100 mm Hg (fig. 3/9). This wave passes from striated to smooth muscle without interruption or irregularity. Because it occludes the lumen and drives food and fluid beyond the lower sphincter, which closes after it, swallowing against gravity in the upside down position is easy. This

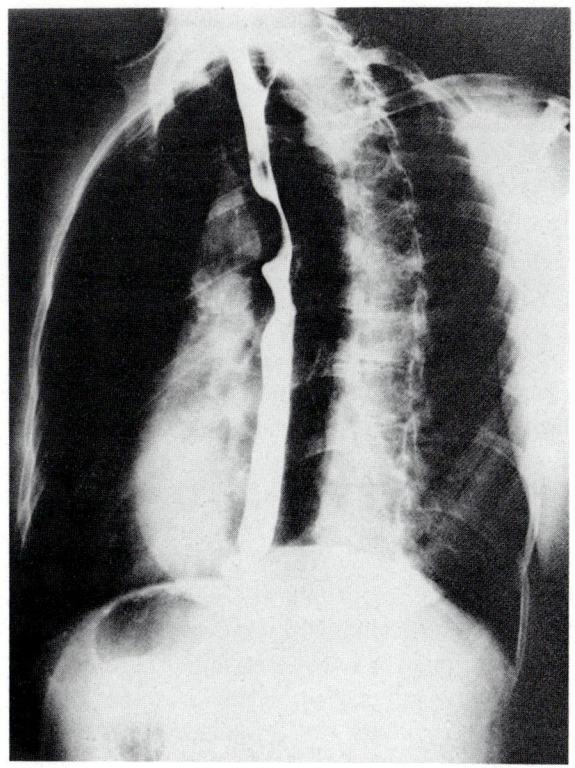

Fig. 3/10. X-ray left anterior oblique projection of normal esophagus showing aortic and bronchial indentations barium swallow. Reproduced, with permission, from *Magee* [14].

propagated wave is dependent on the vagi. If both are cut, the wave will not continue over the denervated muscle. The progressive nature of the wave over the upper striated muscle part depends entirely on sequential firing of progressively lower vagal fibers. This sequence, together with that of the pharynx, depends on the medullary swallowing center. A French physiologist who sutured the central end of the cut vagus to the peripheral end of the cut nerve to sternomastoid noted sequential contraction of that muscle on swallowing when the animal recovered.

Fluids can be swallowed more than once in the time taken for a peristaltic wave to reach the stomach but two primary peristaltic waves cannot exist at once along the length of the esophagus. This is not because there are inhibitory fibers to the striated muscle but simply that the new swallow abolishes all previous activity in the swallowing center and thus activity in the vagal motor efferents ceases. This means that as long as the rate of swallowing exceeds the rate at which primary wave can reach the stomach, the only wave which will make an unimpeded transit of the whole esophagus will be the last one. This presents no difficulty in the upright position as gravity will take fluid through the relaxed lower sphincter, but against gravity rapid swallowing of this sort is impossible. It would be very difficult indeed to chew and swallow solid food fast enough to exceed the esophageal transit time.

There are afferents from the mucosa of the esophagus in both the striated and smooth muscle parts of the esophagus. These, when stimulated mechanically, can induce a peristaltic contraction which will travel from its point of origin to the stomach. This is a *secondary wave;* it also is abolished from above by a swallow. This mechanism is not one which comes into play only if one swallows a balloon, or a large bolus sticks on the way down or something regurgitates from the stomach. It has been shown in cats that this secondary reflex mechanism reinforces each normal swallow. A normally descending bolus will activate the responsible afferents as it passes and, in turn, will increase the impulses travelling down the vagal efferents. In cats local anesthetization of the upper or lower esophageal mucosa has been shown to result in a progressive attenuation of the force of the peristaltic wave. In addition to causing a secondary wave, esophageal distension causes relaxation of the lower esophageal sphincter and a sharp contraction of the upper one. Oftentimes, secondary waves consequent to balloon distension follow one another so rapidly that they do not reach the lower sphincter because they are abolished by the new contraction higher up, just as we saw in the case of rapid water swallowing.

The complication in the story of the lower esophagus comes in that since it is smooth muscle with its own intrinsic nervous system (Auerbach's plexus) it can be shown to have most of the properties of intestinal smooth muscle [17]. It can contract in response to stretch without vagal innervation. This has been shown in the cat esophagus and in the opossum esophagus in vitro. It shows directional polarity in that such contractions pass downwards. An indication or perhaps explanation for this is that the latency for response gets longer as the lower esophageal sphincter is approached. The big question is, does sequential vagal stimulation cease as soon as the bolus reaches the smooth lower esophagus to be replaced by the local response to stretch which the bolus would stimulate or does one get sequential vagal firing, secondary enhancement and the local smooth muscle response to stretch? Stimulation of the central end of a cut vagus will cause peristalic contraction of the smooth muscle in situ. Inhibition of activity in the lower third of the esophagus is abolished if the vagi are cut higher up and the lower sphincter will not relax without the vagi. It seems clear, therefore, that intact vagi are necessary for inhibition of the smooth muscle of the lower third, that it can move a bolus downwards without vagal innervation, but that it will contract sequentially in response to stimuli originating in the swallowing center. It should be clear, of course, that even if the physiological response of the smooth muscle esophagus is independent of extrinsic innervation, without the vagally-dependent upper esophageal muscle no distending bolus could reach it.

Without the intrinsic Auerbach's plexus the lower third of the esophagus is capable of weak and uncoordinated nonpropulsive contractions and the lower esophageal sphincter cannot relax. This is seen in the human malady achalasia of the esophagus in which Auerbach's plexus degenerates.

Lower Esophageal Sphincter [6, 9, 16]. Just as the pharyngoesophageal junction (upper esophageal sphincter) is a zone of increased tonicity, the terminal portion of the esophagus serves as a *gastroesophageal* or *lower esophageal sphincter*. Anatomically, there is no apparent thickening or change in orientation of the musculature in these regions. Physiologically, however, the zones of increased intraluminal pressure are clear. This delineation can be made by slowly pulling a pressure-recording catheter from the stomach into and up the esophagus. Radiological studies also demonstrate sphincter action, showing that swallowed barium may be temporarily slowed in the lower, gastric end of the esophagus prior to entry into the stomach. The length of the inferior high intraluminal pressure zone is two to four centimeters, extending above and below the esophageal hiatus of the diaphragm in normal people. The point at which the reversal of the respiratory pressure occurs is thus in the middle of the high pressure zone. The gastroesophageal sphincter is normally quite efficient in preventing reflux of gastric contents into the esophagus. A common clinical technique to assess the competence of this sphincter is to measure the force required to pull a standard Teflon ball across it from the stomach.

The lower esophageal sphincter at rest is contracted [12]. That this keeps gastric content from regurgitating into the esophagus is proved by the frequency of this unpleasant condition when the sphincter is defective. The esophageal mucosa has no strong intrinsic defenses against gastric juice which can leak into the esophagus and cause pain and eventually esophageal ulceration. As expected, this is most likely to occur in the recumbent position.

When the swallowing reflex starts both upper and lower esophageal sphincters relax. Vagal stimulation in most animals also causes relaxation. The only pharmacological agents, however, which have been shown to prevent this are tetrodotoxin or local anesthetics, both of which block the nerves themselves and not the neurotransmitters, the identity of which remains unknown.

The nature of the tonic contraction, by exclusion, has been labelled intrinsic because no neurohumoral or neuronal blocking agents including tetrodotoxin alter its tonically contracted state. The force of contraction is not constant but can be increased by increasing the intragastric pressure; moreover, it has been contended that a variety of gastrointestinal polypeptide hormones do the same. Positive results have been obtained with gastrin, pancreatic polypeptide, motilin and bombesin. These, indeed, may influence the tone, but there are doubts. They cannot be essential for continuous tonic contraction since in several animals this is maintained even in vitro. Agents which act at muscarinic, α-adrenergic and H_1 receptors increase tone in strips of sphincter muscle while others acting at β-adrenergic, H_2, and dopamine receptors cause relaxation, but antagonists to these agents are without effect on either the isolated or intact sphincter.

Functional Disorders. From the foregoing discussion of the physiology of this organ, one can predict the nature of many functional disorders. For example, dysphagia involving the striated muscle part of the esophagus and the pharynx is most likely to be due not to failure of the upper esophageal sphincter to relax, but to weak propulsive forces since there are no inhibitory fibers to these muscles. Diseases affecting the motor nerves, the muscle or their cells of origin can only weaken contraction. This is borne out by clinical experience. The smooth muscle lower third, on the other hand, since it has both motor and inhibitory fibers might suffer from failure of either one of these. Again, this is the case. In *achalasia,* in which the Auerbach's plexus degenerates, the body of the organ cannot contract nor can the lower esophageal sphincter relax (fig. 3/11). The opposite condition is *diffuse esophageal spasm;* in this the smooth muscle of the esophagus and sphincter contract as unit, and the progressing esophageal wave is not seen. No underlying pathogenesis is known.

If the tonic closing pressure of the lower esophageal sphincter is decreased, reflux of gastric content up into the esophagus can occur espe-

3   Swallowing: Pharynx and Esophagus   73

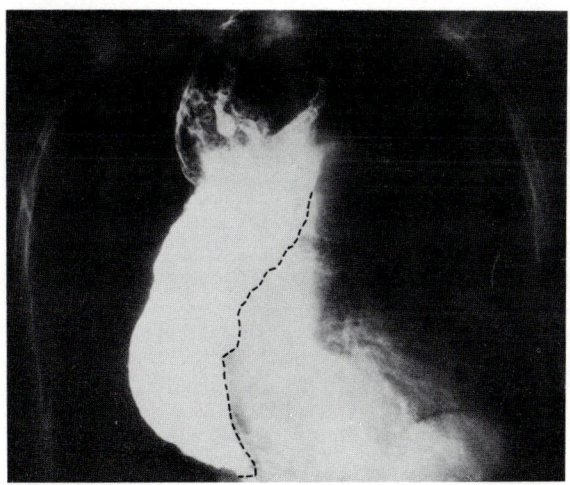

Fig. 3/11. Esophagus of a patient with achalasia. Compare distended esophagus and pinched off lower esophageal sphincter with figure 3/10. Reproduced, with permission, from *Magee* [14].

cially in the recumbent position. This has already been mentioned. Brief episodes of reflux are common and harmless, e.g. during vomiting or with eructation. The latter may produce pain, described as heart burn, which usually yields to antacids. The most important structure guarding against reflux is undoubtedly the sphincter; others, such as the esophagogastric angle and the right crux of the diaphragm assist, but are no substitutes for a competent sphincter.

The esophagus is concerned with material coming up as well as going down. In vomiting, it and its sphincters behave as a flaccid tube. A forceful increase in intra-abdominal pressure, due to contraction of the abdominal muscles and the descent of the diaphragm (the latter lowers the intraesophageal pressure) expels food up the esophagus. This is assisted by strong contraction of the pylorus. Afterwards, a swallow or a secondary peristaltic wave will deliver residual vomitus back to the stomach. The mechanism for belching is essentially the same without the feeling of nausea and salivation. Again, the residual esophageal gas initiates a secondary wave which sends it back to the stomach. There are people who, at will, can suppress the swallowing mechanism and allow fluid to run down under gravity. This is akin to vomiting and belching. Both the upper and lower sphincters remain open. Some Spanish wine drinkers cultivate this trick.

Some individuals can swallow air and eructate at will. This is the basis of some esophageal speech. The air does not enter the stomach since the lower sphincter remains closed. The air is forcibly expelled against the mucosal folds over the cricopharyngeus and a low pitched comprehensible speech can be produced.

References

1 Bossy, J.; Vidic, B.: Existe-il une innervation proprioceptive des muscles du pharynx chez l'homme? Archs Anat. Histol. Embryol. *50:* 273–284 (1967).
2 Basmajian, J.: Grant's method of anatomy; 10th ed. (Williams & Wilkins, Baltimore 1980).
3 Becker, R.; Wilson, J.; Gehweiler, J.: The anatomical basis of medical practice (Williams & Wilkins, Baltimore 1971).
4 Bloom, W.; Fawcett, D.: A textbook of histology; 10th ed. (Saunders, Philadelphia 1975).
5 Brantigan, O.C.: Clinical anatomy (McGraw Hill, New York 1963).
6 Christensen, J.; Carle, D. de: Esophageal motility; in Scientific basis of gastroenterology, chap. 15 (Churchill-Livingstone, New York 1979).
7 Code, C.; Creamer, B.; Schlegel, J.; Olsen, A.; Donaghue, F.; Andersen, H.: An atlas of esophageal motility in health and disease (Thomas, Springfield 1958).
8 Elias, H.; Paula, J.; Burns, R.: Histology and human microanatomy; 4th ed., p. 290 (Wiley, New York 1978).
9 Goyal, R.; Cobb, B.: Motility of pharynx, esophagus and esophageal sphincters; in Johnson: Physiology of gastrointestinal tract (Raven Press, New York 1981).
10 Hammersen, F.: Sobotta/Hammersen histology. A color atlas of cytology, histology and microscopic anatomy; 2nd ed. (Urban & Schwarzenberg, Baltimore 1980).
11 Hwang, K.; Mechanism of transportation of the content of the esophagus. J. appl. Physiol. *6:* 781–796 (1954).
12 Kaye, M.; Showalter, J.: Manometric configuration of the lower esophageal sphincter in normal human subjects. Gastroenterology *61:* 213–223 (1971).
13 Leeson, C.; Leeson, T.: Histology; 3rd ed., p. 343 (Saunders, Philadelphia 1978).
14 Magee, D.: Gastrointestinal physiology (Thomas, Springfield Il. 1962).
15 Netter, F.: The CIBA collection of medical illustrations, vol. 3. Digestive system, part I. Upper digestive tract (CIBA Pharmaceutical Company, Summit 1959).
16 Pope, C.: Esophageal physiology. Med. Clins. N. Am. *58:* 1181 (1974).
17 Weisbrodt, N.; Christensen, J.: Gradient of contractions in opossum esophagus. Gastroenterology *62:* 1159–1166 (1972).
18 Zaino, C.; et al.: The pharyngoesophageal sphincter (Thomas, Springfield 1970).

4 Stomach: Gastric Motility

The Muscular Coat of the Stomach

The musculature of the stomach consists entirely of smooth muscle fibers. The muscle coat (muscularis externa) is arranged in three layers over much of the stomach rather than the usual two; an outer longitudinal layer, an inner oblique layer and a circular layer (fig. 4/1) between the two. These layers are not all complete and are best seen where the musculature is thickest, i.e., in the pyloric portion of the stomach where this extensive and specially modified muscular coat produces the churning and homogenation of the ingested food as the gastric juices are added to it. Because of the relatively complicated contour of the stomach, the terms longitudinal, circular and oblique are not always very accurate, descriptively speaking. The longitudinal muscle fibers of the stomach are continuous with the longitudinal layer of the esophagus. At the cardia, this layer divides into two bands which run along the curvatures (lesser and greater) of the stomach, so that the middle areas of the anterior and posterior surfaces of the stomach remain free of longitudinal muscle fibers. In the pyloric region the two bands coalesce once again into a complete layer which, for the most part, is continuous with the longitudinal layer of the duodenum. The inner, oblique layer is most strongly developed in the fundic region and diminishes as the pylorus is approached. No oblique fibers run along the lesser curvature, and the longitudinal furrow observed on the internal aspect of the lesser curvature due to the absence of this musculature has been called the 'Magenstrasse'. To some extent, the oblique fibers reinforce the areas skirted by the longitudinal fibers. The middle, circular layer is the most complete and is the strongest layer. It is continuous at the cardia with the circular layer of the esophagus and becomes progressively thicker as the pylorus is approached, ultimately forming the muscular ring of the pyloric sphincter (fig. 4/1C). This stout muscular ring bulges the submucosa and the mucous membrane inward so that a circular fold is formed. This fold, having the thickened

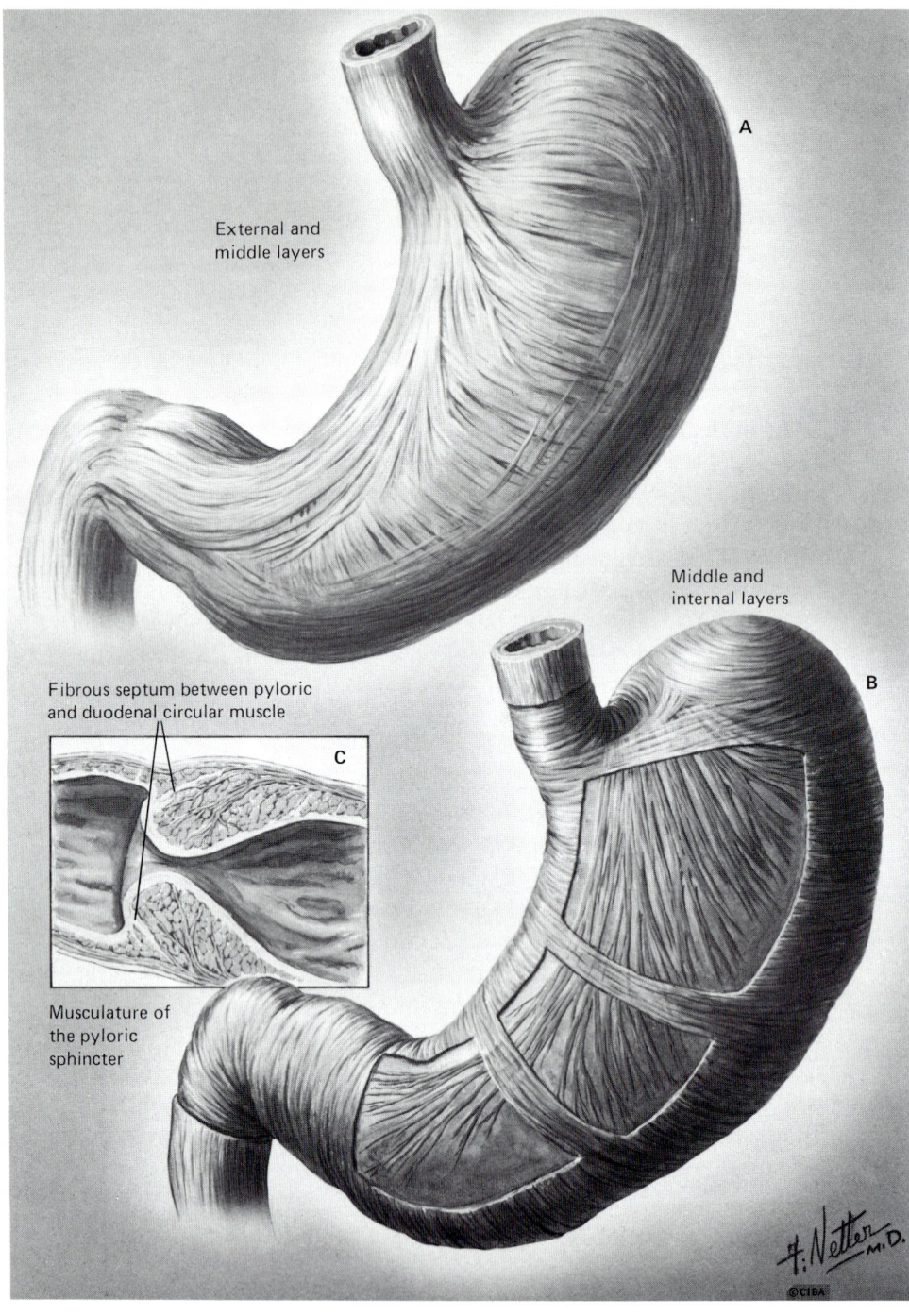

External and middle layers

Middle and internal layers

Fibrous septum between pyloric and duodenal circular muscle

Musculature of the pyloric sphincter

middle coat of the muscularis externa as the chief ingredient of its core, thus differs from most folds of the GI tract which have cores only of submucosa. The lumen of the stomach concurrently narrows as the sphincter is approached. Muscular slips from the outer and inner layers also contribute to the sphincter musculature. The gastric muscle of the pyloric sphincter is not continuous with the circular muscle of the duodenum, but is separated from it by a fibrous septum (fig. 4/1C).

In the past, whether or not the pyloric sphincter was closed or open at rest was a subject for debate. The consensus now seems to be that it is open physiologically, i.e. it offers no noticeable impediment to the flow of fluid into the duodenum [11]. Most investigators have been unable to detect an area of increased pressure on withdrawing pressure transducers from the duodenum to the stomach in the fasting subject. What this means, of course, is that the pyloric sphincter is not a sphincter at all, but has been assumed to be one since it looks like one: a thickened muscle surrounding a narrowed opening. 'Gastroduodenal junction' is, therefore, a better name for this region.

Just as a third layer is added to the usual two layers of the tunica muscularis, a third (i.e. outer circular) layer is added to the usual two layers of the muscularis mucosae over some parts of the stomach. Slips of smooth muscle from the innermost (circular) layer extend into the lamina propria between the mucosal glands, sometimes reaching as far as the gastric epithelium. Thus, in addition to giving additional mobility to the gastric mucosa, the muscularis mucosae probably compresses the mucous membrane to facilitate emptying of the glands of the gastric mucosa.

Vagal Innervation of the Stomach

Nearly all of the stomach (cardia, fundus, body and even proximal antrum) is innervated via the gastric branches of the vagi; however, few fibers from the gastric branches, if any, reach the gastroduodenal junction (fig. 4/2). The four or more gastric branches of the anterior (left) vagal trunk arise at the gastroesophageal junction and are distributed mostly to the

---

Fig. 4/1. The muscularis externa of the stomach. A Longitudinal (external) and circular (middle) layers. B Circular (middle) and oblique (internal) layers. C Musculature at the gastroduodenal junction. Reproduced, with permission, from *Netter* [19].

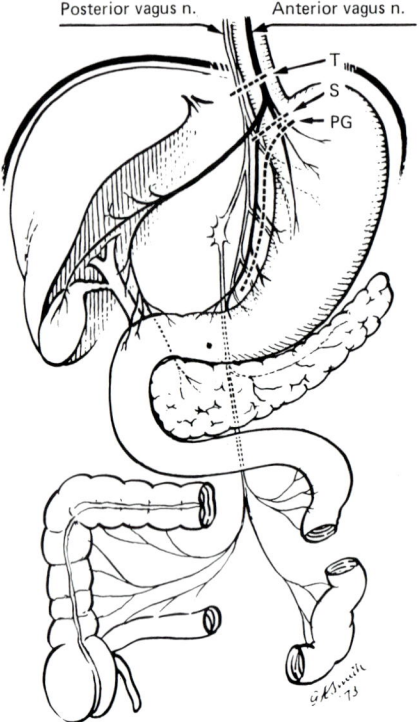

Fig. 4/2. The abdominal distribution of the vagi and the site of the 3 types of vagotomy. T = Truncal; S = selective; PG = proximal gastric or parietal cell. Anterior vagal trunk, black; posterior trunk, white. The nerve of Laterjet lies immediately left of dotted line PG. Reproduced, with permission, from *Kennedy* [14].

cardiac end of the stomach. One branch (the nerve of Laterjet), frequently larger than the rest, often runs along the lesser curvature to within several centimeters of the gastroduodenal junction (but apparently never to the 'sphincter' itself). The gastric branches of the posterior trunk are similarly distributed. The anterior vagal trunk is, nonetheless, the source of the parasympathetic innervation of the gastroduodenal junction, but these fibers arrive via a different route. At the lower end of the esophagus, one or more hepatic branches arise from the anterior vagal trunk which course through the lesser omentum to the porta hepatis. While some fibers pass to the liver at this site others descend, running mainly along the hepatic artery, to the gastroduodenal junction (and pancreas) [8]. This separate routing of the parasympathetic fibers to the gastroduodenal junction is significant to the surgeon attempting to reduce acid secretion by means of vagotomy: truncal vagotomy, with its consequent denervation of the pylorus (and entire lower GI tract) generally results in gastric stasis and thus a drainage procedure is required; selective or proximal gastric vagotomy does not denervate the pylorus, so normal gastric emptying of solids is maintained and drainage procedures are not required (fig. 4/2) [21].

Storage Function

The bolus of food has now passed the lower esophageal sphincter and enters the stomach, an organ concerned both with storage and digestion. The bolus is not fluid at this stage, but in the form of a soft, semisolid mass surrounded by the mucous picked up in the mouth and esophagus. Initially, this mass remains in the upper part of the stomach and enlarges as more swallowed boluses are added to it. As a reservoir for the accumulation of food in intermittent feeders, the capacity of the stomach is considerable. The luminal volume of the empty stomach is only about 50 ml; its cavity is of a caliber not much larger than that of the intestine. The intragastric pressure rises only slightly, however, as food is added to the stomach. If food is added continuously, there is slight initial rise to a plateau and then, after the physiological capacity has been exceeded, an abrupt rise which may be followed by vomiting if ingestion continues (fig. 4/3).

Receptive relaxation [5], part of the swallowing reflex, keeps the stomach from behaving like an inflated tyre or balloon in which the pressure is proportional to the volume of air pumped in. Receptive relaxation is a property of the proximal stomach only and depends on the vagus nerves. The pyloric part of the stomach, which does not function as storage, does not demonstrate this. The transmitter for this inhibitory action of the vagus is unknown: it has, however, been known for a long time that the action of both vagal stimulation and cholinomimetic drugs on the stomach depends on the stomach's activity at the time. Relaxed stomachs are contracted, and vice versa. Experiments employing stimulation of either the parasympathetic vagal nerves or the sympathetic splanchnic nerves to the stomach are complicated by the fact that neither of these is purely cholinergic or adrenergic; both contain a few fibers of the other division of the autonomic nervous system.

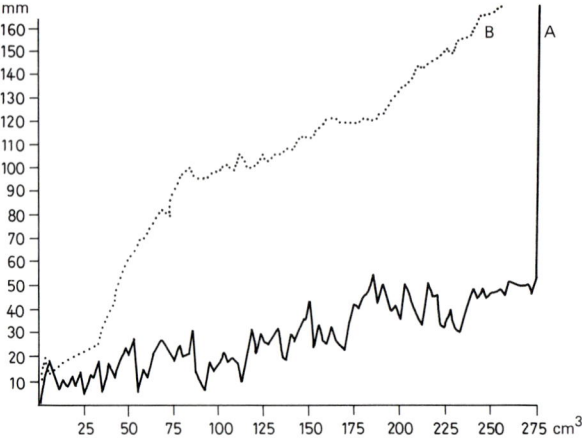

Fig. 4/3. Curve of intragastric pressure in a living rabbit with its abdomen closed (A). Curve in a rabbit stomach in vitro (B). (In both cases saline was added at the rate of 1 ml/min.) Reproduced, with permission, from *Grey* [7].

Digestive and Mixing Activity

In the stomach, then, one has a storing proximal end with a distensible, potentially capacious lumen and a digestive and mixing distal end with a progressively narrowing lumen and thickening musculature as the duodenum is approached. The digestive processing of the food mass that occurs in the distal stomach is in part mechanical and in part chemical. The distal stomach, however, does not secrete hydrochloric acid and pepsin, which are the digestive secretions of the stomach. These are, strangely enough, secreted by the mucosa at the proximal, storage end. They flow around the pultaceous semisolid bolus occupying the lumen there after meals, into the distal end. This is also the fate of fluids consumed with or after meals. The insignificant extent of penetration of the bolus by the gastric juice is shown in the accompanying figure (fig. 4/4). Appreciable digestion of starch can take place only if acid has not penetrated into the mass, since salivary amylase is inactive at a pH below 4.5. The slight peptic digestion of protein, which can occur only at low pH, indicated in the figure probably represents pyloric rather than fundic gastric contents since in these experiments the subjects were induced to vomit the entire contents of their stomachs by the administration of an emetic. Pepsin is active only in a medium more acidic than pH 4.5.

4 Stomach: Gastric Motility

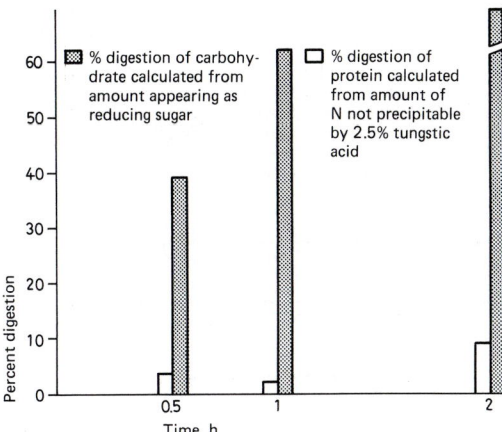

Fig. 4/4. Extent of gastric and salivary digestion of a solid meal in the human stomach. At the times indicated after eating; vomiting was induced with apomorphine. Plotted from data reported by *Beazell* [2].

Proximal Stomach Activity

In keeping with the structural and functional differences between the two ends of the stomach, there are marked differences in activity. After meals, the proximal end of the stomach exhibits slow and sustained tonic contractions lasting from 2 to 6 min and recurring at intervals of 2–4 min [16]. They do not move towards the pylorus but can increase the intragastric pressure to 10–50 cm H_2O. The entire muscular coat contracts. Superimposed on this are quick phasic contractions, these last seconds and increase pressure only by 5–10 cm H_2O. These again do not move caudally. The smooth muscle of other parts of the stomach and of the small intestine exhibits spontaneous, cyclical, electrical activity, but neither this nor the spikes characteristic of such muscular activity are seen here. This must mean that there is no intrinsic muscular activity and that all contraction is induced from without by direct stretching (caused by contained food for example) or by nervous activity; however, there is no clear evidence that vagotomy depresses any mechanical activity in this part of the stomach except for receptive relaxation. The rather steady pressure exerted by the fundic muscle on its contents is a major influence in advancing the food mass progressively into the distal stomach where its free edge can be eroded by the combined actions of gastric juice and muscular activity.

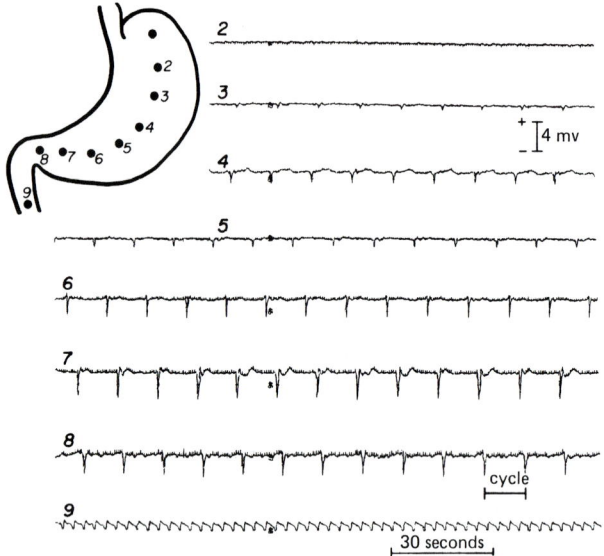

Fig. 4/5. Electrical activity in the walls of the stomach and duodenum in a normal dog. Note that the BER or pacesetter potential (PP) is absent from site 2, but present and of the same frequency at sites 3–8 inclusive. The rate at 9, the duodenum, is much faster. Reproduced, with permission, from *Kelly and Code* [12].

Distal Stomach Activity

The spontaneous electrical activity seen in the distal stomach resembles pacemaker activity in the heart except that in the heart every prepotential reaches threshold and fires. In the stomach, surface electrodes pick up a complex consisting of a positive wave immediately followed by a negative wave, the whole lasting two to three seconds. The size of each of these is 1–2 mV (fig. 4/5). These recur regularly three times per minute in man and 4–5 times/min in dogs. These cycles originate on the greater curvature at the junction of the proximal and distal parts of the organ (fig. 4/6) [22]. In fact, their presence is the label by which one distinguishes between the two parts. This spot [12] like the SA node, dominates the organ because its intrinsic basal electrical rhythm (BER), as it is called, is the most rapid. Should it be destroyed, another area, which will now have the fastest rate, will become dominant. Any part of the distal stomach has pacemaker potential and will assume this as soon as areas with faster rhythm are ablated or inactivated. From the pacemaker region these waves pass circumferentially around the

4   Stomach: Gastric Motility   83

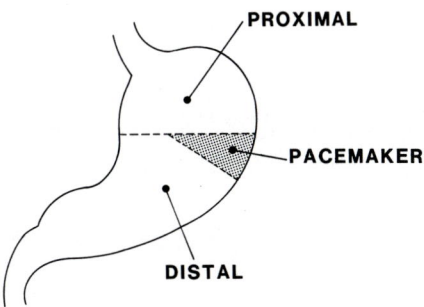

Fig. 4/6. The gastric pacemaker and the dynamic divisions of the stomach.

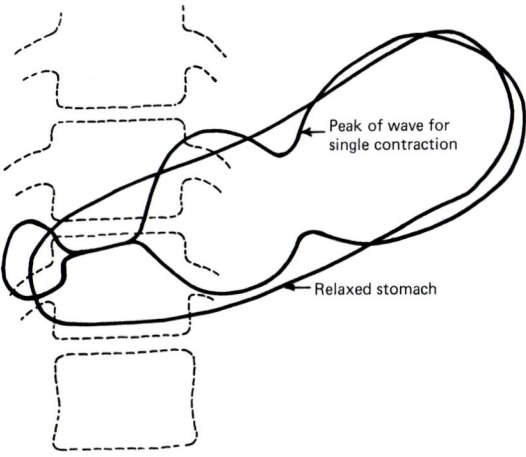

Fig. 4/7. Superimposed tracings from X-rays of the empty stomach. Reproduced, with permission, from *Martin and Rogers* [17].

stomach and towards the pylorus gathering speed as they go. When the BER waves reach threshold spiking occurs and the muscle contracts. The contraction follows exactly the route of the BER seen in the resting organ [3]. Because of the gain in speed as the pylorus is approached, the last three to four cms of the pylorus contract as a unit, but because the lumen of the pyloric canal is smaller it usually closes first (fig. 4/7). This means that not all of the fluid carried before the wave will pass into the duodenum. Some of it will be squirted back into the stomach. The speed with which contractions recur cannot exceed that of the BER. This pattern of events makes it appear fairly certain that the muscle has not within itself the means to bring its rhythmic wave to the firing threshold as does the SA node. Something external is needed, e.g. distension of the stomach. Local anesthetization of the pyloric mucosa will reduce the strength and frequency of contractions. Vagal impulses [18], the composition of the gastric content and circulating hormones all increase gastric motility. During normal digestion, all of these are undoubtedly in action.

Gastric Emptying

Once gastric content has been discharged into the duodenum, the pyloric muscle, now in a contracted state, prevents regurgitation back into the stomach as contraction commences in the duodenum. The gastric BER does not cross into the duodenum. The duodenum has its own intrinsic rate which is not in phase with that of the stomach, but since it is very much faster than the gastric rate during digestion, duodenal contractions start before the pyloric canal muscle has relaxed.

Large particles do not pass into the duodenum. A diameter of 2 mm seems to be the limit. This is much less than the diameter of the pyloric canal and is, therefore, difficult to explain. Indigestible solids with a diameter of more than 1 cm, in dogs, do not empty from the gastric lumen. In our experience, pieces of gastric cannulae and plastic tubing have remained in the stomachs of dogs for months. However, in the fasted patient, tubes tipped with a metal weight up to 1 cm in diameter can fairly, routinely be induced to enter the duodenum. It is likely that only particles in suspension can pass through postprandially because normal postprandial contractions do not completely occlude the gastric lumen and force material through the sphincter. They simply increase the pressure differential between the stomach and duodenum. Particles not in suspension, therefore, do not flow onwards. The pyloric sphincter does not seem to restrain or influence the emptying of fluids. The emptying rate of water is unaltered in dogs if the pyloric canal is held open by a metal pipe [6]. If the pyloric part of the stomach is removed, then the emptying of solids will be disturbed. Large particles will gain entrance to the duodenum and the emptying of mixed meals will be accelerated. Complete gastric vagotomy reduces the amplitude of the contraction waves in the distal stomach and thus delays emptying. To compensate for this surgeons have done drainage operations either by cutting the pyloric muscle longitudinally or by constructing a gastroenterostomy. This gives rise to the opposite difficulty, i.e. precipitous emptying,

4   Stomach: Gastric Motility

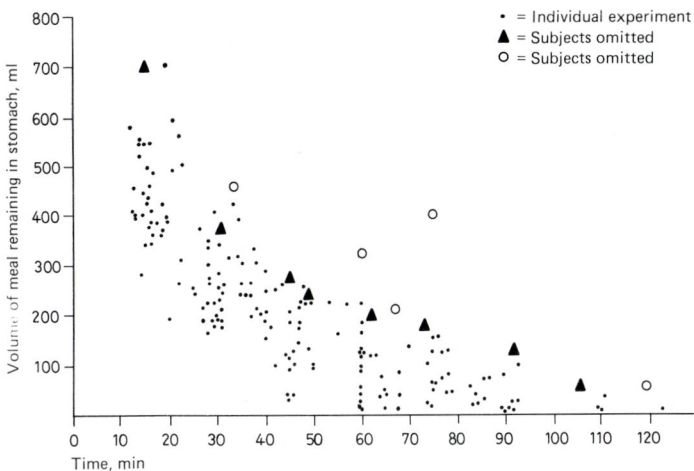

Fig. 4/8. The volume of a meal remaining in the stomach plotted against time. The meal used in the experiments was liquid. Results of 190 experiments on 21 subjects. Reproduced, with permission, from *Hunt and Spurrell* [9].

at least in the initial stages. This probably results from the inability of the pylorus to retain large particles plus the loss of receptive relaxation, consequent to the vagotomy. These difficulties have been greatly reduced by the modern operation which spares the nerves to the pylorus and thus obviates the need for the drainage operation. Receptive relaxation is still abolished. This may cause a feeling of fullness after meals and may accelerate the initial emptying of fluids (fig. 4/2).

The composition of the gastric contents can influence emptying greatly [15]; fluids are more rapidly emptied than solids or partial solids. The relation between the amount of saline in the stomach and the rate at which it leaves is in an exponential one. Fluids containing substances with a caloric value either in solution or suspension empty much more slowly initially. Their rate in relation to the volume in the stomach is linear (fig. 4/8).

The more acid the contents of the stomach, the more slowly they will empty. In part, this is a consequence of duodenal, hormonal inhibitory mechanisms which will be discussed later on, but also it involves gastric mechanisms since fluids confined to fundic pouches reduce motility in

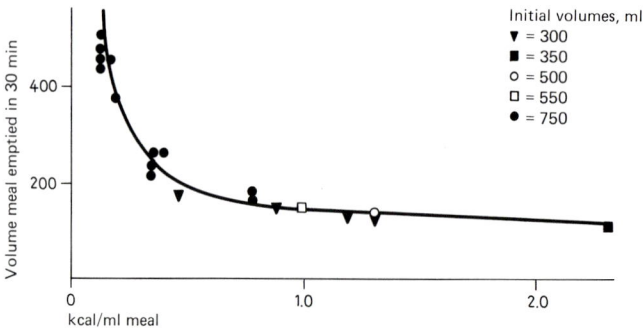

Fig. 4/9. The relationship between the volumes of a meal leaving the stomach per 0.5 h and its caloric density. Reproduced, with permission, from *Hunt and Stubbs* [10].

proportion to the intraluminal [H⁺] [20]. In these same experiments in dogs, increasing concentrations of NaCl increased motility in fundic pouches. In human volunteers, the converse has been found, no doubt due to the influence of the duodenum. The site of these osmoreceptors appears to be the duodenum but the nature of the message or messenger is unknown.

It is argued that the emptying of solid meals is such that the rate of caloric delivery to the duodenum is constant. This again may be a consequence of the release of hormones from the duodenum which depress gastric motility. During normal digestion, this brake on emptying is constant since these hormones are always released into the blood stream. The amount varies with the nature of the food eaten and on a mixed diet roughly with the caloric content. High caloric foods are usually high in fat and fats are powerful liberators of at least one hormone which depresses gastric motility and secretion. Fats with chain lengths of fourteen carbon atoms are the most powerful brake activators and unsaturated fats are more potent than saturated (fig. 4/9).

In the fasting state, the stomach is not completely inactive. At 100- to 110-min intervals, strong spontaneous contractions occur which last ten do twenty minutes. These have been given the cumbersome name of *interdigestive migrating myoelectrical complexes.* Since these involve the whole gut and are accompanied by periodic changes in secretion, they will be dealt with in a separate section.

References

1 Alvarez, W.: An introduction to gastroenterology, chap. 15–19 (Hoeber, New York 1940).
2 Beazell, J.: A reexamination of the role of the stomach in the digestion of carbohydrate and protein. Am. J. Physiol. *132:* 42–50 (1941).
3 Cannon, W.: The nature of gastric peristalsis. Am. J. Physiol. *29:* 250–266 (1911).
4 Cannon, W.: The mechanical factors of digestion (Arnold, London 1911).
5 Cannon, W.; Lieb, C.: The receptive relaxation of the stomach. Am. J. Physiol. *29:* 267–273 (1911).
6 Crider, J.; Thomas, J.: A study of gastric emptying with the pylorus open. Am. J. dig. Dis. *4:* 295–300 (1937).
7 Grey, E.: Observations on the postural activity of the stomach. Am. J. Physiol. *45:* 272 (1918).
8 Hollinshead, W.: Anatomy for surgeons, vol. 2. The thorax, abdomen and pelvis, pp. 398–399 (Harper & Row, New York 1971).
9 Hunt, J.; Spurrell, W.: The pattern of emptying of the human stomach. J. Physiol. *113:* 157–168 (1951).
10 Hunt, J.; Stubbs, D.: The volume and energy content of meals as determinants of gastric emptying. J. Physiol. *245:* 209–225 (1975).
11 Kaye, M.; Mehta, S.; Showalter, J.: Manometric studies of the human pylorus. Gastroenterology *70:* 477–480 (1976).
12 Kelly, K.; Code, C.: Canine gastric pacemaker. Am. J. Physiol. *220:* 112–118 (1971).
13 Kelly, K.; Code, C.: Effects of transthoracic vagotomy on canine gastric electrical activity. Gastroenterology *47:* 51–58 (1969).
14 Kennedy, T.: The vagus and the consequences of vagotomy. Med. Clins N. Am. *58:* 1231–1246 (1974).
15 Kondo, T.; Magee, D.: Effect of variations in intraluminal osmolality and pH on gastric motility and pepsin secretion. Itl. J. Gastroent. *12:* 6–9 (1980).
16 Linde, J.; Duthie, H.; Schlegel, J.; Code, C.: Motility of the gastric fundus. Am. J. Physiol. *201:* 197–202 (1961).
17 Martin, C.; Rogers, F.: Hunger pain. Am. J. Roentg. *17:* 222–227 (1927).
18 Martinson, J.: Studies on the efferent vagal control of the stomach. Acta physiol. scand. *255:* suppl., pp. 1–24 (1965).
19 Netter, F.: The CIBA collection of Medical Illustrations, vol. 3. Digestive systems, part I. Upper digestive tract (CIBA Pharmaceutical Company, Summit 1959).
20 Quigley, J.; Read, M.; Radzow, K.; Meschan, J.; Werle, J.: The effect of hydrochloric acid on the pyloric sphincter and adjacent portions of the digestive tract and on the process of gastric evacuation. Am. J. Physiol. *137:* 153–159 (1942).
21 Shackelford, R.; Zuidema, G.: Surgery of the alimentary tract; 2nd ed., p. 368 (Saunders, Philadelphia 1981).
22 Weber, J.; Kohatsu, S.: Pacemaker localization and electrical conducting patterns in the canine stomach. Gastroenterology *59:* 717–726 (1970).

5 Stomach: Gastric Secretion

The Gastric Mucosa

The mucosa of the stomach is a pale, grayish-pink color in the living. In the empty (and therefore contracted) stomach, it is thrown into numerous folds or ridges known as *rugae* which run mainly in a longitudinal direction (fig. 5/1A). The height and number of rugae depends on the level of gastric filling; both diminish with filling and they disappear, for the most part, when one is 'full' (see fig. 4/1). Along the minor curvature of the stomach, where the mucosa is more firmly attached to the muscular layer and the oblique layer of the muscularis externa is absent, a furrow between two of these rugae is the so-called 'Magenstrasse' ('stomach street'). Along this furrow saliva and very small quantities of fluid are said to flow from esophagus to duodenum.

Besides these coarse folds, the gastric mucosa is subdivided by shallow furrows which produce a mosaic of finer surface elevations (areae gastricae) which do not disappear with filling. When the surface of the gastric area is magnified, as with a hand lens or under scanning electron microscopy (see fig. 5/2), it can be seen that the gastric mucosa is further characterized by the presence of minute invaginations of the surface epithelium. These are the *gastric pits* (foveolae). They extend into the lamina propria (fig. 5/1B–E). There are approximately three million such pits in the gastric mucosa. At the bottoms of these pits are the orifices of the multicellular glands of the mucosa. The shape and depth of the gastric pits, and the types of epithelial cells comprising the glands associated with them, vary in different regions of the stomach (fig. 5/3). Accordingly, three regions of the stomach are distinguished: (a) a narrow, ring-shaped zone around the cardia, called the *cardiac area* and containing the cardiac glands; (b) a much larger area including the *fundus and body* (proximal two-thirds or more) of the stomach which contains the gastric glands (proper), also called the fundic glands by some, and (c) a distal region (somewhat less than the distal third of the stomach), extending more proximally on the lesser than on the greater curvature, called the *pyloric region* and containing the pyloric glands (fig. 5/1A).

Surface Epithelium (Gastric Lining Epithelium)

Although the conformation of the gastric pits and the cellular composition of the associated glands varies from region to region, the epithelium lining the luminal surface (including the gastric pits) of all three regions of the stomach is the same. At the gastroesophageal junction, there is an abrupt transition from the stratified epithelium of the esophagus to a columnar epithelium that will line the remainder of the GI tract as far as the anus (see fig. 2/3). The columnar epithelium lining the stomach is composed of tall (20–40 µm) mucous-secreting cells (*surface mucous cells,* fig. 5/1F) which all look – and are – alike. This enables one to easily distinguish gastric mucosa from that of the small and large bowel where the lining epithelium is composed of alternating cell types (goblet cells alternating with nonmucous-secreting columnar cells). The gastric surface mucous cells are structurally typical of mucous-secreting cells (see 'Glandular Epithelial Tissue', chapter 1), except that their nuclei, though somewhat basally-located, are rounded and not flattened against the base (fig. 5/1F, 5/4). Further, the mucigen occupying the apical secretory granules is of a peculiar type which does not stain with many mucin-specific dyes. Unlike the mucous secreted into the oral cavity, the gastric mucous is a neutral polysaccharide which is not precipitated by acid. The mucous is released from the cells in such a way that it forms continuous sheets which line the stomach, lubricating and protecting it. In view of the detrimental phenomena consequent to the stomach's function, including mechanical and chemical insults from a wide range of extrinsic (ingested) substances as well as from intrinsic substances such as HCl, digestive enzymes and refluxed duodenal contents, this latter task is a formidable one. When the mucous sheet – and hence the underlying cells – are damaged by certain types of food and drink (e.g. alcohol), areas of epithelium may slough into the lumen. Even under normal physiological conditions with a healthy diet, the mortality rate of the surface epithelial cells is high. The cells continuously desquamate (approximately 1.5% of the population each hour) so that the entire epithelial lining is replaced every 3–4 days. This is compensated for by a constant migration of new cells arising in the isthmuses of the gastric pits. Even when extensive wounds of

Structure of the Gastric Mucosa

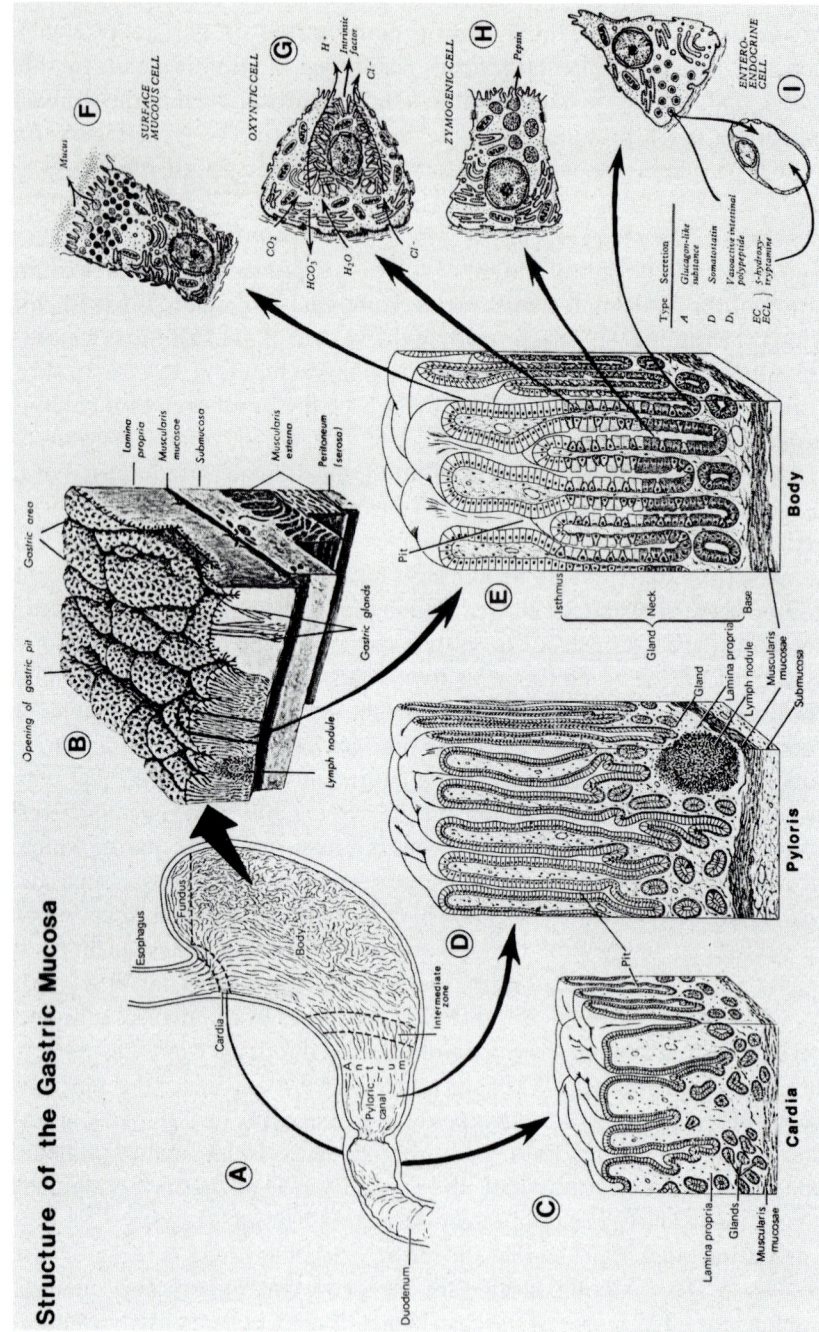

the stomach are inflicted, as by surgery, they normally heal quite rapidly, presumably because of the normally high proliferation rate of new cells [14].

Lamina propria

The zone between the lining epithelium and the muscularis mucosa is so interrupted by the gastric pits and, deep to them, so densely occupied by the mucosal glands, that it is difficult to recognize it as lamina propria (see fig. 5/1C–E). Indeed, it exists only as a matrix filling the gaps between the gastric pits and, below the bases of the pits, the even narrower spaces between the mucosal glands. Except for the glandular elements which protrude into and occupy most of it, the presence (especially early in life) of small collections of lymphoid tissue called *gastric lymphatic follicles,* or *nodules* (fig. 5/1D) and the fact that it is almost devoid of elastic elements, it is typical of lamina propria in composition (see 'General Structure of the Gastrointestinal Canal', chapter 1).

Cardiac Glands. The glands in the cardiac area of the lamina propria of the stomach closely resemble the esophageal cardiac glands in appearance (fig. 5/1C). The gastric *cardiac glands* are coiled, tubular glands – either simple or compound – the often large lumina of which are lined by cuboid, mucous-secreting cells. Several such glands open directly into the base of a single gastric pit. The gastric pits in this region are relatively short and have small, round openings. The ratio of depth of the gastric pit to the thickness of

---

Fig. 5/1. Structure of the gastric mucosa. **A** Diagrammatic representation of the stomach, demonstrating the rugae on its internal surface. Regions of the stomach are indicated. **B** Semidiagrammatic reconstruction of an approximately 3.5 mm tissue block cut from the wall of the gastric body as indicated in **A**. Luminal surface of tissue block demonstrates 'gastric areas' on the surface of which the openings of the gastric pits are evident. **C, D** Diagrammatic representations of tissue blocks from the cardia and pylorus, as indicated in **A**. **E** Diagrammatic representation of a tissue block cut from **B** as indicated. **F–I** Four of the five cell types associated with the gastric pits and glands, as indicated in **E**; mucous neck cells are not demonstrated. **A, C–E** adapted from *Junquiera and Carneiro* [26]. **B** is adapted from *Copenhaver, W.; Bunge, R.; Bunge, M.:* Bailey's textbook of histology; 16th ed. (Williams & Wilkins, Baltimore 1971), in which it appeared slightly modified from *Braus*. **F–I** from *Williams, P.; Warwick, R.:* Gray's anatomy; 36th British ed. (Saunders, Philadelphia 1980).

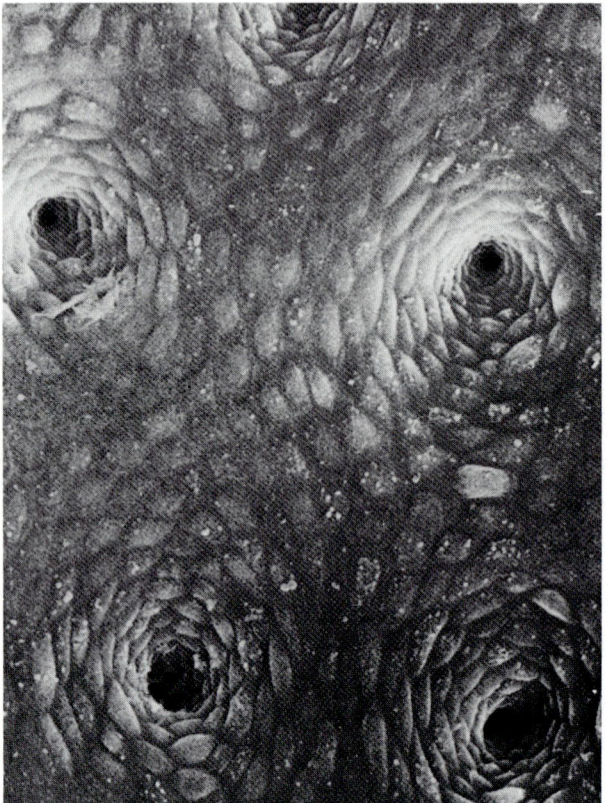

Fig. 5/2. 585 × enlargement (via scanning electron microscopy) of the gastric luminal surface demonstrating the round orifices of the gastric pits. Reproduced, with permission, from *Leeson and Leeson* [32].

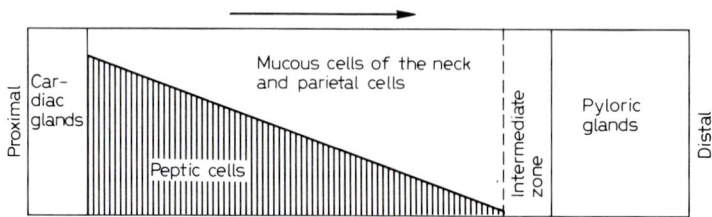

Fig. 5/3. Graphic demonstration of the distribution of epithelial cell types along the length of the stomach. Slightly modified from *Greep, R.; Weiss, L.*: Histology; 3rd ed. (McGraw-Hill, New York 1973), where it appeared on p. 572 modified from *Babkin, B.*: Secretory mechanisms of the digestive glands; 2nd ed. (Hoeber, New York 1950).

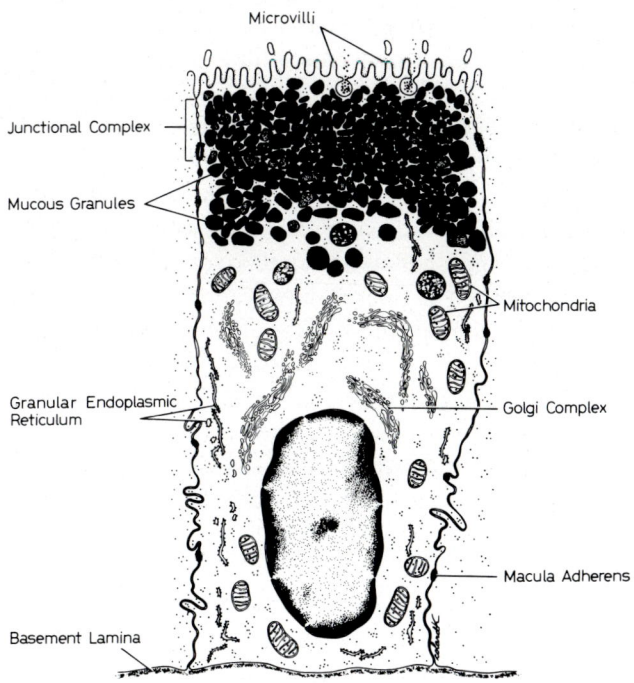

Fig. 5/4. Schematic representation of a gastric surface mucous cell. From *Bloom and Fawcett* [6].

the gland-occupied lamina propria is approximately 1:1. The gastric cardiac glands secrete little enzyme and, their functional significance not being known, are considered to be of little practical importance.

Gastric Glands. The *gastric* (proper) or oxyntic glands of the fundus and body are the most abundant glands of the mucosa (there are approximately 15 million) and are the most important contributors to the secretion of gastric juice, producing nearly all the gastric enzymes and hydrochloric acid (as well as some of the mucous) secreted into the stomach. They are mainly branched, tubular glands in humans, densely packed within the lamina propria (fig. 5/1E). Although they may be coiled or bent where they abut against the muscularis mucosa, for the main part they are straight glands with their narrow lumina oriented perpendicularly to the mucosal surface. Hence, they are almost always sectioned longitudinally in preparations cut perpendicularly to the surface. They are described [57] as having a *base* or *body* (the basal third, roughly), a *neck* (the middle third) and an *isthmus* (upper third) by which they are continuous with a gastric pit (see fig. 5/1E). Recall that the pits are not parts of the glands, but merely inpocketings of the mucosal surface which are linked with the surface epithelium. One to several glands open into the bottom of each gastric pit. Although their orifices are still small and round (fig. 5/2) – and hence appear the same from the surface – the gastric pits are somewhat shallower than those of the cardiac area. The gland-occupied lamina propria is thicker, however, so that the ratio of the depth of the gastric pits to the thickness of the glandular lamina propria is approximately 1:4.

Cytologically, the gastric glands proper are the most highly differentiated of the glands of the stomach's mucosa. At least five distinct cell types contribute to these glands, four of which are exocrine in function; but these are not evenly distributed over the different segments of the glands. The various cell types are adequately distinguished in PAS and hematoxylin-stained sections.

The *isthmus* contains two cell types. *Immature*, relatively undifferentiated, *columnar cells* located here undergo mitotic division to maintain the population of surface epithelial cells – and probably the other cell types of the gastric glands as well (although these details are not yet certain). New surface epithelial cells migrate from the isthmuses into the gastric pits, where division ceases, and from there to the free surface from which they will eventually be shed into the lumen. Only a few mucin granules are found in the apical cytoplasm of the immature or partly mature surface epithelial cells of the isthmus. The amount of mucous will increase progressively as the cells ascend from the glandular isthmuses into the pits and then to the mucosal surface. Distributed among the immature and partly mature cells – but occurring only at intervals – are large, rounded *parietal* (oxyntic = 'acid-forming') cells (fig. 5/1G). These appear to have a clear cytoplasm with a marked affinity for acid dyes such as eosin (although – as will be explained under 'Ultrastructure of the Parietal (Oxyntic) Cell', this chapter, further on – they are not truly 'acidophilic') and dark, centrally-placed nuclei. As if from compression by the surrounding cells, their basal aspects bulge into the adjacent lamina propria, giving the glandular epithelium a 'beaded' appearance with some cells appearing to be completely crowded away from the glandular lumen. The ultrastructure of these important and interesting cells, believed to be responsible for the production of stomach acid, will be described in detail further on.

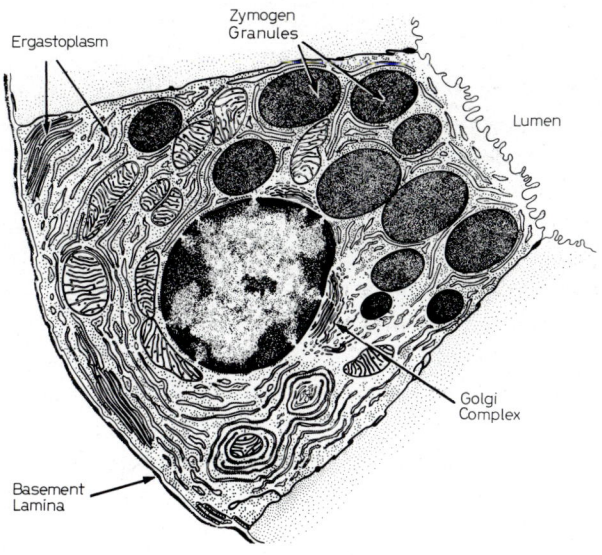

Fig. 5/5. Schematic representation of a chief (zymogenic) cell. From *Bloom and Fawcett* [6].

The *neck portion* of the glands is composed primarily of *mucous neck cells* [4]. Located only in this site, they are relatively few in number and are literally stuffed with mucous, giving them a foamy appearance. The mucous is, however, histochemically distinct from that of the surface epithelial cells in that it is an acidic rather than a neutral polysaccharide. Otherwise, these cells are typical in structure of mucous-secreting cells (see 'Glandular Epithelial Tissue', chapter 1). Individual parietal cells are distributed among groups of mucous neck cells, their roundness tending to deform the adjacent neck cells so that they are characteristically irregular in shape. Occasionally, mitotic figures will be seen among the mucous neck cells which, having a lifespan of about one week, are replaced at about half the rate of the surface epithelial cells.

In the *base* or *body* of the glands, the mucous neck cells are abruptly replaced by *chief* (zymogenic) *cells* which comprise most of this portion of the glands (fig. 5/1H, 5/5). These cuboidal or low columnar cells have all the characteristics of protein-synthesizing cells (see 'Glandular Epithelial Tissue', chapter 1) and bear close similarity to the salivary serous cells and pancreatic acinar cells. The chief cells produce the enzymes of the gastric secretion; their supranuclear secretory granules contain *pepsinogen,* the antecedent of the enzyme *pepsin.* In conformity with this, granules disappear from the cells when the stomach is stimulated to produce its secretion. Parietal cells are found here, too, sprinkled singly among the chief cells, but are less frequent than in the isthmus and neck regions of the glands. Other cell types found in the bases of the glands are undifferentiated columnar cells (present in smaller numbers than in the isthmuses) and enteroendocrine cells.

Enteroendocrine cells (components of the gastroenteropancreatic endocrine system, to be dealt with in detail later – see chapter 12) are found scattered singly and infrequently between the chief cells and the basement membranes of the bases of the gastric glands (fig. 5/1I). They are characterized by the presence of irregularly infolded nuclei surrounded by cytoplasmic granules which *accumulate at the basal aspect* of the cells. This polarization, plus the fact that the 'apical' portion of the cells rarely reaches the glandular lumen, suggests that the substance within the granules is released into the bloodstream (or into the extracellular space for passage into the circulation) rather than into the lumina of the gastric glands.

At least two types of enteroendocrine cells are found in the mucosa of the stomach. Most of those associated with the bases of the gastric glands are the strongly argentaffin *EC cells,* which synthesize and store *serotonin* (5-hydroxytryptamine). Another type associated mainly with the pyloric glands is the argyrophilic *G cell* which secretes the peptide hormone *gastrin* [13]. These substances are discussed elsewhere (see chapter 12).

Pyloric Glands. In some species (koala, wombat, beaver, pangolin, grasshopper, mouse) the gastric glands are organized into large, complex glandular structures, set further away from the gastric lumen but communicating with it either by a single or a limited number of large orifices [30].

The small, round orifices of the gastric pits of the cardiac, fundus and body of the stomach are replaced by oblong slits in its pyloric region. The pits themselves are considerably deeper than elsewhere (fig. 5/1D). Further, the gland-occupied lamina propria is much shallower or thinner than in body and fundus, so that the ratio of the depth of the gastric pits to the thickness of the glandular lamina propria is again 1:1, although the total thickness of the mucosa is greater here than in the cardiac area where the ratio was the same. The *pyloric glands* which open into the pits are shorter, have larger lumina and are sufficiently convoluted that one rarely sees a gland sectioned along its length. They are also less densely packed within the lamina propria and their basal portions are more branched than the gastric glands. At the pylorus itself the glands are lengthened and enlarged and may penetrate through the muscularis mucosae – which is deficient here – into the submucosa.

The pyloric glands are, generally speaking, lined by a single type of cell which resembles and, in fact, has been said by some investigators to be identical with the mucous neck cells of the gastric glands proper and the cells of the cardiac glands. However, they are selectively stained with cresyl violet and the Giesma mixture of dyes [6]. Occasional enteroendocrine cells – primarily the gastrin-producing G cells – are seen. Parietal cells, though scarce, are apparently invariably present in both fetal and postnatal pyloric glands – especially in the sphincteric region. (In adults they may also appear in the proximal part of the duodenum, near the pylorus [31].) For the main part, however, the pyloric glands – like the cardiac glands – produce very little enzyme or acid but mostly mucous.

Intermediate Glands. An *intermediate zone,* some millimeters in width in humans, between the body and pyloric regions (fig. 5/1A) is described by some authors as a transitional zone in which a fourth type of gland, the *intermediate glands,* are found. These glands share characteristics of gastric and pyloric glands. In the dog – an animal widely used for physiological experimentation, the intermediate zone is unusually well developed, being some 1–1.8 cm in width [6].

Acid and Pepsin Producing Cells

Thus, the mucosa of the entire stomach secretes, but for all practical purposes only that of the body and fundus produces either acid or pepsin, for that is where the parietal and chief cells are concentrated. There are other cells which produce pepsin, however, as we will see later. The earliest convincing evidence that these two cells had the roles which have been ascribed to them was provided by *Linderstrom-Lang* et al. [34], who sliced serial sections of fresh-frozen gastric mucosa parallel to the mucosal surface. They found the pH of the sections which included the necks of the glands (where the parietal cell count is highest) to be the lowest (fig. 5/6). In the sections which included the bases of the glands, peptic activity and chief cell count was highest. The superficial gastric mucosa and the entire pyloric antrum – essentially devoid of both these cell types – were not acid and

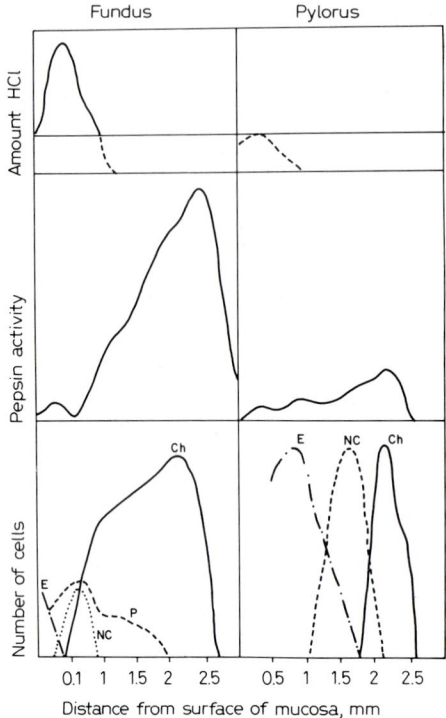

Fig. 5/6. The relationship between the number of cells of a given type and the production of HCl and pepsin. The cells were counted in serial sections cut parallel to the luminal surface. E = Surface epithelial cells; NC = mucosal neck cells; P = parietal cells; Ch = chief cells. Reproduced, with permission, from *Linderstrom-Lang* et al. [34].

demonstrated very little peptic activity. This simply confirmed what had been suspected for a very long time.

Secretory granules can now be isolated and identified, but acid production by isolated parietal cells, for reasons which will be evident later, can still only be surmised. Among the more compelling evidences indicting the parietal cell as the actual source of HCl are the observations that acid is found only in stomachs having parietal cells as an element of the mucosa, and that the larger number of parietal cells present, the greater the acid production. (In human disease, the number of parietal cells is closely correlated with the acid-producing capacity of the stomach. In atrophic gastritis, for example, both parietal and chief cells are greatly

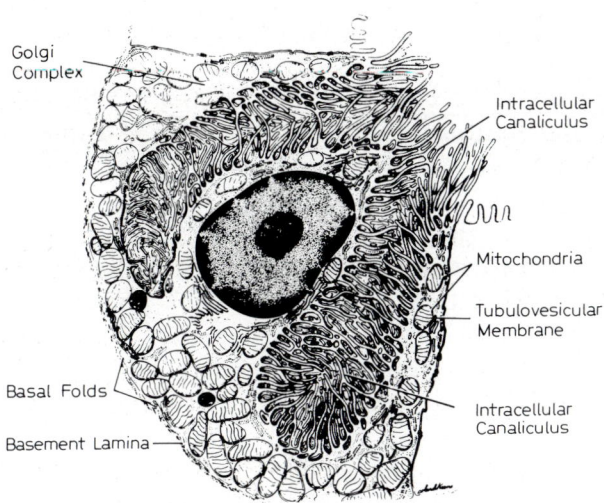

Fig. 5/7. Schematic representation of a parietal (oxyntic) cell. The cell is depicted during the actively-secreting phase, during which the intracellular canaliculi are prominent, although nearly occluded with microvilli, and there is a relative paucity of tubulovesicular structures. From *Ito* [21].

reduced in number; consequently, the acid and pepsin activity of the gastric juice is reduced proportionately.) Further, acid is initially detected in the fetal stomach immediately following the differentiation of parietal cells [21].

Radioautographic studies performed with labeled vitamin B_{12} strongly suggest that the parietal cells are, in humans, the site of production of a glycoprotein, *gastric intrinsic factor,* necessary for vitamin B_{12} absorption. In other species, however, this substance may be present in other cells.

Ultrastructure of the Parietal (Oxyntic) Cell

Being the first cells of gastric glands to be described, these cells have retained the nondistinct name 'parietal', although the more descriptive term 'oxyntic' (= 'acid-forming') has been more recently applied.

These cells are usually larger than their surrounding chief or mucous neck cells and are ovoid or pyramidal in shape (fig. 5/7). Their centrally-placed nuclei are spherical, and occasionally binucleate or even multi-nucleate cells are seen. Mitotic figures among parietal cells are a rare occurrence. This is due in part to the fact that parietal cells (as well as chief cells) are relatively long-lived. However, recent evidence indicates that parietal cells are formed primarily from dedifferentiation of other cell types (mucous neck cells in particular). This at least appears to be the case in circumstances which cause a marked increase in the number of parietal cells, as in surgical resection of the intestine [61].

Foremost among the features which characterize parietal cells is the presence of a deep, circular cytoplasmic invagination extending into the cell from its apex (fig. 5/7). The invagination is commonly referred to as an *intracellular* (or *secretory*) *canaliculus,* because on section it does simulate the intracellular canaliculi of other cells. However, electron microscopy reveals that these 'canaliculi' are not actually within the cytoplasm (at least during the active secretory phase of the cell) but are complex trenches or circular grooves formed by involution of the cell membrane of the apical surface. Thus the contents of the so-called intracellular canaliculi are really extracellular. The invaginations extend deeply into the cell from the apex toward – and sometimes nearly reaching – the base of the cell, often completely surrounding the nucleus in such a way that, when sectioned, they are seen on all sides of the nucleus, except basally (see fig. 5/7). During the nonsecretory of 'resting' phase, however, the canaliculus may become internalized, with the cell closing off its apical opening to the lumen of the gastric gland, the canalicular contents thus becoming truly intracellular (fig. 5/8) [23].

Microvilli. The cell membrane lining these channels – and that covering what there is of an apical or luminal surface of the cell – forms an amazing abundance of microvilli which project into (and nearly occlude, in large part) the lumen of the canaliculi and, apically, into the lumen of the gastric gland. The microvilli display an inconstant arrangement and shape, their number and length varying according to the secretory activity of the cell. The involution of the cell membrane to form the secretory canaliculus and the formation of microvilli combine to give the parietal cell a tremendously increased luminal surface area – in spite of the fact that the cell may appear, under light microscopy, to have been almost 'squeezed out' and away from the lumen of the gastric gland by adjacent cells to the point where it apparently has almost no apical or luminal surface at all. The involuted membrane studded with microvilli provides a vast area for membrane phenomena – a fact of great importance to the production of hydrochloric acid by the parietal cells.

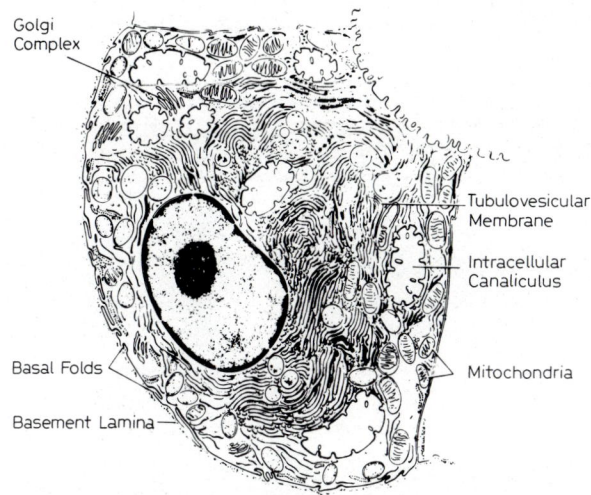

Fig. 5/8. Schematic representation of a parietal (oxyntic) cell. The cell is depicted here during the nonsecreting phase. An abundance of tubulovesicular structures occupy much of the cytoplasm. The intracellular canaliculi are 'sealed off' from the gastric lumen (i.e. they have become 'internalized'). They appear distended but largely devoid of microvilli. From *Ito* [21].

Free acid is not found within the parietal cell, nor are membrane-bound, acid-containing secretory granules found. Studies on living gastric mucosa using a variety of indicator dyes such as neutral red have shown that the cytoplasm of the parietal cell is, in fact, somewhat alkaline. (The apparent acidophil of these cells under light microscopy will be explained shortly.) However, the secretory canaliculus and the lumen of the gastric gland contain free acid. Thus, the cell membrane is a highly selective structure which both segregates and secretes the constituents of the acid. The membrane of the parietal cell microvilli lacks the prominent morphological coating (the filamentous glycocalyx) conspicuous on the other cells in the gastric gland – and in the GI tract in general. Instead, experiments have shown that their membranes stain positively for neutral glycoconjugate, an unusual carbohydrate cell coating that has not been found on any other cell type. It has been suggested that this feature of parietal cell microvilli may be of special significance in binding water and thus provide for the proton transport believed to take place at this site [53, 10 and 56 in ref. 21].

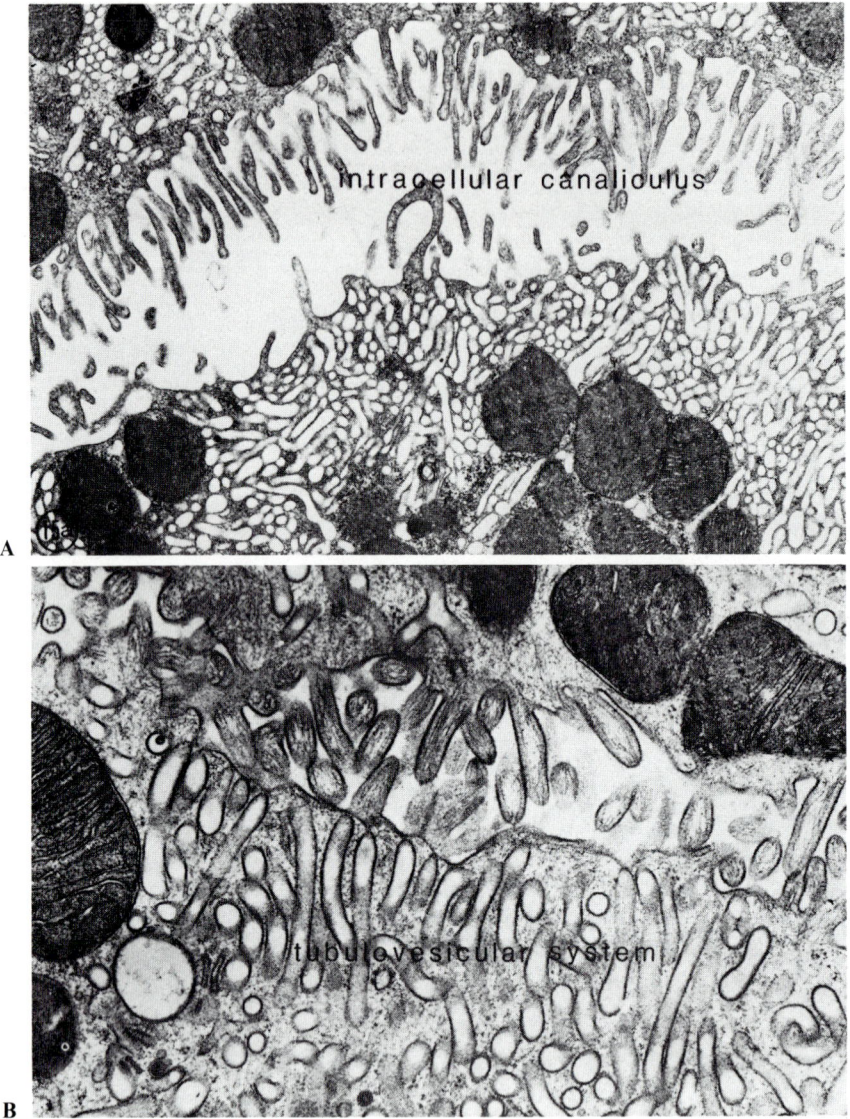

Fig. 5/9. Electron micrographs of portions of mouse parietal cells. **A** Although microvilli are moderately abundant in the intracellular canaliculus, the abundance of tubulovesicular membranes suggests intermediate acid secretory activity. Although the tubulovesicular membranes closely approximate the plasma membrane, attempts to show continuity between the two have been unsuccessful. × 19,000. **B** This tissue was quick-frozen in the living state and fixed to ensure minimal artifact. The predominantly tubular-shaped tubulovesicular system does not appear to open into the lumen of the intracellular canaliculus. × 28,000. From *Ito* [21].

Tubulovesicular System. The cytoplasm just deep to the cell membrane of the secretory canaliculus is permeated by an extensive system of smooth-surface, minute, convoluted tubules and vesicles – the *tubulovesicular system* (fig. 5/7–9). These elements are considered to be distinct from smooth endoplasmic reticulum. Just as there is variation in the degree of development of the microvilli with the cell's secretory activity, there is considerable variation in the abundance of the tubulovesicular system. During active acid production (such as that in response to electrical vagal stimulation or that stimulated by secretogogues such as histamine, carbachol, gastrin, insulin or reserpine) there is a rapid depletion of the membrane (fig. 5/7), the total reduction of the tubulovesicular system being as great as 90% [51]. Concurrently, the canalicular microvilli increase in size and number, becoming prominent and luxuriant, occluding the canalicular lumen. Stereological measurements have shown increases of fourfold and greater in the microvillar surface of parietal cells in mice during such stimulation [19, 51, 61, 63]. Conversely, in nonsecreting cells ('resting cells' or cells inhibited by acetazolamide), the tubulovesicular system becomes very extensive while the canaliculi appear dilated, there being relatively few microvilli and those of shorter length (fig. 5/8). Further, some investigators [23, 51] noted that during such times the intracellular canaliculus became completely internalized, closing off its apical opening to the lumen of the gastric gland – although this was not a constant feature of all parietal cells in the nonsecretory configuration.

Little is known about the process by which the cell reverts from one morphological state to the other. An obvious process which would explain the rapid loss of tubulovesicular membrane and concurrent increase in microvillar membrane would be direct continuity, the membrane becoming exteriorized by membrane flow or exchange. This has been suggested by several investigators. Actin filaments observed within the microvilli [60] and microtubules observed within the apical cytoplasm [11] have been suggested as a possible means of withdrawing the microvillar membrane from the surface and into the cytoplasm to reconstitute the tubulovesicular membrane. However, this process of membrane exchange has not been demonstrated with tracers and there is no evidence to date that such transposition occurs in mammalian parietal cells [22]. Further, some distinct differences in the microvilli and tubulovesicular system have been noted. Microvilli show positive histochemical reactions for ATPase and *p*-nitrophenylphosphatase; the tubulovesicular system does not [47]. Carbonic anhydrase appears to be localized in the microvillar cores and on the cytoplasmic surface of the cell membrane, but is not consistently associated with the tubulovesicular system [58]. Attempts to show continuity between the tubulovesicular system and the plasma membrane of parietal cells have been unsuccessful (fig. 5/9) [22].

It has been widely speculated, but never shown, that the tubulovesicular system plays some role in the transport of chloride and hydrogen ions across the cell. *Sedar* [52] has suggested that water transport might be a function of the tubulovesicular system since the rate of water transport by the cell is enormous during hydrochloric acid secretion.

Mitochondria. The secretory canaliculus, as an involution of the cell membrane into the cytoplasm, encroaches upon the cytoplasm to occupy much of the volume of the parietal cell. The remaining cytoplasm is literally stuffed (30–40% occupied) with plump mitochondria which appear spherical or elongated on section (and therefore are probably sausage shaped) [18, 22]. These mitochondria, in turn, are densely packed with elaborate cristae and display numerous dense granules in the intercristal matrix. The presence of such abundant and complex mitochondria in the parietal cell suggests that the metabolic processes in these cells are highly energy consuming. In fact, the parietal cell exhibits histochemical peculiarities which characterize it as one of the cells with the highest observable energy metabolism [26].

The gastric mucosa has long been known to be a tissue of exceptionally high oxidative activity. Studies on isolated enriched parietal cell fractions have shown an oxygen consumption rate about five times higher than that of mucous cells [45, 53]. It is the mitochondria which stain strongly with acid dyes such as eosin, giving the parietal cell a strongly acidophilic appearance under light microscopy. Although an early light microscopy study [33] suggested visible changes in the mitochondria associated with secretory activity, most recent studies do not indicate significant changes in mitochondrial size or number in the active or nonsecretory phases of the cell.

Ribosomes. The cytoplasm contains few ribosomes, little rough-surfaced endoplasmic reticulum and no secretory granules. The Golgi complex is frequently located between the nucleus and the cell base, in contrast to the supranuclear position that it occupies in most epithelial cells, but it may be located almost anywhere in the cytoplasm. Several Golgi complexes may be found in the peripheral or basal cytoplasm of a single cell [21].

Basal Border. The basolateral membrane of the parietal cell apparently carries the receptors for histamine, gastrin, acetylcholine, and other substances which stimulate secretion of hydrogen ions into the lumen and bicarbonate ions into the interstitium [42]. The basal border of the parietal cell may be smooth, but more often has uniform basal folds approximately as thick as the diameter of the microvilli. These plications are most prominent on parietal cells and are not a regular feature of gastric epithelial cells. The functional significance of this basal specialization is not known, but may represent an amplification of the surface area associated with the release of bicarbonate into the interstitum and circulatory system [21].

Gastric Digestion and Secretion

Food is present in the stomach either as chyme, in the active pyloric region or as a semisolid pultaceous mass in the storage part of the stomach. Fluid drunk runs around the mass, mostly via the lesser curvature (Magenstrasse) and into the pyloric antrum. It is the same with the gastric secretion. Even though the digestive secretion of the stomach originates in the body (the storage part of the stomach) it penetrates into the mass of food there only slightly (see fig. 4/4). Mixing of food with secretion takes places in the lumen of the antrum. Here the semisolid bolus becomes chyme, which is a suspension of food in gastric juice and any swallowed fluid.

The principal digestive constituents of this juice are hydrochloric acid and the proteolytic enzyme pepsin. Hydrochloric acid secretion has been the subject of an immense amount of study and pepsin of very much less. No doubt this is because the secretion of such large amounts of a highly concentrated mineral acid is a unique biological phenomenon and, because the acid of stomach bears the blame for the fairly common occurrence of peptic ulceration of the stomach, duodenum or esophagus, and because acid is easy to measure; peptic activity is a little more difficult to study.

Nonparietal Secretion (table 5/I)

The acidity of the gastric contents is variable, depending on the amount of acid secreted, on the buffering by food and on the other secretions of the stomach which are not acid. These other secretions are known as *nonparietal.* The acidity of parietal cell secretion is constant, but the volume secreted is variable. The maximal amount which can be secreted, as might be imagined, is a measure of the number of active parietal cells or of the

Table 5/I. Composition of nonparietal secretion concentration (mEq/l) [human data from Makhlouf et al., 41]

	Na^+	K^+	HCO_3^-	Cl^-	All cations	All anions
Human	136.7	6.4	25.0	117.8	143.1	142.8
Dog	140.8	7.6	14.3	134.9	148.4	149.2

Dog: mean of various sources.

parietal cell population, as it is usually called. Persons who hypersecrete acid have been found to have a higher than normal maximal stimulated output and a proportionately larger parietal cell population.

Most glands produce isotonic secretions. The stomach is no exception, but the parietal cell secretion contains significant concentrations of only H^+, K^+ and Cl^-. The $[K^+]$ is 1/20 that of $[H^+]$. This means that this isotonic secretion must be highly acid indeed at all rates of secretion. In fact, in man the $[H^+]$ is about 150 mEq/l and the $[K^+]$ about 7 or a little higher, as expected, than in the plasma. When parietal secretion is small, even though the concentration is constant, less acid will enter the stomach and its neutralization by food and by nonparietal secretion will be significant. Neutralization becomes less significant as the rate of parietal acid secretion increases.

Nonparietal secretion appears to be constant in amount and continuous at rest and during active digestion. It contains mucin, which will be discussed later. It is the sole source of Na^+ and of HCO_3^- in gastric juice, but it also contains Cl^- and K^+ both of which of course are constituents of the parietal secretion. Its pH and its HCO_3^- concentrations are close to those of plasma in man, therefore, it is slightly alkaline in reaction. The importance of this will be seen later on (table 5/I). This secretion has been the subject of a great deal of study in an effort to explain why the secreting side of the gastric mucosa is more electrically negative than the nutritive side (−45 mV). The answer is that Cl^- is actively secreted into the gastric lumen [20]. There is evidence now that HCO_3^- is also, but it cannot be as important, because if the Cl^- transporting machinery is abolished the transmucosal potential difference virtually disappears. At one time it was proposed that this electrical energy contributed to the secretion of HCl; this is unlikely since the transmucosal potential difference is seen throughout the stomach in both acid secreting and nonsecreting mucosae.

Gastric Acid (Parietal Secretion)

When active acid secretion starts the transmucosal potential difference (PD) in the parietal cell region falls since now a cation is being actively secreted as well. In mucosae in which the active transport of Cl^- has been blocked and which in consequence has no resting transmucosal PD, a positive PD is seen with the onset of gastric acid secretion.

Active secretion is not the sole source of nonparietal Cl^-; HCO_3^- can as elsewhere exchange with Cl^- (the surface epithelial cells contain carbonic anhydrase), but this is not likely to be of great importance since the gastric mucosa is not a very leaky one.

H^+ Secretion. The source of this remarkable production of H^+ has been a puzzle. The most widely discussed hypotheses in recent times have been the carbonic acid idea which held that the dissociation of H_2CO_3 provided the H^+. This gained support from the evidence of abundant carbonic anhydrase in and around the gastric glands. The idea was abandoned when it was found that the then available carbonic anhydrase antagonists did not abolish stimulated gastric acid secretion [7]. The redox theory held that dehydrogenation reactions, which took place in the course of glycolysis and oxidation cycles, provided the H^+. In other words, H^+ secretion was linked to oxygen consumption. It is very true that acid secretion can easily be reduced by anoxia, but when the energy expended, calculated from the acid produced, is compared with that calculated from the increment in oxygen consumption resulting from secretion, clearly ridiculous secretory efficiencies of several hundred per cent are obtained.

The currently favored theory is a modification of the carbonic acid idea which avoids the difficulty with carbonic anhydrase antagonists. It started with the work of *Davies and Edelman* [7], who found, in frogs, that although the antagonists did not block acid secretion, in the course of time the mucosa underwent destruction consequent to the accumulation of OH^-. This would happen if water were the origin of the H^+, which having been secreted, would leave a OH^- behind. Carbonic acid, it is felt, provides the H^+ to neutralize this OH^-, the HCO_3^- entering the blood stream in exchange for the Cl^-, which then enters the lumen of the stomach with the acid. The source of the carbonic acid is CO_2 and water, both of which enter from the plasma (fig. 5/10).

There is a one to one relationship between bicarbonate leaving the cell to enter the blood stream and H^+ or Cl^- leaving the parietal cell for the ducts of the gastric glands. Thus, in carrying out experiments with isolated mucosa

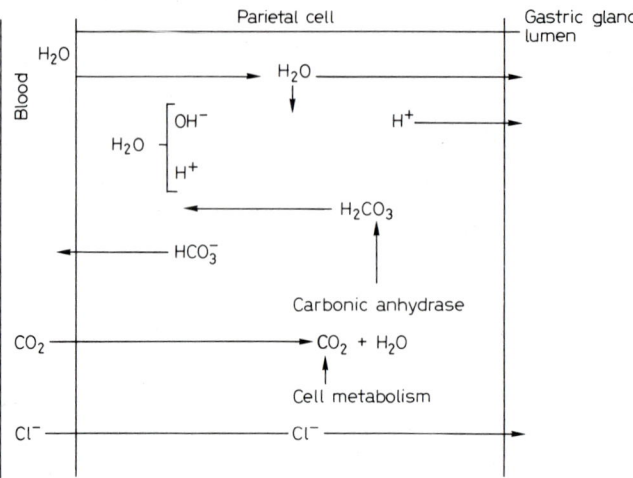

Fig. 5/10. A hypothesis to explain the formation of HCl by gastric parietal cell.

a sheet must be used which completely separates the secretory from the nutrient surface otherwise the acid will be completely neutralized by the bicarbonate. This is why acid secretion cannot be measured directly in isolated cells. In the intact animals or man, when the stomach is secreting vigorously following meals, the venous blood draining it reflects its bicarbonate load by becoming more alkaline than usual. This is known as the *alkaline tide*. Even the urine formed during vigorous gastric acid secretion becomes more alkaline in reaction.

In the above scheme (see fig. 5/10) the only energy requiring movement of ions is the secretion of the H^+. The concentration gradient against which it is secreted is very large since the $[H^+]$ in the cell is very low and in the lumina of the gastric glands very high; Cl^- and HCO_3^-, on the other hand, move passively. The Cl^- going into the stomach follows the actively secreted H^+. Between plasma and parietal cell, Cl^- and HCO_3^- simply follow concentration gradients just as they do in and out of the erythrocytes. It has been argued, however, that this chloride shift is too rapid to be entirely dependent on passive forces. It is felt now that it is assisted by a HCO_3^- halogen-activated ATPase [27]. If halogens, in addition to Cl^-, are present in the plasma these will be secreted with H^+ in proportion to their plasma concentrations.

5   Stomach: Gastric Secretion   109

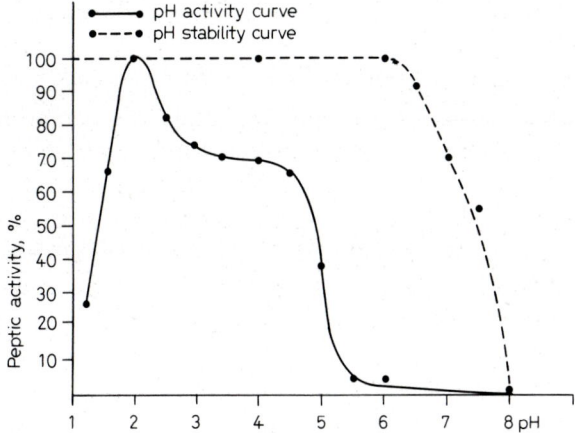

Fig. 5/11. The pH stability and activity of pepsin. Reproduced, with permission, from *Piper and Fenton* [44].

Pepsin [59]

Because acid gastric juice always contains pepsin, the notion has arisen that acid and pepsin are always secreted together. We have seen that acid and pepsin originate in separate cells, but only the parietal cell seems to produce water. Pepsin cannot be collected or measured unless it is in solution. If, however, water is introduced into gastric pouches in dogs it is quite easy to demonstrate that under certain circumstances (to be mentioned later) pepsin can be secreted copiously in the absence of acid secretion.

Pepsin is a proteolytic enzyme active only in a highly acid pH. It is present in the granules of its cells of origin, the chief cells, as an inactive zymogen or precursor. The precursor, pepsinogen, is a molecule with a molecular weight of 42,000. This is activated almost instantaneously in solution at pH 2, but progressively more slowly as the pH rises. Above pH 6 it is not activated at all (fig. 5/11). Activation results from the cleavage of six basic peptides with a combined molecular weight of 7,000, from the parent molecule. One of these can recombine reversibly with pepsin, thus inactivating it again if the pH rises above 5.5. Above pH 5.5 pepsin is without proteolytic actively. Indeed, by pH 4 the activity has started a precipitous fall. The pH optimum for pepsins is in the neighborhood of 2. It is difficult to be more precise because it is possible electrophoretically to separate pepsin and pepsinogen into several distinct entities with pH optima ranging from 1.6 to 3.6. Various terminologies have been used to describe these [49, 50,

59]. The simplest, although it does not follow international convention, is that of *Samloff* [49, 50]. Pepsins which are found only in the chief cells he calls *pepsinogen I*. Those found elsewhere by immunofluorescence he calls *pepsinogen II*. Group I is made up of 5 electrophoretically separable molecules the ratios of which are genetically determined and vary from person to person. Group II has only two components. We have not referred, heretofore, to pepsin as occurring anywhere except in the body of the stomach where there are chief cells, but some peptic activity can be detected in the superficial mucosa of the whole stomach, in the pyloric glands and in the mucosa of the first part of the duodenum. This is all of the group II type.

Pepsinogens of both types can be found in the blood of normal people. In normal urine pepsinogen I can be detected. Here it is called *uropepsinogen*. Leakage of digestive enzymes back into the blood stream is a common occurrence and can be of diagnostic significance.

Pepsins will split most naturally occurring proteins including mucins. Pepsins have a special affinity for bonds involving phenylalanine, tyrosine or leucine. They are *endopeptidases,* i.e. they act within the molecule rather than at its ends. Pepsins clot milk. The acid itself denatures the protein causing it to precipitate, but pepsins remove a large peptide. The remainder precipitates as a calcium salt known as *paracasein*. In the abomasum of young ruminant animals, which corresponds to the stomach of the nonruminant, a milk-clotting enzyme renin is found. This is structurally distinct from pepsins, but its enzymic activity is identical.

Mucus [1]

This is the lubricant and the agent which helps to keep mucosa damp throughout the gastrointestinal tract. We have already seen its production by salivary glands (major and minor) by esophageal glands, by mucus glands and by surface epithelial cells in the stomach. We will encounter it also in the small intestine and throughout the colon. Since the stomach is inclined to digest itself and the duodenum, and not infrequently does, and since mucus secretion seems to be one of its major protective mechanisms we will deal with its basic nature here [1].

Properties of Mucus. Mucus unlike the other digestive secretions is not completely water soluble. Even in a great excess of water it forms a gel. The higher its concentration the more viscous the gel. In other words, the large branches of its complicated molecules are sticky. Water can enter between

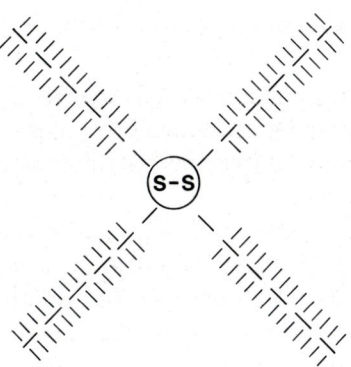

Fig. 5/12. Diagrammatic representation of the structure of gastric mucins showing the central protein (MW 70,000). The protein backbones hold to it by S-S bonds and the carbohydrate side chains attach to the backbone. Modified from *Allen* [1].

them (fig. 5/12). The molecules separate to an extent, but never completely, when they are intact. The fully hydrated mucus gel is in the region of 95% water plus its dissolved electrolytes. This electrolyte solution is, in the stomach, the nonparietal secretion, which as we have seen has a high pH and a relatively high concentration of HCO_3^-. Not only does mucus adhere to itself, but it adheres also to the mucosa, probably to the cells there which are producing it. Thus, it provides a barrier between the lumen of the stomach and the mucosa [56]. It functions rather like a thermal insulating material only its extensive interstices are filled with a slightly alkaline fluid rather than with air. It provides what is now called an *unstirred layer,* i.e. one in which there is no mixing. Mucus will protect a mucosa only if it is a barrier to the agents which are harmful, in the case of the stomach H^+. If it is a barrier, it is effective only as long as it remains intact.

Mucus is not an impenetrable barrier to the back diffusion of H^+ but this is of no consequence as long as HCO_3^- secretion keeps pace with back diffusion of H^+. Much more H^+ than HCO_3^- is secreted during digestion but H^+ mixes freely in the entire gastric lumen whereas the HCO_3^- enters and is entrapped in the mucus layers. Microelectrical studies have substantiated a pH gradient across the mucus layer from pH 2 on the luminal surface in the antrum to pH 7 on the mucosal side. Mucus itself is a poor buffer but its contained HCO_3^- is a good one [16, 17].

Not only is mucus not an impenetrable barrier to H^+, but it does not resist proteolysis by pepsin [43]. It can, therefore, exist as a barrier only if its secretion can keep up with its destruction. Proteolysis results in solubilisation of the molecule. In other words, its side chains are removed and it loses its stickiness. Proteolytic enzymes further down the gut do likewise. Compared with proteolytic action on food proteins the action on mucus is slow since optimal proteolysis occurs in solution and the mucus gel slows diffusion of the large proteolytic enzyme molecules. This diffusion barrier also conceivably protects the mucosa from the action of toxins and bacteria.

The mucus layer obviously cannot cover the mouths of the gastric glands. Either the cells of the neck have special properties which prevent the adherence of mucus or more simply and therefore, in my view, more likely, the flow from the glandular orifices simply washes the mucus out of the way.

Mucus is not a chemically homogenous substance but all varieties have general features in common. Each consists of a protein backbone with carbohydrate side chains (fig. 5/12). These side chains are attached to short peptide leaders which are in turn anchored via disulphide bridges to a large protein core. Disruption of these bridges destroys the gel. The side chain sugars are N-acetylgalactosamine, galactose, fucose and various sialic acids. The proportion of these is variable and indeed not all of them are invariably present. Related glucopolysaccharides in connective tissue and joints contain uronic acid, but gut mucus does not. Seventy per cent of the weight of the mucus molecule is carbohydrate. The number of sugar molecules per side chain may vary from 2 to 22. It is these side chains which are responsible for the stickiness of the molecule and it is these which are severed from the protein core during proteolysis after which both they and the naked protein will then go into solution and the gel will be disrupted.

An intriguing property of many of the mucus glycoproteins is that they are antigenic to the AB0 blood groups. They can therefore interfere, if present, with the erythrocyte agglutination which is the basis of blood typing. The structure of the carbohydrate chain is the determining feature. Group A activity is found in molecules with carbohydrate chains ending with N-acetylgalactosamine; group B chains end with galactose. About 20% of people secrete neither and have a terminal fucose. Whether one's glycoproteins have either A or B antigenic activity or none is genetically determined and has been related to susceptibility to peptic ulcer disease.

Mucus Secretion. Mucus exists preformed in vesicle in the cells which produce it. The vesicles fuse with the plasma membranes and then disgorge their contents a vesicle at a time. This occurs in response to local stimuli such as pressure, distension, and irritation and to stimulation via nerves or hormones. Strong acids and irritants such as mustard, eugenol, and acetic acid have been used to stimulate secretion, but with powerful stimuli of this sort rapid release may be the result of disruption of the whole apex of the cell or even of exfoliation of surface epithelial cells.

Under physiological conditions food and any sort of physiological irritant such as gastric acid in contact with a mucous membrane will cause secretion. The mucous membranes of the gut and those glands, such as the salivary glands, which produce it, are all stimulated to secrete by cholinergic stimulation. In the stomach vagal stimulation or cholinomimetic drugs given either intravenously or topically will increase secretion and all are blocked by atropine. A logical presumption is that any physiological process in the gastrointestinal canal which increases a glandular secretion will increase mucus along with the secretion unique to it. The hormonal phase of gastric secretion for example which is known to stimulate acid and pepsin secretion will also increase mucus. Secretin, the duodenal hormone which stimulates the secretion of water and bicarbonate by the pancreas, is a potent stimulant of gastric mucus and pepsin, but not of acid secretion. This is intriguing since an acid duodenum is the most potent stimulant for secretin release. Certain prostaglandins (which have achieved some prominence as possible therapeutic agents in peptic ulcer disease because they depress acid secretion) also stimulate gastric mucus production. Nearly all of the anti-inflammatory and nonopiate analgesic agents such as salicylates, butazolidine, glucocorticoids and ACTH depress mucus production. Gastric ulceration has been associated with prolonged therapy with many of these.

Intrinsic Factor [9]

An important glycoprotein in gastric juice is intrinsic factor. This is necessary for the absorption of vitamin B_{12}, important for the maturation of red blood cells in bone marrow. Unless the vitamin is bound to this protein it cannot be absorbed from its specific site, the distal ileum. Intrinsic factor differs a good deal from the mucus glycoproteins. It is only 15% carbohydrate and it appears to be produced in the parietal cells. Atrophy of the gastric mucosa very often spares the antrum, but such patients nevertheless will require parenteral vitamin B_{12} to prevent pernicious anemia, because in the absence of the acid secreting mucosa they lack intrinsic factor. As would be suspected from its parietal cell origin anything which increases acid secretion will also increase intrinsic factor production. This large B_{12} intrinsic factor complex in turn binds to the absorption site in the lower ileum. Gastric acid facilitates the release of vitamin B_{12} from protein complexes in the food. Patients with impaired pancreatic secretion also show impaired B_{12} absorption. The probable explanation for this is the presence of the plasma B_{12} transporting protein in gastric juice and saliva, here called *R protein*. This is a globulin glycoprotein which binds B_{12} avidly in the duodenum. Pancreatic proteases split the B_{12} from it thus permitting the vitamin to combine with intrinsic factor which is normally in great excess. B_{12} combined with R protein is not absorbed.

Regulation of Secretions (fig. 5/13)

Cephalic Phase

Before breakfast the stomach is empty. It contains only nonparietal secretion which is usually slightly alkaline and may contain regurgitated duodenal juice. The empty stomach becomes active for brief periods (see below) regularly at 100- to 110-min intervals when migrating myoelectrical complexes start. At these peaks nonparietal secretion increases, acid may; it

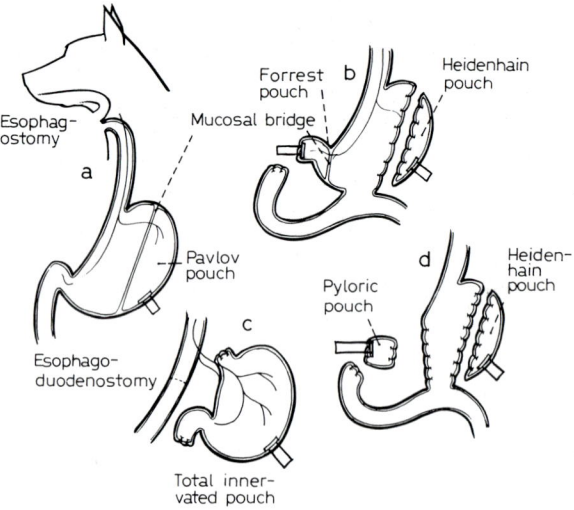

Fig. 5/13. Various preparations commonly used in experimental gastrointestinal physiology. In both the Pavlov pouch and the innervated antral pouch (Forrest pouch) the pouch is separated from the main stomach only by two layers of mucosa back to back. Reproduced, with permission, from *Magee* [35].

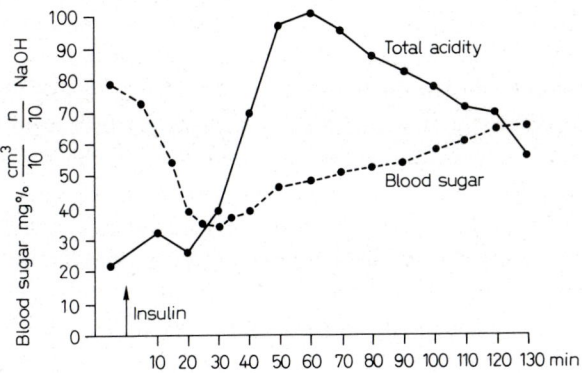

Fig. 5/14. The effect of hypoglycemia on the secretion of acid from an innervated fundic pouch. Reproduced, with permission, from *Roholm* [45].

is at this time that bile-stained duodenal juice may regurgitate to the stomach. Once food is placed in the mouth in man, or is seen, smelt or suggested in most experimental animals, a copious secretion of acid gastric juice high in pepsin will start. This is the cephalic phase of gastric secretion, so called because its stimuli all originate in the head. After successful bilateral vagal section between the head and the stomach this phase of gastric secretion is no longer seen. It can be duplicated by injecting acetylcholine into the arteries supplying the acid and pepsin secreting mucosa. Juice stimulated in this way, like the usual cephalically stimulated secretion, is characteristically high in pepsin. This phase of gastric secretion can be tested in a physiological manner in animals and man with intact vagi by sham feeding. In experimental animals an opening can be made in the esophagus such that swallowed food falls to the floor rather than entering the stomach. In man food can be chewed and then spat out. The former produces copious secretion. The latter in man for obvious reasons is not as satisfactory. Fortunately, there is an alternative which is used frequently in practice. Reduction of the glucose uptake by the hypothalamus duplicates the cephalic phase. In patients, therefore, following attempted bilateral vagotomy, if a low blood glucose, produced by injecting insulin, results in a copious secretion of acid and pepsin the attempted vagal section must be judged a failure. This is the *Hollander test* (fig. 5/14). A variation used experimentally employs 2-deoxy-*D*-glucose which blocks the intracellular metabolism of glucose and has the same effect as low blood glucose.

It is evident that vagal fibers supply and stimulate parietal and chief cells directly and activate them cholinergically, but this is not all. A small increase in secretion can be seen during sham feeding from vagally denervated gastric pouches. The explanation for this is that vagal fibers supply the G cells in the pyloric antrum and cause them to liberate the secretory hormone gastrin. This hormone, by its presence in the blood stream, causes the vagally denervated pouch to secrete acid even in the absence of its secretomotor nerves. Removal of the antrum or the employment of a number of the manipulations which depress the G cells, to be dealt with later, will abolish this. This indirectly stimulated juice is low in pepsin. The existence of this indirect mechanism, deduced as mentioned above, has now been substantiated by direct measurement of circulating gastrin. The cephalic phase and all the procedures used to duplicate it, sham feeding, direct vagal stimulation, insulin, 2-deoxy-*D*-glucose, raise blood gastrin levels. Atropine in sufficient doses will prevent this and of course vagotomy will abolish it. Following successful vagotomy the cephalic phase is abolished for 6 months or so. After this it often recovers, but rarely completely. The reasons for the partial recovery are still only guesswork.

Gastric Phase

Once the meal has been swallowed the cephalic phase is over, but secretion continues unabated until the stomach is empty. This is the gastric phase. The factors responsible for this are gastric, i.e. the food in the stomach itself acting on the mucosa keeps the secretion going. Simple mechanical pressure of the food bolus against the mucosa causes acid and pepsin secretion. This is true even of vagally denervated mucosa but is much more obvious if the vagi are intact. The mechanism is vagovagal, i.e. the afferents and efferents both travel in the vagus nerves (fig. 5/15). Claims have been made that a variety of chemicals including amino acids stimulate the mucosa directly. These are in dispute at the moment as absorption of the amino acids into the bloodstream and subsequent stimulation via gastrin has not been eliminated. The secretion which results from direct stimulation, although obvious, is small. Mechanical stimulation, particularly distension of the pyloric antrum, which produces no acid or pepsin itself, results in copious secretion of both. The amount of acid and pepsin produced is directly proportional to the pressure, within physiological limits, in the pyloric antrum. The juice so produced is typical hormonally stimulated low pepsin juice. Distension of from 10 to 20 cm H_2O stimulates the antral G cells and raises the levels of blood gastrin which, in turn, via the circu-

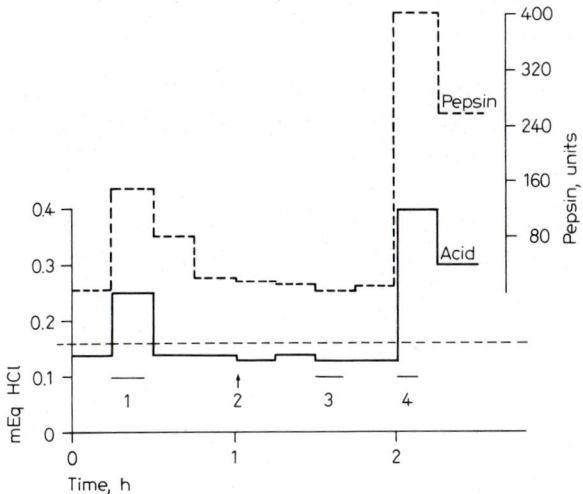

Fig. 5/15. At 1 the central end of the branch to the right dorsal vagus trunk was stimulated. At 2 both vagal trunks were sectioned and the stimulus was repeated ineffectively at 3. At 4 the peripheral end of the central vagal trunk was stimulated. Reproduced, with permission, from *Harper* et al. [15].

lation, stimulates the secretory mechanisms of the gastric glands and mucous cells. Degrees of antral distension insufficient to produce significant elevations in blood gastrin level can nevertheless still cause good gastric secretion. This may mean only that current assay methods are not sensitive enough (fig. 5/16). At higher pressures blood gastrin levels off and eventually declines. Distention is an important component of all physiological stimuli which result in gastrin release. Every food bolus consumed produces some distension and contains in addition potent chemical stimuli to the G cells. The important ones are digestion products of protein, especially the products of peptic digestion: the amino acid B-alanine, which is a constituent of carnosine present in meat; ethanol, caffeine and calcium ions also are potent gastrin releasers. Calcium, ethanol and some of the amino acids stimulate acid secretion after intravenous administration also. In the case of calcium and ethanol this almost certainly is the result of gastrin release via the blood stream. There are other factors not yet recognized, for example, decaffeinated coffee is just as strong a stimulant of acid secretion as are the ordinary varieties. An important but unphysiological local stimulant of gastrin

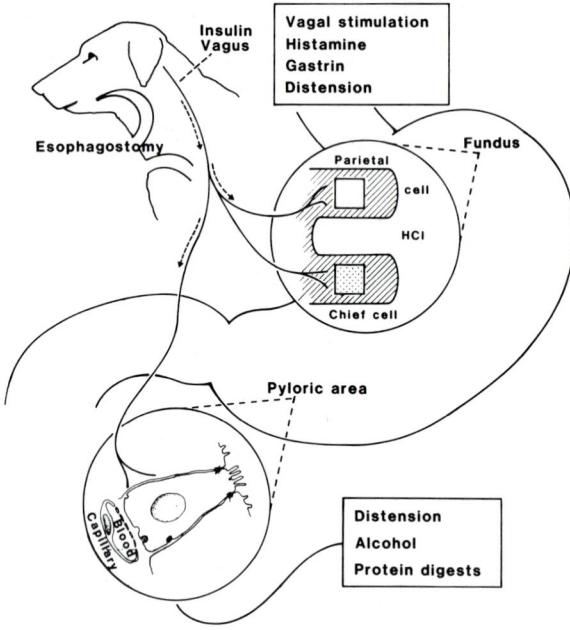

Fig. 5/16. Diagram to illustrate the regulation of gastric acid secretion. The vagus is shown innervating the chief and parietal cells directly and also the G cell in the pyloric area. Modified from *Magee* in *Ruch and Patton* [36].

release by G cells is acetylcholine. Lavage of antral pouches with this will increase acid secretion greatly from denervated gastric pouches.

For years before the development of immunofluorescence techniques (which made possible the identification of hormone-secreting cells, amongst them the G cells), it was felt that stimulation of G cells by distension or chemical stimuli in the antrum was via a mucosal nervous intermediate. The G cell, however, is close to the surface and since it is an open type enteroendocrine cell has microvilli which actually protrude into the lumen so that they can evidently be stimulated directly. However, the pharmacological evidence which gave rise to this nervous intermediate idea remains unchallenged. Acetylcholine applied topically releases gastrin; topically applied local anesthetics prevent gastrin release. Ganglionic blocking agents prevent gastrin release as measured indirectly by Heidenhain pouch secretion; circulating hormone levels do not seem to have been measured in these experiments. Finally, large doses of atropine depress gastrin production. Vagal stimulation is of course capable of causing an increase in blood gastrin levels, but those, and there are many, who propose the idea that acetylcholine is without effect on the G cell argue that even this is a non cholinergic action of the vagi. Acetylcholine has been outlawed by some investigators in an effort to explain why it is that the low basal levels of plasma gastrin seen in the fasting state are raised by small doses of atropine and by vagotomy. This paradox brings us to the action of gastrin on the parietal cell.

The orthodox view is that this is a direct action on specific parietal cell receptors. There are a few difficulties with this view: one is that the H_2 antihistamine agents block the action of gastrin, and another is that very small doses of atropine also block its action on acid secretion. After bilateral vagotomy the responsiveness of the parietal cell is depressed, but both small doses of atropine and vagal section raise resting plasma gastrin levels. The orthodox explanation for this is that the cholinergic fibres in the vagi to the G cells are inhibitory. This explanation is unsatisfactory because after the common operation now called parietal cell vagotomy (see fig. 4/2), in which the nerve supply to the pylorus remains intact, the resting plasma gastrin levels are still elevated. In addition, doses of atropine large enough to produce unequivocal muscarinic blockade depress gastrin levels.

Virtually every procedure, drug and disease, which reduces the ability of the parietal cells to respond to gastrin results in elevated resting plasma gastrin levels. It is generally observed that endocrine glands, which is what the G cells are, increase hormone production when the responsiveness of their target organ falls. An extreme example of this is seen in pernicious anemia. Such patients cannot secrete acid at all, but their blood gastrin levels are very high. If the relationship between gastrin and parietal cell were a simple stimulus and response one, the high gastrin levels would always mean copious secretion, but this is not always the case. It is true that in normal people and animals feeding raises the gastrin levels and secretion. Graded distension as seen above will raise the gastrin levels, but there are many circumstances under which gastrin levels are high and secretion low or vice versa. The secreting mucosa must be able to signal the G cells [40]. Cooling the antrum increases the response of a Heidenhain pouch to gastrin and the cooling of a pouch of secreting mucosa will increase the secretion of an uncooled pouch [28]. Both of these reactions require an innervated pyloric antrum and an intact sympathetic nerve supply (β-adrenergic agents depress gastrin-stimulated secretion [39]).

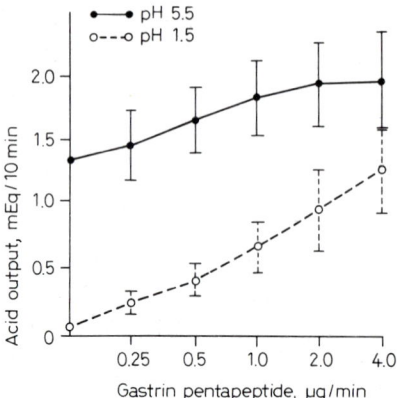

Fig. 5/17. Dose-response curves for continuous intravenous gastrin pentapeptide and acid secretion with antral acetylcholine at pH 5.5 and 1.5. Each point is the mean of eight experiments in eight dogs, the vertical bars represent the standard errors of the mean. Reproduced, with permission, from *Magee and Nakajima* [37].

Antral Inhibition

The G cells of the antrum are easy to stimulate (see above) except when the luminal pH falls below 2. Ten percent ethanol in the antrum at pH 5 is a good stimulant of secretion from a Heidenhain pouch, but not at pH 2 (fig. 5/17). Plasma gastrin levels are reduced by acidification of antral stimulants. The sensitivity of the G cells to stimuli falls progressively as antral pH falls from about 5.5 to become completely insensitive below pH 2. For many years an inhibitory hormone was suspected. It is now clear that there is no such hormone; however, the experiments done and misinterpreted as evidence for such a hormone still stand, begging interpretion. It has been a tacit assumption that if the secreting stomach were vagally denervated that is was totally denervated. Sympathetics have been admitted to blood vessels and such like, but not to parietal and chief cells. The paucity of experiments on sympathetic influences on gastric secretion is an indication of this attitude. There is now evidence that the pylorus via β-adrenergic fibres exercises tonic depression on the secreting mucosa. The old experiments taken as proof of a pyloric inhibitory hormone in which removal of the pyloric antrum increased Heidenhain pouch secretion can be interpreted as interruption of this sympathetic inhibitory pathway [40].

The orthodox explanation for the elevated fasting plasma gastrin in patients incapable of secreting acid (achlorhydriacs) is that acid inhibition is

5 Stomach: Gastric Secretion

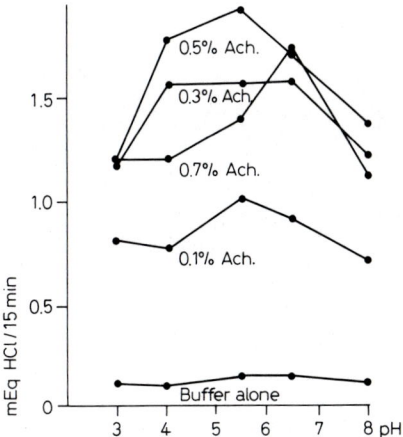

Fig. 5/18. The effect on Heidenhain pouch acid secretion of bathing the antral pouch with acetylcholine for 15 min at various concentrations and pH. From *Antinone* et al. [3].

no longer present. Acid inhibition is a brake upon gastrin release; therefore, it cannot occur except during active gastrin release, i.e. after meals. During fasting, however, there is no intragastric stimulation of gastrin release; therefore, the explanation must be false. Alkaline solutions do not by virtue of their pH stimulate gastrin release, in fact pH's much above 7 start to depress release (fig. 5/18).

Acid inhibition of gastrin release is a feedback mechanism which prevents the stomach from becoming excessively acid and is one of the stomach's important protective mechanism.

Intestinal Phase (fig. 5/19)

Pavlov's school found that the presence of food in the proximal part of the small intestine could cause acid secretion from a vagally denervated gastric pouch. Patients who have had their pyloric antra removed have low fasting plasma gastrin as might be expected, but they nevertheless show elevations following meals due, undoubtedly, to the entry of food into the intestine. The maximal secretory response from this source is only about 10% of that from the antrum. The stimuli including distension are exactly the same as for the antral G cells and gastrin is the hormone released.

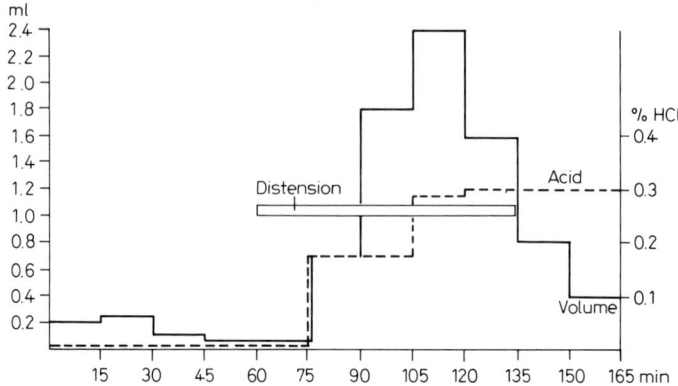

Fig. 5/19. The effect of distension of an intestinal loop on acid secretion. Reproduced, with permission, from *Sircus* [51].

Gastrin

Since 1911 it has been established that the gastric phase of gastric secretion is hormonal and more recently the intestinal phase also. This certainty was based entirely on indirect but classical methodology. Transplanted pouches of fundic mucosa secreted following food. Removal of the pyloric antrum abolished this. The effect could be duplicated by intravenous injections of extracts of antral mucosa, but unfortunately extracts of all sorts of other tissues also could stimulate secretion. The case for an antral hormone was not, therefore, a watertight one. The fly in the ointment was histamine, which is a potent gastric secretory stimulant that can be extracted from any tissue in the body. When this was recognized there were many who held that histamine itself was the secretory hormone. The isolation of several nonhistamine acid stimulants from antral mucosa and the final determination of the amino acid sequence of a potent acid and pepsin stimulant from the antrum, by *Gregory* [12] and his colleagues, did not end the debate. To do this it had to be shown, as it was, that *Gregory's* [12] gastrin (which became accepted as *the* gastrin) was not a histamine liberator. Histamine has had a comeback from oblivion in that a new class of antihistamine drug (H_2 blockers) depresses not only the action of histamine on gastric secretion, but also that of gastrin and cholinomimetics. It is not very good against pepsin secretion. Two explanations have been advanced to deal with this new difficulty: one, the final common mediator idea, states that intracellular histamine is the ultimate stimulus for the parietal cell, whether gastrin, the vagi or cholinomimetics are actually acting on it. The other notion states that the parietal cell has specific receptors for gastrin, histamine and cholinomimetics and that optimal secretion is obtained only when all three are active. H_2 blockade would eliminate the histamine receptor. This hypothesis would mean that H_2 blockers could reduce, but not abolish gastric acid secretion, but in dogs this is not so.

Pepsin secretion, which is stimulated by all three stimuli, has not been included in this debate. Its stimulation by gastrin and cholinomimetics can easily be diminished by atropine, but unlike gastrin-stimulated acid secretion, also by ganglionic blockade [38]. Thus the stimulation of pepsin secretion by gastrin seems to need a ganglion with nicotinic transmission and a postganglionic muscarinic site for optimal action. The behaviour of the parietal cell is different in that ganglionic blockade does not diminish the action of gastrin; atropine does in very small doses in dogs. In man the action of cholinomimetic agents and of atropine are much less obvious than in dogs.

After vagotomy or atropine both parietal and chief cells are much less sensitive to gastrin than before. This, according to the unorthodox view expressed earlier, explains the elevated resting blood gastrin levels and is also the rationale behind vagotomy as a treatment for patients thought to secrete excessively (duodenal ulcer).

Gastrin as originally extracted by *Gregory* [12] is a straight chain polypeptide made up of 17 amino acids (see fig. 13/3). The C-terminal 4 were found to contain all the activity of the whole molecule, but this tetrapeptide is not as potent as the whole molecule. No other fragment nor anything smaller than the tetrapeptide was found to be active.

In the course of time 34 and 13 amino acid gastrins were isolated from the plasma. Inactive peptides from the N terminus corresponding to G-17 and 13 have been isolated also. Both G-34 and G-17 are found in G cells. It seems that the large gastrin is actually formed within the cell and cleaved there to produce G-17 and the inert N-terminus peptide. A further gastrin known as BBG (big, big gastrin) seems to be an association with plasma protein. It is a significant constituent of fasting plasma gastrin.

After a meal both G-17 and G-34 in the plasma rise, the latter higher and later. This seems to mean that G-17 is stored gastrin; 75% of the gastrin obtained from G cells is G-17 while the 34 is de novo synthesis which has not yet been cleaved. The intrinsic potency of G-17 is about 5 times that of G-34,

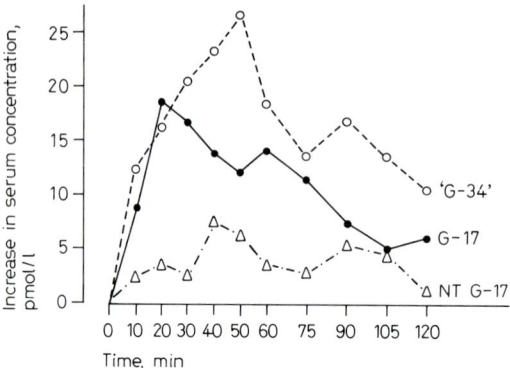

Fig. 5/20. Increase in gastrin concentrations in blood following food (mean of 17 normal subjects). NT G-17 is the inactive N-terminal polypeptide resulting from the splitting of G-34 to G-17. Reproduced, with permission, from *Dockray and Fisher* [8].

but since the latter last 6 times as long in the circulation their action on the parietal cell is about equal (fig. 5/20).

The G cells of the antrum and duodenum are typical APUD cells containing large granules and with superficial villi. More will be said about this system of cells later.

A remarkable feature of gastrin is that the active terminal peptide is identical with that of cholecystokinin (CCK), the hormone which stimulates pancreatic enzyme secretion and contracts the gallbladder. As might be expected gastrin also has these actions to a slighter extent than has CCK. CCK will cause the quiescent stomach to secrete acid and pepsin and augment both from the cholinergically stimulated stomach. However, if gastrin and CCK are given together they are mutually antagonistic on the stomach, but strangely not on the gallbladder or pancreas. Here they seem to be additive.

As is the case with any endocrine gland-target cell relationship, removal of the gland, in this case the G cells, results in atrophy of the target cells, the secretory mucosa. There is nothing unusual or special, therefore, about what is called the 'trophic' action of gastrin.

We have seen that circulating gastrin is raised in conditions which depress the secreting mucosa. It is also raised in persons with impaired kidneys. This evidently is the main catabolic site for the hormone. The small bowel seems to be another site of catabolism. Extensive resections of small bowel are sometimes followed by hypergastrinemia and gastric hypersecretion. Preparations of the C-terminal tetrapeptide, now commercially available for gastric analysis, by contrast, seem to be catabolised mostly by the liver.

Another cause for greatly elevated blood gastrin levels is gastrinoma, which is a tumor of gastrin-producing cells. This can, in theory, occur wherever there are G cells, but in actuality occurs most frequently in the D cells of the pancreas, which do not normally produce gastrin. This tumour, which is usually malignant, causes enormously elevated gastrin levels, huge gastric secretion, duodenal ulceration and often diarrhea, which can be cured only by removal or destruction of the neoplasm and its metastases. In patients with these tumors both G-34 and G-17 are almost equally increased in the blood and the cells themselves seem to lack the usual storage ability. A diagnostic feature of these tumors is that an injection of the pancreatic hormone secretin paradoxically raises blood gastrin levels further. While hypergastrinemia from G cell hyperplasia is diminished by secretin. This latter is the only exception to the rule stated earlier that things which depress the secreting mucosa elevate plasma gastrin. Secretin depresses gastric acid secretion.

Peptic Ulcer

The stomach protects itself, we have seen, with its mucous and non-parietal secretions and also by way of acid inhibition of the G cells. There are also duodenal mechanisms which will be discussed later which regulate the acidity of the duodenum. These mechanisms at times fail and according to the orthodox view, peptic ulceration of the duodenum or stomach is the result. The acid secretion in the gastric ulcer is usually below normal while in the duodenal ulcer it is higher than normal. In the former case the mucosal gastric defenses are evidently diminished. In the latter there is too much acid for normal duodenal mucosa. The duodenal ulcer patient has a larger parietal cell population than normal, his maximal capacity to secrete acid is above normal and his stomach, even when empty, continues to secrete. It is at times like this, when the acid is unbuffered by food, that pain is felt, i.e. in the small hours of the morning. In these patients fasting plasma gastrin levels are not elevated. Indeed many investigators have claimed that they are diminished. In fact, every single contention mentioned above has been denied by someone. To be strictly honest one must admit that we really do not know why duodenal ulcers occur. Acid is logical, but far from incontestable. People without acid do not get duodenal ulcers and in those people who have ulcers with symptoms (not all do) procedures or medicines which reduce gastric acid alleviate discomfort. Acid seems therefore to have something to do at least with the symptomatology [62].

The treatment of both duodenal and gastric ulcer is to neutralize or reduce acid secretion. This can be done medically with antacid therapy or with H_2 blockers (cimetidine or ranatidine) or, if these are unsatisfactory, surgically.

Operative Therapy

The objective of the surgeon is to remove as much of the acid-secreting machinery of the stomach as possible without interfering with its emptying or storage. The operation most favored at the moment is one in which the vagal branches supplying the secreting mucosa are cut leaving those to the pylorus intact (see fig. 4/2). Thus, the cephalic nervous phase is abolished, but vagal release of gastrin from the antrum can still occur. The depressed action of gastrin on the parietal cells plus normal emptying are considered more than adequate compensation for unimpaired G cells.

After this operation the resting acid secretion is reduced by 80% and the maximal response to gastrin by 50%. The recurrence rate, which is between 5 and 10%, can be reduced if antrectomy is added but the advantages of low plasma gastrin are offset by emptying problems graphically known as the *dumping syndrome*. In this, rapid emptying, often of hypertonic gastric contents, produces distension, pain, weakness and diarrhea.

A good deal of gastric physiology can be learned from early attempts to treat duodenal ulcer (fig. 5/21). Subtotal gastrectomy aimed at removing the antrum and as much as possible of the secreting mucosa. This operation had a high operative mortality and the reduced capacity of the stomach plus rapid emptying produced post-prandial distress. The Billroth I type produced less distress than the Billroth II. As one might expect the gastroenterostomy distal to the ampulla of Vater impairs digestion and also bypasses the gastric inhibitory duodenal acid feedback mechanism. In addition to all of this blind loops tend to become a breeding ground for bacteria. Ulcers at the anastomosis are to be expected in any gastrojejunostomy. The longer the blind loop the greater the risk of anastomatic ulcer and all the other post-prandial symptoms. Some have argued that the jejunal mucosa is more susceptible to damage by gastric juice than is the duodenal. Others feel that the duodenal mucosa has no special intrinsic resistance but is protected by the secretions, notably pancreatic juice, which pour into it.

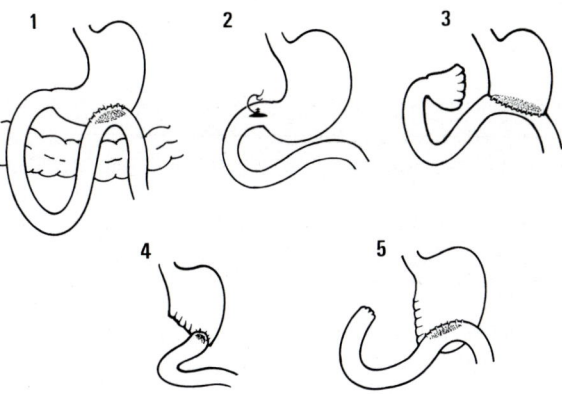

Fig. 5/21. Diagrams of operative procedures which have been used to treat duodenal and gastric ulcer. For vagotomy see figure 4/2. 1 = Anterior gastrojejunostomy (gastroenterostomy); 2 = pyloroplasty (to increase the diameter of the pyloric canal); 3 = pyloric exclusion operation (no stomach removed); 4 = Billroth I gastric resection; 5 = Billroth II gastric resection. From *Ivy* et al. [24].

A simple gastrojejunostomy (fig. 5/21) one might imagine, by reducing the time a meal spends in the antrum, would reduce gastric secretion. On the contrary, it increases it, because duodenal contents can regurgitate back into the empty stomach and keep the production of acid going after it should have stopped.

Perhaps the most disastrous operation intended to reduce gastric secretion was that which separated the pylorus from the body of the stomach, but left it attached to the duodenum (fig. 5/21). Continuity was reestablished by gastrojejeunostomy. This increased gastric secretion from the parietal cell area enormously. This operation is now used as a method to produce experimental peptic ulcers in dogs. What is happening here is that gas and regurgitation of duodenal contents stimulate the G cells to unrestrained production of gastrin, unrestrained because there is no acid brake on the mucosa, and also in the eye of those who believe in it there is no longer any inhibitory innervation to or from the antrum and secretory mucosa. Several groups of workers have transplanted the antrum as a pouch into the colon in dogs and as expected copious gastric secretion has been the result. Today every operation for peptic ulcer includes vagotomy, whatever else may be done. The consequences of vagotomy, which does not spare the pyloric innervation, have been dealt with in the previous chapter.

References

1. Allen, A.: Structure and function of gastrointestinal mucus; in Johnson, Physiology of the gastrointestinal tract, vol. 1 (Raven Press, New York 1981).
2. Alonso, D.; Durbin, R.: Gastric secretion cellular aspects; in Duthie, H.; Wormsley, K.; Scientific basis of gastroenterology (Churchill-Livingstone, Edinburgh 1979).
3. Antinone, R.; Bluvas, R.; Magee, D.: Antral acidity and gastric secretion. Ann. Surg. *166:* 990–994 (1967).
4. Bensley, R.: The gastric glands; in Cowdry, Special cytology, vol. 1, pp. 198–230 (Hoeber, New York 1932).
5. Bloom, G.: Gut hormones (Churchill-Livingstone, Edinburgh 1978).
6. Bloom, W.; Fawcett, D.: A textbook of histology; 10th ed. (Saunders, Philadelphia 1975).
7. Davies, R.; Edelman, J.: The function of carbonic anhydrase in the stomach. Biochem. J. *50:* 190–194 (1951).
8. Dockray, G.; Taylor, I.: Heptadecapeptide gastrin. Measurement in blood by specific radioimmunoassay. Gastroenterology *71:* 971–977 (1976).
9. Donaldson, R.: Intrinsic factor and the transport of cobalamin; in Johnson, Physiology of the gastrointestinal tract, vol. 1, chap. 22 (Raven Press, New York 1981).
10. Forte, T.; Forte, J.: Histochemical staining characterization of glycoproteins in acid-secreting cells of frog stomach. J. Cell Biol. *47:* 437–453 (1970).
11. Forte, T.; Machen, T.; Forte, J.: Ultrastructural changes in oxyntic cells associated with secretory function. A membrane-recycling hypothesis. Gastroenterology *73:* 941–955 (1975).
12. Gregory, R.: The isolation and chemistry of gastrin; in Code, Handbook of Physiol. Soc. Washington, sect. 6, vol. II, chap. 46 (American Physiological Society, Washington 1967).
13. Greider, M.; Steinberg, V.; McGuigan, J.: Electron microscopic identification of the gastrin cell of the human antral mucosa by means of immunocytochemistry. Gastroenterology *63:* 572–583 (1972).
14. Ham, A.: Histology; 7th ed. (Lippincott, Philadelphia 1974).
15. Harper, A.; Kidd, C.; Scratcherd, J.: Vago-vagal reflex effects on gastric and pancreatic secretion and gastro-intestinal motility. J. Physiol. *148:* 417–436 (1959).
16. Heatley, N.: Some experiments of partially purified gastrointestinal mucosubstance. Gastroenterology *37:* 304–312 (1959).
17. Heatley, N.: Mucosubstance as an barrier to diffusion. Gastroenterology *37:* 313–318 (1959).
18. Helander, H.; Hirschowitz, B.: Quantitative ultrastructural studies on gastric parietal cells. Gastroenterology *63:* 951–961 (1972).
19. Helander, H.; Hirschowitz, B.: Quantitative ultrastructural studies on inhibited and on partly stimulated gastric parietal cells. Gastroenterology *67:* 447–452 (1974).
20. Hogben, C.: Transport of chloride by isolated frog gastric epithelium. Origin of the gastric mucosal potential. Am. J. Physiol. *180:* 641–669 (1955).
21. Ito, S.: Functional gastric morphology, chap. 17; in Johnson, Physiology of the gastrointestinal tract, vol. 1, pp. 517–550 (Raven Press, New York 1981).
22. Ito, S.; Schofield, G.: Studies on the depletion and accumulation of microvilli and

changes in the tubulovesicular compartment of mouse parietal cells in relation to gastric acid secretion. J. Cell Biol. *63:* 364–382 (1974).
23 Ito, S.; Munro, D.; Schofield, G.: Morphology of the isolated mouse oxyntic cell and some physiological parameters. Gastroenterology *73:* 887–898 (1977).
24 Ivy, A.; Grossman, M.; Bachrach, W.: Peptic ulcer (Blakiston, Philadelphia 1950).
25 Jorpes, J.; Mutt, V.: In Jorpes, Mutt, Secretin, cholecystokinin, pancreozymin and gastrin, chap. 1 (Springer, Berlin 1973).
26 Junqueira, L.; Carneiro, J.; Contopoulos, A.: Basic histology; 4th ed. (Lange, Los Altos 1983).
27 Kasbekar, D.; Durbin, R.: An adenosine triphosphatase from frog gastric mucosa. Biochim. biophys. Acta *105:* 472–482 (1965).
28 Kondo, T.; Magee, D.: Evidence for antral inhibition of pentagastrin from experiments using mucosal cooling. J. Physiol. *270:* 37–50 (1977).
29 Konturek, S.: Gastric secretion physiological aspects; in Duthie, Wormsley, Scientific basis of gastroenterology (Churchill-Livingstone, Edinburgh 1979).
30 Krause, W.; Cutts, J.: Concise text of histology (Williams & Wilkins, Baltimore 1981).
31 Leela, K.; Kanagasuntheram, R.: A microscopic study of the human pyloroduodenal junction and proximal duodenum. Acta anat. *71:* 1–12 (1968).
32 Leeson, R.; Leeson, T.: Histology; 3rd ed., p. 347 (Saunders, Philadelphia 1976).
33 Lim, R.; Ma, W.: Mitochondrial changes in the cell of the gastric glands in relation to activity. Q. Jl exp. Physiol. *16:* 87–110 (1926).
34 Linderstrom-Lang, K.; Holter, J.; Speborg, H.; Ohlsen, W.: Beiträge zur enzymatischen Histochemie. XIII. Die Enzymverteilung im Schweinemagen als Funktion seines histologischen Aufbaus. Arch. Pharm. Chem. *227:* 1–15 (1936).
35 Magee, D.: Gastro-intestinal physiology (Thomas, Springfield 1962).
36 Magee, D.: In Ruch, Patton, Physiology and biophysics (Saunders, Philadelphia 1965).
37 Magee, D.; Nakajima, S.: The effects of antral acidification on the gastric secretion stimulated by endogenous and exogenous gastrin. J. Physiol. *196:* 713–721 (1968).
38 Magee, D.; Dutt, B.: Effect of ganglionic and muscarinic blockade on the mechanism of pepsin secretion stimulation. Am. J. Physiol. *227:* 1178–1180 (1974).
39 Magee, D.: Adrenergic activity and gastric secretion. Proc. Soc. exp. Biol. Med. *151:* 659–662 (1976).
40 Magee, D.: Gastric fundo-pyloric relationships. A discussion. Mt Sinai J. Med. *49:* 1–6 (1982).
41 Makhlouf, G.; McManus, J.; Card, W.: A quantitative statement of the two component hypothesis of gastric secretion. Gastroenterology *51:* 149–171 (1966).
42 Neutra, M.; Padykula, H.: The gastrointestinal tract, chap. 19; in Weiss, Histology: cell and tissue biology; 5th ed., pp. 658–706 (Elsevier Biomedical, New York 1983).
43 Pearson, J.; Allen, A.; Venables, C.: Gastric mucus: isolation and polymeric structure of the undegraded glycoprotein; its breakdown by pepsin. Gastroenterology *78:* 709–715 (1980).
44 Piper, D.; Fenton, B.: pH stability and activity curves of pepsin with special reference to their clinical importance. Gut *6:* 506–508 (1965).

45 Romrell, L.; Coppe, M.; Munro, D.; Ito, S.: Isolation and separation of highly enriched fractions of viable mouse gastric parietal cells by velocity sedimentation. J. Cell Biol. *65:* 428–438 (1975).

46 Roholm, K.: Clinical investigation into the effects of intravenous injection of insulin. Acta med. scand. *73:* 472–492 (1930).

47 Rubin, W.; Allasgharpour, A.: Demonstration of a cytochemical difference between the tubulovesicles and plasmalemma of gastric parietal cells by ATPase and NPPase reactions. Anat. Rec. *184:* 251–264 (1970).

48 Sachs, G.; Spenney, G.; Rehm, W.: Gastric secretion; in Crane, International review of physiology gastrointestinal physiology, vol. 11, chap. 5 (Union Park Press, Baltimore 1977).

49 Samloff, I.: Slow moving protease and the seven pepsinogens. Electrophoretic demonstration of the existence of light proteolytic factions in the human gastric mucosa. Gastroenterology *57:* 659–660 (1969).

50 Samloff, I.: Pepsinogens, pepsins and pepsin inhibitors. Gastroenterol. *60:* 586–604 (1971).

51 Schofield, G.; Ito, S.; Bolander, R.: Changes in membrane surface areas in mouse parietal cells in relation to high levels of acid secretion. J. Anat. *128:* 669–692 (1979).

52 Sedar, A.: The fine structure of oxyntic cells in relation to functional activity of the stomach. Ann. N.Y. Acad. Sci. *99:* 9–29 (1962).

53 Sedar, A.: Electron-microscopic demonstration of polysaccharides associated with acid-secreting cells of the stomach after 'inert dehydration'. J. ultrastruct. Res. *28:* 112–124 (1969).

54 Sircus, W.: The intestinal phase of gastric secretion. Q. Jl exp. Physiol. *38:* 91–99 (1953).

55 Soll, A.: The action of secretogogues on oxygen uptake by isolated mammalian parietal cells. J. clin. Invest. *61:* 370–380 (1978).

56 Spicer, S.; Sun, D.: Carbohydrate histochemistry of gastric epithelial secretions in dog. II. The role of the mucous barrier in the defense of the stomach vs. peptic ulceration. Ann. N.Y. Acad. Sci. *140:* 762–783 (1967).

57 Stevens, C.; Leblond, C.: Renewal of the mucous cells in the gastric mucosa of the rat. Anat. Rec. *115:* 231–245 (1953).

58 Sugai, N.; Ito, S.: Carbonic anhydrase, ultrastructural localization in the mouse gastric mucosa and improvements in the technique. J. Histochem. Cytochem. *28:* 511–525 (1980).

59 Taylor, W.: Biochemistry of pepsins; in Code, Handbook of physiology, vol. V, sect. 6, chap. 120 (American Physiological Society, Washington 1973).

60 Vial, J.; Orrego, H.: Actin-like filaments and membrane rearrangements in oxyntic cells. Proc. natn. Acad. Sci. USA *73:* 4032–4036 (1976).

61 Winborn, W.; Seelig, L.; Nakayama, H.; Weser, E.: Hyperplasia of the gastric glands after small bowel resection in the rat. Gastroenterology *66:* 384–395 (1974).

62 Wormsley, K.: Duodenal ulcer: does pathophysiology equal aetiology? Gut *24:* 775–780 (1983).

63 Zalewsky, C.; Moody, F.: Stereological analysis of the parietal cell during acid secretion and inhibition. Gastroenterology *73:* 66–74 (1977).

6 The Biliary Tract

Gross Structure and General Function

The biliary tract is the excretory apparatus of the liver, draining its exocrine secretion into the GI tract (duodenum) (fig. 6/1A). It consists of: (1) the *common hepatic duct,* formed by the junction of the *right* and *left hepatic ducts* which leave the liver at the porta hepatis; (2) the *gallbladder,* a slate-blue, piriform sac some 7–10 cm in length which serves as a reservoir for the bile; (3) the *cystic duct,* draining the gallbladder; (4) the *common bile duct,* formed by the junction of the common hepatic and cystic ducts, and (5) the *hepatopancreatic ampulla* (of Vater) wherein the common bile duct and pancreatic ducts unite and empty into the descending part of the duodenum (fig. 6/1B).

The liver, in its function as an excretory pathway for the breakdown products of hemoglobin and bilirubin, secretes continuously. In its function as a producer of substances which aid in digestion (bile salts), it need only secrete during digestion in animals that are intermittent eaters and digesters such as humans and carnivores. However, even in these it secretes continuously. Upon laparotomy in such animals following the usual overnight fast, the gallbladder is found to be full; bile drawn from it is seen to be almost black in color and very viscous. In contrast, bile draining directly from the liver via the hepatic ducts is seen to be a golden-brown color and highly fluid, even after a fast. Many animals that are continuous digesters have no gallbladder at all, e.g. rats, horses, and some ruminants such as deer. In them bile dribbles constantly into the duodenum. There are some continuous digesters with gallbladders, but the bile in these animals is never viscous; it remains dilute. A characteristic of the continuously digesting animal is that the sphincter of the hepatopancreatic ampulla (sphincter of Oddi) is weak. It may, in fact, be absent in animals without gallbladders. Intermittent digesters by contrast have strong sphincters which remain closed except when the

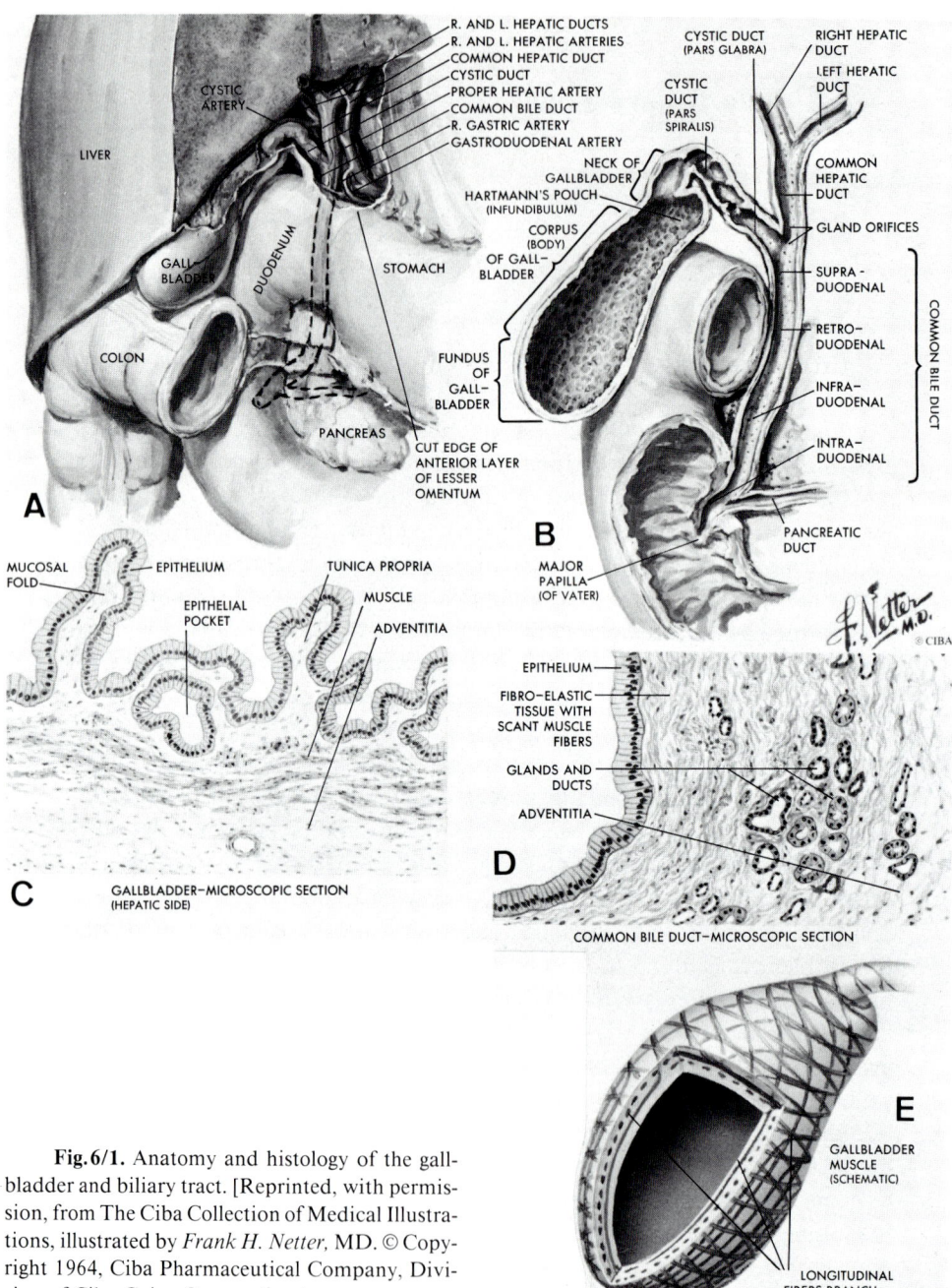

Fig. 6/1. Anatomy and histology of the gallbladder and biliary tract. [Reprinted, with permission, from The Ciba Collection of Medical Illustrations, illustrated by *Frank H. Netter*, MD. © Copyright 1964, Ciba Pharmaceutical Company, Division of Ciba-Geigy Corporation.]

stomach is discharging into the duodenum and digestion is proceeding. We may surmise then that the biliary system functions to deliver – and in intermittent digesters, to store and concentrate – the continuously produced secretions of the liver, i.e. it provides synchrony of supply and demand. In humans, when food enters the duodenum the sphincter of Oddi relaxes and the gallbladder contracts, pushing its viscous contents into the duodenum. This is usually a slow process, taking up to 30 min for the normal gallbladder to empty.

Microstructure

Biliary Tract vs. GI Canal
Although they develop as part of a diverticulum of the embryonic foregut, the gallbladder and most of the biliary pathway (with the exception, perhaps, of the middle and lower portions of the bile duct) vary morphologically from what has been described as 'the general structure of the gastrointestinal canal' (chapter 1). Here the wall consists only of a surface epithelium, a lamina propria, a fibromuscular layer and a serosa or adventitia (fig. 6/1C). It has been commonly accepted that the smooth muscle-containing layer on which the mucous membrane rests is comparable (except in thickness) with the muscularis externa of the GI canal, and that the gallbladder and closely-associated ducts are characterized by the absence of a muscularis mucosae [8, 23]. However, it is actually a stratum of fibrous (reticular and elastic) tissue in which a comparatively skimpy amount of smooth muscle is arranged in loose bundles. Rather than having an outer longitudinal and an inner circular orientation, the bundles interlace, being disposed in all directions (including longitudinal and circular) but mainly oblique, so that they appear to spiral around the lumen (fig. 6/1E). Further, the myocytes are smaller in size, more irregular in sectional profile and less densely packed, and have fewer intercellular junctions (nexuses) than those of the intestinal muscularis externa. In these features, the fibromuscular layer more closely resembles the muscularis mucosae of the gut [16]. Thus, on purely descriptive grounds, it has recently been suggested [1] that 'the muscularis of most of the bile duct is equivalent to – and probably derived from – the muscularis externa of the intestine. However, in the rest of the biliary pathways and in the gallbladder, the muscularis is equivalent to the muscularis mucosa and may, therefore, be a derivative of the intestinal mucosa.' If this is correct, it is actually the lack of

a muscularis externa which characterizes the gallbladder. Recent studies of the innervation of the gallbladder and biliary pathways support this view [2]. In the lower bile duct where there clearly is a muscularis externa, autonomic plexuses occur on both sides of it (i.e. there are interconnected, ganglionated submucosal and subserosal plexuses which seem to correspond to – and be continuous with – the submucosal and myenteric plexuses, respectively, of the gut). The occurrence of ganglion cells, undoubtedly parasympathetic, has been confirmed in the intrinsic plexuses of the biliary pathways and the gallbladder of primates [24, 25]. This plexus has been described as being intramuscular, but in the guinea pig at least it actually lies at the outer surface of the fibromuscular layer. If the proposals of *Cai and Gabella* [1, 2] are correct, the subserosal tissue of the gallbladder is the equivalent of the submucosa of the gut, and the ganglionated plexus of the gallbladder is the equivalent of and is in continuity with the submucosal plexus of the intestine and not the myenteric plexus as generally assumed.

Both presynaptic parasympathetic fibres (derived from both vagi but primarily the anterior (l.) vagal trunk) and postsynaptic sympathetic fibres extend from the celiac plexus (along branches of the hepatic artery) to reach the gallbladder. Afferent fibres course with the splanchnic nerves as well as with the right phrenic nerve, as evidenced by the pain referred to the right shoulder in gallbladder disease.

Gallbladder

When the gallbladder is contracted, the mucous membrane loosely connected to the fibromuscular layer is thrown into a great number of folds or wrinkles; the elevations thus created take the form of elongated and decussating minute rugae which, in aggregate, confer a honeycomb appearance. Thus, on section, the irregular mucosa might appear to be riddled with glands (fig. 6/1C). However, there are no glands in the mucosa of the gallbladder except near its neck, where small, simple, mucous-secreting tubuloalveolar glands open on the luminal surface. When the gallbladder is distended, most (but not all) of its mucosal folds disappear.

The mucosal epithelium of the gallbladder is composed of tall, columnar cells, each cell resembling the one beside it (fig. 6/1C). In this respect, its appearance is similar to the luminal epithelium of the stomach, but the cells themselves more closely resemble the absorptive cells of the intestine and, like them, are provided with microvilli (fig. 6/2). Secretory (mucous) granules are present in the apical halves of some cells, particularly in the neck

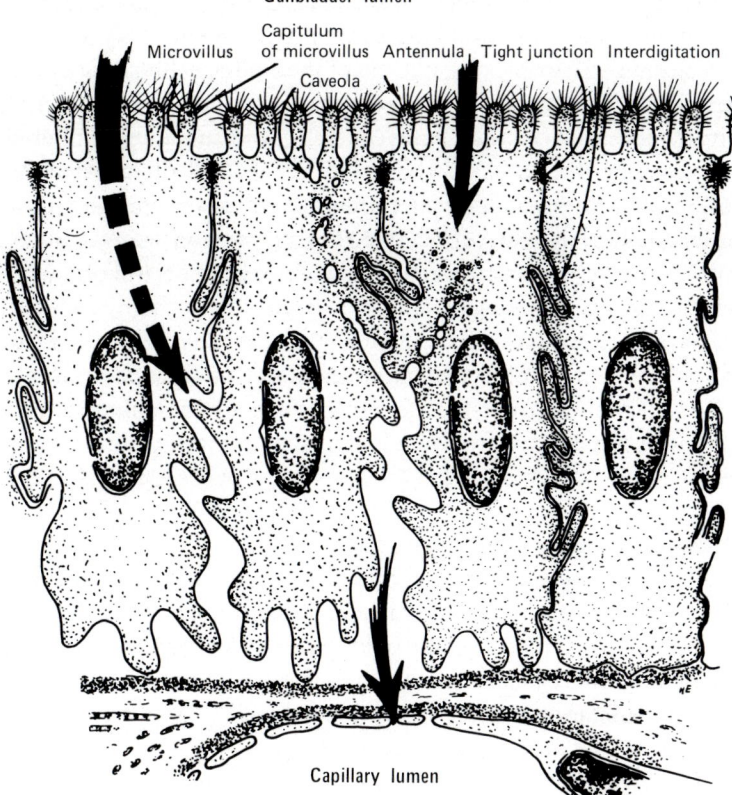

Fig. 6/2. Diagrammatic representation of the structure and function of the mucosal epithelium of the gallbladder. On the apical aspect of the columnar cells, note the microvilli and the pinocytotic activity at their bases. Both narrow and widely-dilated basal intercellular spaces are depicted to demonstrate the mucosa during inactivity and during active absorption of water, respectively. Reproduced, with permission, from *Elias* et al. [5], based on the work of Kaye, Lane, Wheeler and Whitlock: Anat. Rec. *151:* 369 (1965).

region; these are secreted into the lumen [13, 20]. However, the primary function of the lining cells is absorptive rather than secretory [8].

Bile Concentration. The reason the bile in the gallbladder of the fasted dog, man or other intermittent feeder is thick and viscous is that when the strong choledochal sphincter is closed and bile cannot enter the duodenum, the continuous secretion of the liver enters the gallbladder where it is accommodated by being concentrated. The daily secretion of the liver in man is about 1 liter; the gallbladder can accommodate only about 50 ml.

Electron microscopic observation of the canine gallbladder epithelium [3] demonstrated regularly-arranged microvilli apically with pinocytotic activity evident between their bases (fig. 6/2). Additionally, the intercellular spaces between the bases of the columnar cells demonstrated marked dilatation with an abundance of capillaries in close approximation [15]. These features are indicative of active absorption of water and solutes from the bile; hence the concentration of bile and reduction in volume. Under normal circumstances the gallbladder does not reabsorb any of the constituents of bile except Na^+, Cl^- and HCO_3^-. The mechanism of bile concentration appears to involve the active transport of these substances, thus creating an osmotic gradient from the lumen of the gallbladder to the capillaries of the lamina propria, along which the water is drawn (fig. 6/2, arrows).

Hepatopancreatic Junction

The junction of the common bile duct, pancreatic duct, and the duodenum is an anatomic area of medical and physiological importance. It is here that, in man at least, the flow of both bile and pancreatic enzymes into the duodenum are regulated and the gallbladder is controlled. In the adult human, the associated common bile duct and major pancreatic duct obliquely perforate the circular muscle of the duodenum (fig. 6/3). The ducts jointly empty their contents into the hepatopancreatic ampulla (Vater), which then pass through an orifice in the ampulla into the duodenum. In most nonruminant animals, however, there are separate openings for the biliary and pancreatic ducts. In such animals (carnivores, for example) the embryological ventral pancreatic duct (the duct of Wirsung – the major secretory pancreatic duct in man) persists only as a small accessory duct, which may or may not open in common with the bile duct. The embryological dorsal pancreatic duct (the duct of Santorini) develops into the main pacreatic duct, with its own opening into the duodenum a few centimeters distal to that of the common bile duct.

Fig. 6/3. The choledochoduodenal junction and the microanatomy of the sphincter of the hepatopancreatic ampulla (sphincter of Oddi). [Reprinted, with permission, from The Ciba Collection of Medical Illustrations, illustrated by *Frank H. Netter*, MD. © Copyright 1964, Ciba Pharmaceutical Company, Division of Ciba-Geigy Corporation.]

6 The Biliary Tract

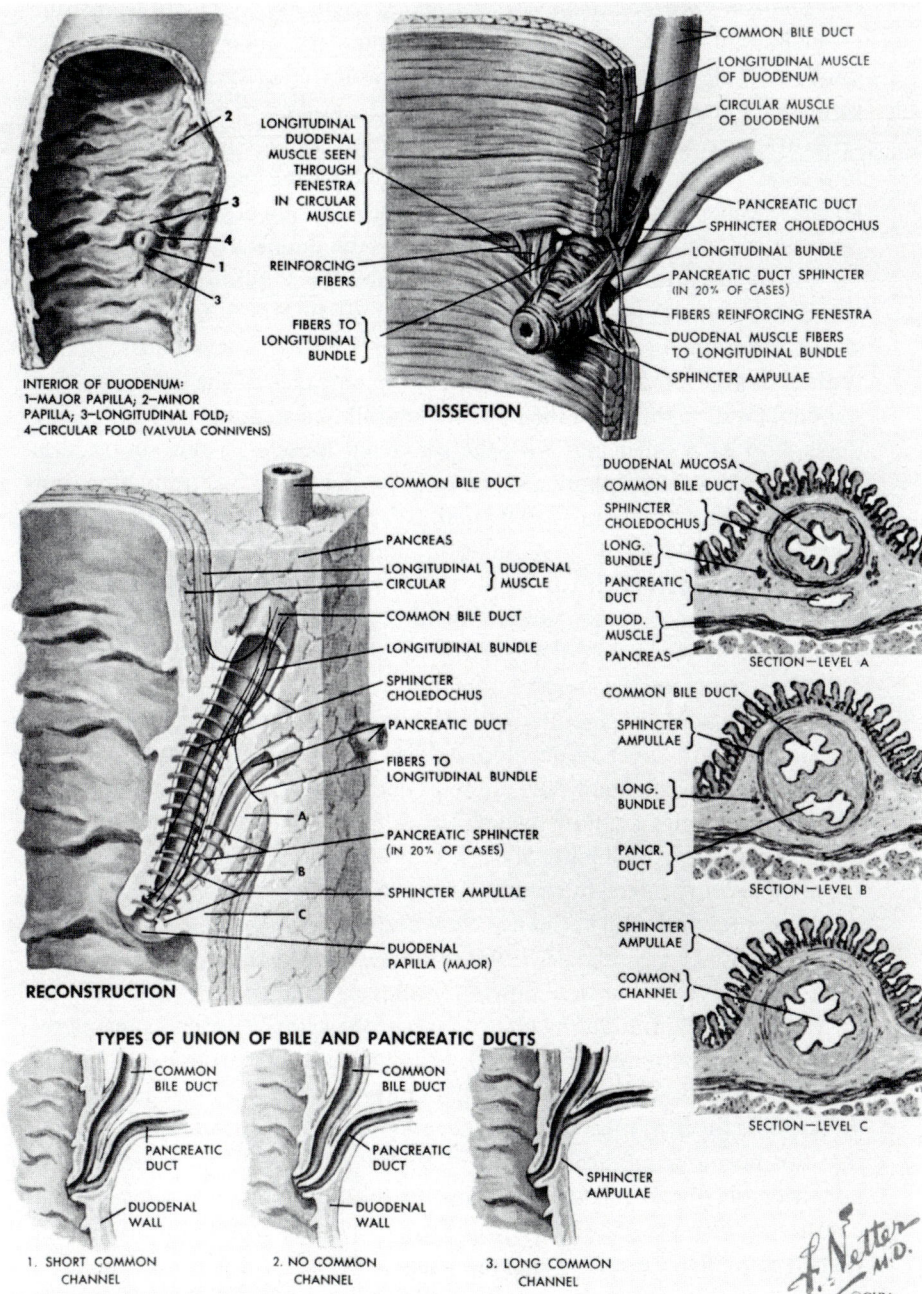

Sphincter of Hepatopancreatic Ampulla (Sphincter of Oddi). Returning to the human, as the associated bile and pancreatic ducts pass through the duodenal wall, they are invested by a common ring of smooth muscle, the sphincter of the hepatopancreatic ampulla (sphincter of Oddi). The structure varies greatly from individual to individual, but the paradigm includes four subdivisions (fig. 6/3). (1) The *sphincter choledochus* is a strong, annular sheath surrounding the termination of the common bile duct beginning a centimeter or less before the duct penetrates the duodenal wall and extending to its junction with the pancreatic duct. This sphincter is the most consistent portion; its contraction regulates the flow of bile and, retrogressively, the filling of the gallbladder. (2) The *fasciculi longitudinales* (longitudinal bundles in fig. 6/3) extend from the entrance of the ducts into the duodenal wall to the tip of the duodenal papilla, in so doing connecting the ducts with each other and with the duodenal muscles. Their contraction retracts and erects the papilla, shortening the ducts and facilitating the flow of bile into the duodenum. (3) The *sphincter pancreaticus* is present only in a few individuals (20% of the population) and is highly variable in its development when present. It surrounds the intraduodenal portion of the pancreatic duct, prior to its junction with the ampulla. (4) The *sphincter ampullae,* a continuation of the sphincter choledochus and, if present, the sphincter pancreaticus beyond the union of the ducts, is the terminal musculature actually surrounding the common channel, the hepatopancreatic ampulla (Vater). Its development is variable, but it is virtually always considerably weaker than the sphincter choledochus.

These muscles are now recognized to be both anatomically and functionally distinct from duodenal muscle. It is of course true that, because of the oblique course taken by the two ducts through the duodenal wall, that duodenal contraction will influence flow and pressure in the common duct when its sphincter is relaxed. If the sphincter is contracted, however, duodenal activity cannot initiate flow. Physiological concentrations of CCK induce bile flow by stimulating contraction of the gallbladder and causing relaxation of the sphincter of Oddi [10, 17]. Pharmacological doses can put the sphincter into spastic contraction. This, combined with the evidence that the sphincter is the only muscle caused to relax by CCK, leads to the conjecture that this hormone has an indirect action on the muscle. The sphincter has also been shown to relax in response to the hormone even when out of continuity with the gallbladder; thus, its relaxation is not conditional on increases in biliary pressure caused by the action of CCK in the gallbladder. There is no substantial description of the nerve supply to the region.

Gallbladder Emptying

A good deal of attention has been paid to pressures in the biliary tract: e.g. the maximal secretory pressure, the maximal pressure generated by gallbladder contraction and the resistance of the sphincter. The interest in these was the result of the idea that acute pancreatitis was caused by regurgitation of bile into the pancreas due to gallbladder contraction against a closed sphincter. In the postabsorptive state the resistance of the human sphincter has been estimated at 12–15 cm water. It is stated that gallbladder pressure is less than this, thus the gallbladder cannot force the sphincter. The secretory pressure of the liver in man rises to 29–39 cm bile. These arguments have lost much of their meaning today since it is clear that the major sphincter is not the ampullary one which surrounds the common channel for both secretions, but is the choledochal one which guards the common bile duct alone; and again it seems clear today that the presence of bile in the pancreatic duct system cannot activate pancreatic proteolytic enzymes, thereby setting off acute pancreatitis.

As stated previously, when food enters the intestine, the sphincter relaxes and the gallbladder contracts. Nowadays, the hormone CCK is considered to be a sufficient explanation for this. The presence in the duodenal lumen of partically digested fats and proteins brings about liberation of this hormone.

When there are no digesta in the upper small intestine no CCK is liberated from the mucosa; the choledochal sphincter is therefore contracted and the bile now has nowhere to go but into the relaxed gallbladder. Until a few years ago it was felt that two distinct hormones were required to bring about gallbladder contraction on one hand and the secretion of enzymes by the pancreas on the other. *Jorpes and Mutt* [14] recognized that these two polypeptides were chemically identical. Their physiological identity had been suspected for some time. Since the need for a gallbladder hormone was recognized before that for a pancreatic one, and since this hypothetical gallbladder contracting material had been named cholecystokinin in 1928, *Jorpes* proposed that this name be retained and that the name pancreozymin be abandoned. *Ivy and Oldberg* [12] extracted a material from pigs' upper small intestine which, on injection into dogs, caused gallbladder contraction. In cross-circulation experiments (carotid to carotid) they demonstrated that the introduction of HCl into the duodenum of one dog caused contraction of the gallbladders of both. Others used viviperfused gallbladders (i.e. acutely transplanted into the neck) and found the same thing.

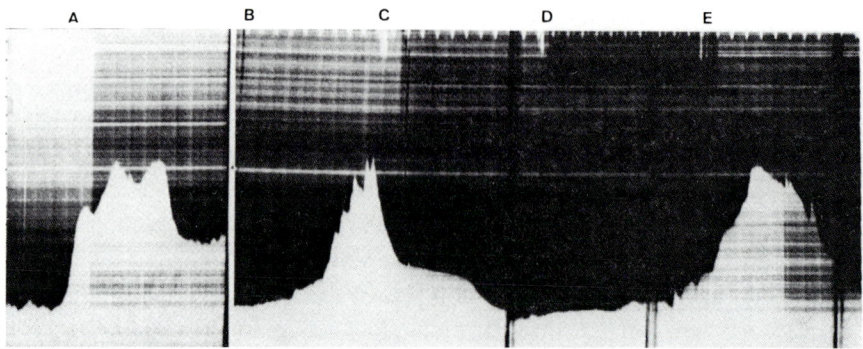

Fig. 6/4. The effect of CCK and acidification of the duodenum on gallbladder pressure in a conscious dog. **A** CCK is given intravenously. **B** N/10 HCl in the duodenal loop. **C** Ganglionic blockade with hexamethonium. **D** N/10 HCl again. **E** Intravenous CCK again. Reproduced, with permission, from *Hong* et al. [10].

The need for a pancreatic hormone was not recognized early, because vagal stimulation caused an increase in pancreatic enzymes and was considered a sufficient explanation until 1948 when the ability of duodenal mucosal extracts to increase pancreatic enzymes was recognized. The active material was called pancreozymin but is now known to be CCK. This aspect of the hormone will be dealt with in more detail under pancreas.

Both pancreatic and gallbladder activities were best elicited by the presence of digests of fat and protein in the duodenum and less so by acid, but prior administration of ganglionic blocking agents or procaine to the duodenal mucosa prevented intraduodenal activation of the gallbladder muscle (fig. 6/4). By contrast, mucosal extracts given intravenously were still active. Indeed pure preparations of CCK and its analogs now available, on intravenous administration, can contract the gallbladder even after tetrodotoxin (which blocks nerve terminals), anticholinergics (atropine) and ganglionic blocking agents. Clearly, therefore, the action of CCK on the gallbladder is independent of extrinsic or intrinsic nerves, but release of the hormone from the intestinal mucosa requires cholinergic nerves and a synapse (ganglion).

The action of extrinsic nerves on the gallbladder has always been equivocal. Some claim contraction following vagal stimulation, others deny this. Acetylcholine can cause contraction of isolated strips of gallbladder muscle, but no one has claimed that either vagal stimulation or cholinomimetics cause delivery of bile into the duodenum.

Many investigators have reported an increase in the size of the X-ray gallbladder shadow following bilateral vagotomy. We, however, found no differences in the sensitivity of the gallbladder to CCK with or without vagi in conscious dogs (but see chapter 9). The modern techniques of selective vagotomy which preserve the hepatic branch are said to preserve normal gallbladder size and tone.

Cholecystokinin

Cholecystokinin (CCK), like gastrin, is a straight chain polypeptide; it contains 39 amino acids at its longest; like gastrin the whole molecule chain is unnecessary for activity or even optimal activity and it occurs naturally in 39, 33 or 8 amino acid forms. The C-terminal octapeptide is much more potent in every respect than the whole molecule. For full CCK activity a sulphated tyrosine in position 7 is necessary. Gastrin by contrast is equally active with or without sulphation of the corresponding tyrosine and both forms occur naturally. The C-terminal tetrapetide which is identical with that of gastrin has much weaker CCK activities than the whole molecule, but without it the remainder of the molecule is inactive. Both the terminal tetrapetide and the whole gastrin molecule will contract the gallbladder but the whole CCK molecule has 20–30 times the molar potency of either. It is doubtful if this action of gastrin or the tetrapeptide has any physiological significance.

The only functionally significant muscular structures in the entire biliary tract are the gallbladder and the sphincter choledochus (Oddi). The common and hepatic ducts are essentially fibroelastic tubes with a columnar epithelial lining. The common bile duct in man has only an occasional smooth fiber which may be oriented in any direction. In one study [18] muscle fibers were detected in only 12 of 100 human common bile ducts. This has been a controversial matter, since radiologists have described peristaltic activity in the common duct and the colicky pain associated with the passage of bile stones down the duct is typical of that from smooth muscle tubes. It is quite likely that movement seen occasionally in the bile duct is passive resulting from gallbladder contractions and rhythmic activity in the sphincter with elastic recoil in the walls of the duct itself.

The conclusion drawn so far is that bile enters the duodenum only in response to meals. This would mean that after a fast of 2–3 days the gallbladder would either fill beyond its capacity and rupture or overflow or else the individual would get jaundice, as happens with any other obstruction to the emptying of the biliary tract. Neither of these, of course, occurs because bile in substantial quantities enters the duodenum regularly every 100–110 min in man (see chapter 9). This is in phase with the periodic interdigestive motility bursts, which pass down the gut. The stimulus for this is unknown. These bursts account for the frequent presence of bile in the fasting human stomach.

Bile

Bile Salts

When bile enters the intestine, the bile salts play an essential role in the absorption of lipids (to be considered later). They are not absorbed with lipids in the upper reaches of the small intestine, but in the ileum where they no longer have a digestive role. The lower ileum unlike the upper intestine harbors a substantial bacterial flora. These organisms are capable of destroying bile salts to the extent that they can no longer be absorbed [7]. This is the fate of about 10% of the bile salt which enters the duodenum. A larger portion (varying from species to species), up to 25% in man, is only slightly modified. The former appears in the feces; the latter is absorbed back into the bloodstream and known as secondary bile salts. The common ones in man are salts of deoxycholic and lithocholic acids. These are derived from cholic and chenodeoxycholic salts, respectively. Cholic and chenodeoxycylic acids are manufactured de novo from cholesterol by the human liver and are known as primary bile acids. Cholic is by far the most abundant primary bile acid in human bile and deoxy the most abundant secondary acid.

In the normal biliary tract and intestine free bile acids do not exist except in the bacterially contaminated ileum and colon. In man all the acids are present as sodium salts of conjugates of both glycine and taurine. Some animal species have only taurine conjugates, some only glycine and some, like man, both. The properties essential for optimal lipid absorption depend on this conjugation. Intestinal bacteria can and do deconjugate bile salts.

Once absorption from the ileum into the portal blood has taken place [9], bile salts primary and secondary will be delivered unchanged by the liver back into the bile; free bile acids will be reconjugated first. As long as digestion is proceeding, these salts will re-enter the duodenum directly. When digestion has ended and no more CCK is released the choledochal sphincter will contract and bile will enter the gallbladder to be concentrated and stored. The initial spurt of bile into the duodenum when a meal starts is

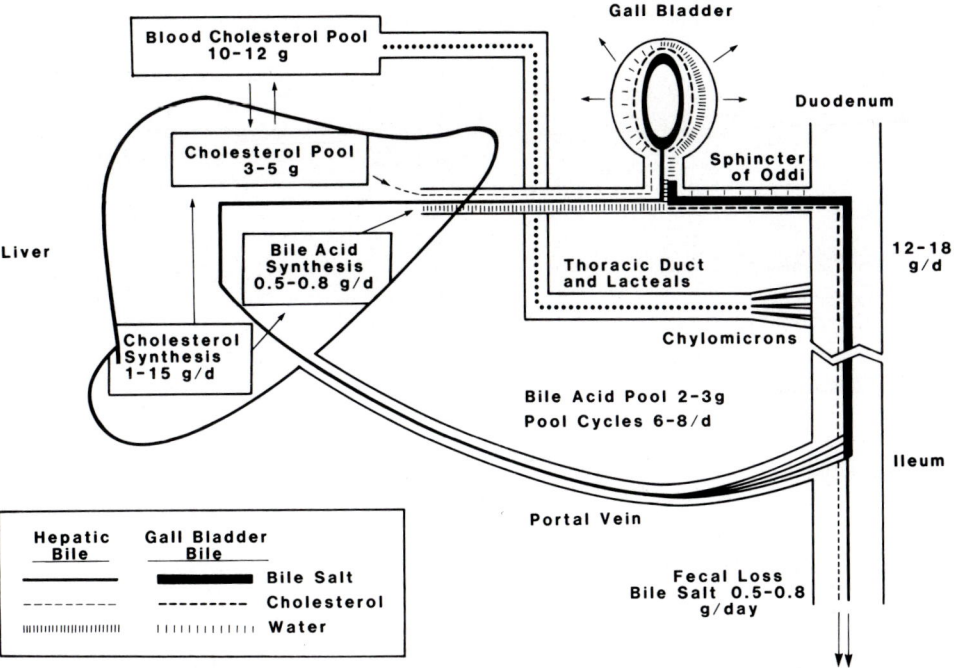

Fig. 6/5. The enterohepatic circuit of bile salts and values for fecal loss, hepatic production and pool size. The lines for bile salt and cholesterol entering the gallbladder are thin and, those leaving, thick to indicate concentration. Note the converse for water.

concentrated, high in bile salt and bilirubin; once this has been discharged, subsequent bile comes straight from the liver and is, therefore, unconcentrated. This circuit of bile salt, from liver to intestine and back again, is known as the *enterohepatic circulation* (fig. 6/5). In the normal human there are about 3–8 complete circuits per day and, as noted above, there is about a 10% loss of bile salt in the feces at each circuit. This amounts to about 0.5 g/day. The new production of bile by the liver exactly replaces this loss. Thus, the total amount of bile salt in the body, or the *bile acid pool* as it is called, is kept constant (table 6/I). How the liver can monitor the size of this pool is a mystery when it is remembered that most of the bile acid pool is either in the intestine or sequestered in the gallbladder. The blood concentration does not, therefore, reflect the size of the pool. If antibiotics are used to decrease the intestinal flora destruction of bile salt in the intestine will be reduced and so will production by the liver. If, on the other hand, bile salt is lost through a draining bile fistula or through failure of the terminal ileum to

Table 6/I. Composition of human bile

	Lower and upper means (mg/dl) from several sources	
	gallbladder	hepatic
Acid	3,790 ± 2,060 9,812 ± 2,940	319 ± 124 2,098 ± 847
Phospholipid	2,030 ± 1,560 4,339 ± 1,872	337 ± 158 1,055 ± 337
Cholesterol	390 ± 245 778 ± 367	129 ± 36 245 ± 86
Bilirubin	208 ± 144 293 ± 194	19 ± 5 65 ± 13

Bile acid pool size (g) in man
Cholic acid   1.04 ± 0.21
Deoxycholic acid   0.42 ± 0.11
Chenodeoxycholic acid   0.81 ± 0.17
Lithocholic acid   0.19 ± 0.02

reabsorb it, the bile acid pool will fall and production of primary salt by the liver will be accelerated.

Water and Electrolytes

The liver secretes a bile salt solution which is isotonic with plasma. In addition, the cells of the bile cannaliculi secrete an isotonic solution containing no bile salt or bilirubin, but which is high in bicarbonate. Secretion of this is stimulated by secretin, another duodenal hormone which acts on the pancreas (see fig. 7/2). In fact, secretin stimulates the secretion of bicarbonate and the water follows to meet the osmotic requirement for isotonicity. During digestion these two solutions enter the intestine. When digestion ends the gallbladder has to deal only with the water associated with bile salts, because then the duodenal production of both secretin and CCK ends.

The normal gallbladder mucosa absorbs only water and electrolytes: its mucosa absorbs both Na^+, Cl^- and HCO_3^- actively [14]; water follows these ions so that the solution absorbed is isotonic with plasma. Virtually all the studies of the mechanisms of gallbladder absorption have been confined to Na$^+$ and Cl$^-$. Absorption of these ions is stated to be coupled and, therefore, electrically neutral, i.e. since an anion and cation are transported actively and together there is no potential difference across the absorbing mucosa. Actually very small potential differences of 3–8 mV have been detected. Little attention has been paid to HCO$_3^-$, except to note that if HCO$_3^-$ is present, water is transported much more rapidly than in its absence. Presumably HCO$_3^-$ also is transported actively.

It was from a consideration of gallbladder absorption that some general hypotheses about the isotonic movement of water were formulated. It was argued that water would only follow an osmotic gradient and that this was established between the intercellular channels and the lumen. These channels are blocked at the luminal surface by tight junctions held to be impervious to water and electrolytes (fig. 6/2). The osmotic gradient, it was proposed, is created because cells having absorbed ions would then pump them into the lateral channels which would, as a result, become hypertonic. Water would then be pulled through the cell by osmotic forces to restore isotonicity.

Validation of this hypothesis awaits the demonstration both of a hypertonic channel fluid and the impermeability of the tight junctions. The evidence at the moment is that, on the contrary, the tight junctions are quite permeable.

Whatever the mechanism for absorption, the gallbladder mucosa absorbs water and electrolytes until such time as the diffusion back into the organ is equal to the absorption and equilibrium results. This back diffusion is likely to be via these same tight junctions and lateral channels.

Bilirubin

Bile pigments are the tetrahydropyrolic end products of the catabolism of hemoglobin and other heme proteins (fig. 6/6). Concentration in hepatic bile is 20–65 mg/100 ml. In the gallbladder it may reach 200–300 mg/100 ml complexed with phospolipid, lipoproteins, bile salts and cholesterol. Bilirubin contains 12 double bonds. Gut bacteria progressively reduce these to a maximum of 6 producing *stercobilinogen,* the most saturated metabolite; others less saturated are *stercobilin* and *urobilinogen* [17]. Of the latter, 100–200 mg/day appears in the stool making it the largest fecal representative of the bile pigments. These pigments are not responsible for the colour of feces since they are present there as the colorless chromogens.

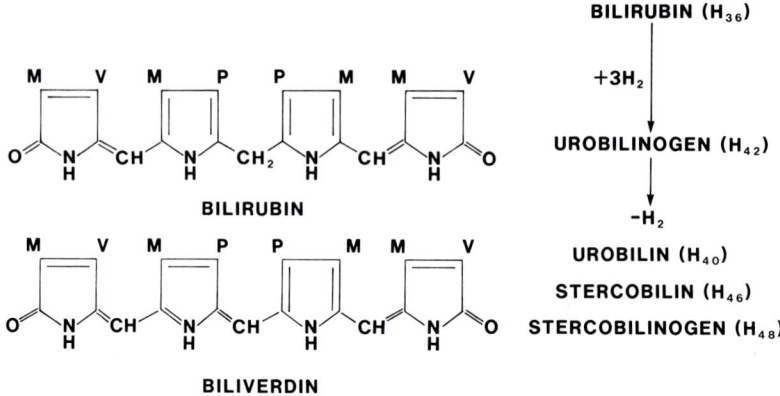

Fig. 6/6. Bilirubin, biliverdin and the intestinal degradation of bile pigments.

Since bile obviously does add colour to feces, bilirubin itself (of which 5–20 mg/day eludes the intestinal flora and appears in the stool), or some as yet uncharacterized pigments derived from it must be responsible. Only about 50% of the daily production of bilirubin can be accounted for by recognized fecal derivatives. Oxidation of urobilinogen and stercobilinogen does result in colored urobilinogen and stercobilin, but these are unusual in feces.

The bile pigment seen sometimes in the fasting stomach and in vomitus is green (biliverdin). It results from the oxidation of bilirubin. In many herbivorous animals and rodents the biliary tract bile is habitually green, but in man it never is.

In the intestine conjugated bilirubin diglucuronide is not absorbed, but unconjugated bilirubin readily can be, and so can the urobilinogen and stercobilinogen. Very little unreduced bilirubin reaches the feces, and bilinogens in small amounts are found in normal urine. They are also found in bile since like bile acids they are subject to enterohepatic circulation. It would seem that reduction takes place before deconjugation and, therefore, there only is little free bilirubin in the gut to be absorbed.

Approximately 40, 30 and 24% of intestinal urobilinogen, stercobilin and free bilirubin are absorbed passively from the intestine, large and small. These normally enter the enterohepatic circulation and are returned to the intestine. If the liver is injured, the usual trace amounts of urobilinogen in the urine will be increased. In obstructive jaundice urobilinogen disappears from the urine, and in hemolytic jaundice when the bilirubin entering the gut is greatly increased, urobilinogen formation and the amount absorbed will increase and so, of course, will the urinary urobilinogen.

Jaundice. There is no absorption of bilirubin, bile salts, cholesterol or biliary lipids from the biliary tract under normal circumstances, but in obstructive jaundice, in which the common duct is blocked, reabsorption from the biliary tract must occur. Such patients become obviously jaundiced and often suffer from cutaneous itching due to the elevated blood levels of bile salt. The mechanism for this absorption is almost unstudied. Unconjugated bilirubin, since it is less highly charged than conjugated, should more readily enter the bloodstream, but in obstructive jaundice the raised plasma bilirubin is conjugated. To enter the bile bilirubin must undergo conjugation in the liver to a water soluble diglucuronide. Very old evidence suggests that most of it is absorbed via the lymphatics and into the thoracic duct. The possibility is that the sustained high pressure in the biliary tract breaks the continuity of the epithelial cells and these large molecules are salvaged from intercellular spaces by the lymph vessels.

Secretion from the biliary mucosa itself is another subject that is almost uninvestigated. If the cystic duct is obstructed gallbladder bile will gradually disappear and in time be replaced by a viscous slimy fluid. This is mostly mucus and is known as *white bile.*

Jaundice or bile pigment staining of the skin can be divided into two types, depending on whether or not the bilirubin which accumulates in the plasma has been secreted into the biliary tract. In obstructive jaundice, mentioned above, bilirubin does enter the plasma from the biliary tract. Bilirubin is produced in the marrow of flat bones from the effected red cells. If the liver is unable to clear the plasma of this, either because it cannot take it up effectively or because red cells are being destroyed too rapidly, jaundice will result. This is, however, different from obstructive jaundice in that in the latter the plasma bilirubin is conjugated mainly (80%) as a diglucuronide, but monoglucuronides and sulphates occur also. These conjugates are highly water soluble. In the hemolytic or prehepatic type of jaundice, the plasma bilirubin is unconjugated and therefore not very water soluble. Unconjugated bilirubin exists in the plasma in association with albumin. The upshot of this is that, although in all forms of jaundice the skin is yellow, only in the obstructive type (posthepatic) is the urine yellow also. Conjugated bilirubin, being water soluble, can enter the urine; the unconjugated type cannot. The prehepatic hemolytic type of jaundice is called, as a result, *acholuric.* The physiological jaundice seen in newborn infants, when the fetal red cell population is undergoing reduction to normal postnatal levels, is of this type. In *hemolytic jaundice* (excessive destruction of red cells, with a normal liver) the amount of conjugated bile pigment entering the intestine will be elevated and, therefore, the feces will be more darkly coloured than normal. In this condition the normal liver is overwhelmed. The glycuronyl transferase mechanism, which is responsible for bilirubin conjugation, is relatively deficient in the newborn, especially in the premature, and the liver as a result is easily overwhelmed. This mechanism may be defective throughout life with continuous jaundice as the outcome (Crigler-Najjar syndrome). In obstructive jaundice and in diseases in which the liver cannot conjugate bilirubin the feces will, of course, contain no bile pigment and become 'clay colored', but in the former there will be pigment in the urine and in the latter condition not. If the liver cannot conjugate bilirubin the pigment cannot enter the bile, but neither does conjugation guarantee entrance. There are congenital conditions in which, despite normal conjugation, bilirubin cannot leave the hepatocytes (Dubin Johnson syndrome). In obstructive jaundice, in contrast to the prehepatic type, bile salts will be absent from the intestine and levels in the plasma will rise.

As usually portrayed (fig. 6/6), one would expect unconjugated bilirubin to be highly water soluble since the two propionic acids (P and P) have free carboxyls. In actual fact, within the bloodstream these carboxyls are masked by hydrogen bonding to the nitrogens and oxygens of the two outer rings respectively; this renders the compound insoluble. Glucuronide conjugation is with these carboxyls, and this prevents internal ring formation and thus preserves the water solubility. In cases of severe prehepatic jaundice, especially in infants, the patient is usually exposed to blue light; this causes rotation of the outside rings, thus disrupting the internal linkages and freeing the carboxyls of the propionic acids. Bilirubin thus becomes water soluble and can enter the bile without conjugation. Photoisomers, however, are unstable and rapidly resume the stable insoluble form.

Gallstones. All bile, even the highly concentrated gallbladder bile, is isotonic with plasma. This can only mean that the concentration of particles must remain constant as water is withdrawn from gallbladder bile. Since the larger particles remain in solution they must coalesce. This is true of the bile salts; possibly the coalescence of bile salts is determined by total solute concentration rather than on the concentration of the bile salts alone. It is possible, then, that bile salt coalescence compensates for the increasing concentration of other molecules and that bile salts coalesce with other large molecules. Since the absorbed water and electrolytes are isotonic, when bile salts coalesce this will leave free water for the solution of the other large organic molecules. This coalescing property of bile salts has been extensively studied. Critical concentrations for coalescence of the various naturally occurring bile salts have been described. Five or more single molecules may coalesce to form a single particle in the gallbladder. Coalescence is not seen in hepatic bile because the concentration of bile salt is much below the critical point.

Bile contains substantial concentrations of cholesterol (table 6/I). Cholesterol is not very water soluble, but it is held in solution by the combined actions of bile salts and lecithin. This becomes especially critical in the gallbladder in which cholesterol concentration rises along with that of the other large molecules. A rise in biliary cholesterol or a fall in either or both of the other two constituents could result in precipitation of cholesterol. This occurs from time to time. The usual cause seems to be a small bile acid pool.

Bile salts are amphipathic substances. The carboxyl and hydroxyl groups are hydrophilic and the large steroid portion is hydrophobic. When the concentration rises these aggregate to form micelles with the hydrophilic end out and the hydrophobic in. This hydrophobic nucleus can dissolve or dissolve in lipid. One hundred molecules of bile salt can hold three of cholesterol in such aggregates, but if lecithin, also amphipathic, is present in addition, then nine cholesterol molecules can be held. Lecithins in bile and monoglycerides in the intestine generally lower the critical micellar concentration of bile salts. In the absence of bile salt-lecithin micelles, or if they are few, cholesterol will crystallize, in other words, the cholesterol solution will become super saturated.

The bile salt concentration of bile is not constant. During digestion it is high because the enterohepatic circulation is active, but during an overnight fast the concentration of bile salt in the hepatic bile will be low. This fasting bile may be super saturated with cholesterol but on entering the gallbladder, where bile salt concentrations are high in normal people, it ceases to be so.

In most animals the secretion of lecithin and of cholesterol varies directly with the output of bile salt (fig. 6/7) but, in man, this relationship is much less evident for cholesterol. An independent source of cholesterol is suggested. Bile acids by mouth, except chenodeoxycholic and ursodeoxycholic, usually increase biliary cholesterol which, of course, is undesirable

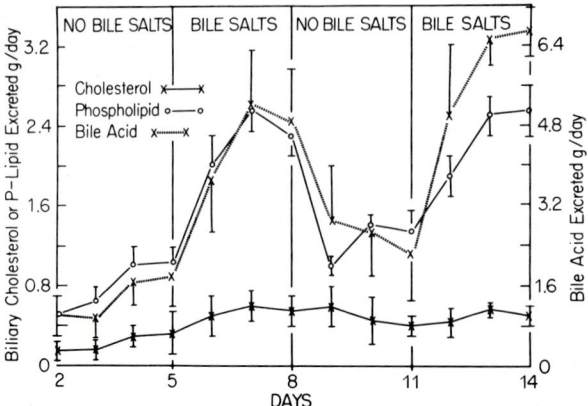

Fig. 6/7. Daily secretion of bile salts, phospholipids and cholesterol by five patients with indwelling biliary T tubes with and without bile salt feeding. Reproduced, with permission, from *Swell* et al. [26].

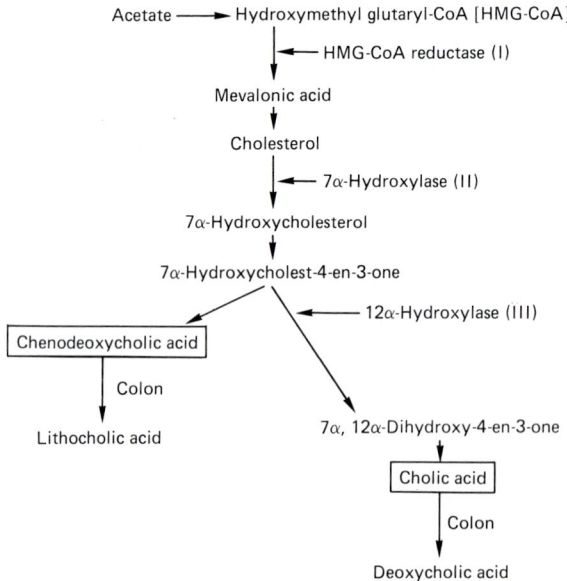

Fig. 6/8. The pathway for cholesterol and bile acid synthesis and the rate limiting steps I, II, and III. Chenodeoxycholate is thought to reduce the enzyme activity at step I. From *Swell* et al. [26].

in people in whom the bile is already super saturated. Human cholesterol stones placed in dog's gallbladders will dissolve slowly. Attempts have been made, by feeding bile acids, to increase the bile salt pool and thus dissolve stones. Stones have indeed been dissolved, but not as originally thought, through an increase in pool size. The two effective acids, chenodeoxycholic and ursodeoxycholic, unlike the others, reduce biliary cholesterol, probably by reducing the activity of the enzyme between acetate and mevalonic acid (HMG Co-A reductase) (fig. 6/8). This is the most plausible explanation for their effectiveness. This step is common to both the synthesis of cholesterol and bile acids. Therapy with bile acids is slow (6 months) and, in addition, it must be continuous, because when it is stopped the patient's endogenous bile is as lithogenic as ever. Surgical removal of the gallbladder eliminates the condition as now the patient is incapable of producing a concentrated bile and the cholesterol can remain in solution.

Following a cholecystectomy, bile must flow continuously into the intestine. Fortunately, after this operation the tone of the choledochoduodenal sphincter usually diminishes [18]. If it does not, severe discomfort results. Many surgeons perform a sphincterotomy as part of the cholecystectomy. A special instrument is passed down the common duct which can incise the duct and overlying sphincter muscle from within.

Following cholecystectomy the daily number of enterohepatic circuits will be greatly increased (from about 3–8 to 12–16). This will mean a bigger loss of bile salt and increased production of primary acids by the liver to keep the pool constant.

References

1 Cai, W.; Gabella, G.: The muculature of gallbladder and biliary pathways in the guinea-pig. J. Anat. *136:* 237–250 (1983a).
2 Cai, W.; Gabella, G.: Innervation of the gallbladder and biliary pathways in the guinea-pig. J. Anat. *136:* 97–109 (1983b).
3 Chapman, G.; Chiardo, A.; Coffey, R.; Wineke, K.: The fine structure of mucosal epithelial cells of a pathological human gallbladder. Anat. Rec. *154:* 579–616 (1966).
4 Eastwood, M; Mitchell, W.: Biliary excretion by the liver; in Duthie, Wormsley, Scientific basis of gastroenterology (Churchill-Livingstone, Edinburgh 1979).
5 Elias, H.; Pauly, J.; Burns, E.: Histology and human microanatomy; 4th ed. (Wiley, New York 1978).
6 Gerolami, A.; Sarles, J.: Biliary secretion and motility; in Crane, Gastrointestinal physiology. II Int. Res. Physiol., vol. 12 (University Park Press, Baltimore 1977).

7   Gray, C.; Nicholson, D.; Quincy, R.: The fate of bile in the bowel; in Code, Handbook of physiology, sect. 6: Alimentary canal, vol. 5 (American Physiological Society, Baltimore 1968).
8   Ham, A.: Histology; 7th ed. (Lippincott, Philadelphia 1974).
9   Hoffman, A.: Enterohepatic circulation of bile acids in man; in Clinics in gastroenterology 6:1 (Saunders, London 1979).
10 Hong, S.; Magee, D.; Crewdson, F.: The physiological regulation of gallbladder evacuation. Gastroenterology *30:* 625–630 (1956).
11 Itoh, Z.; Takahashi, I.: Periodic concentrations of the canine gallbladder during the interdigestive state. Am. J. Physiol. *40:* G183–189 (1981).
12 Ivy, A.; Oldberg, E.: A hormone mechanism for gallbladder contraction and evacuation. Am. J. Physiol. *86:* 599–613 (1928).
13 Johnson, F.; McMinn, R.; Birchenough, R.: The ultrastructure of the gallbladder epithelium of the dog. J. Anat. *96:* 477–487 (1962).
14 Jorpes, J.; Mutt, V.: Secretin, cholecystokinin, pancreozymin and gastrin. Handbook of experimental pharmacology. Part XXXIV (Springer, Berlin 1973).
15 Kaye, G.; Wheeler, H.; Whitlock, R.; Lane, N.: Fluid transport in the rabbit gallbladder. A combined physiological and electron microscopic study. J. Cell Biol. *30:* 237–268 (1966).
16 Lane, B.; Rhodin, J.: Fine structure of the lamina muscularis mucosae. J. Ultrastruct. Res. *10:* 489–497 (1964).
17 Magee, D.: In Taylor, The biliary system, pp. 233–248 (Davis, Philadelphia 1965).
18 Mahour, G.; Wakim, K.; Soule, F.; Ferns D.: The common bile duct after cholecystectomy: comparison of common bile ducts in patients who have intact biliary system with those in patients who have undergone cholecystectomy. Ann. Surg. *166:* 967 (1967).
19 Mueller, J.; Jones, A.; Long, A.: Topographical and subcellular anatomy of the guinea pig gallbladder. Gastroenterology *63:* 856–868 (1972).
20 Netter, F.: The CIBA collection of medical illustrations, vol. 3. The digestive system – part III: Liver, biliary tract and pancreas. (CIBA Pharmaceutical Co., Summit 1964).
21 Polak, J.; Bloom, G.: Hormones of the gastrointestinal tract; in Duthie, Wormsley, Scientific bases of gastroenterology (Churchill-Livingstone, Edinburgh 1979/University Park Press, Baltimore 1977).
22 Simmond, W.: Absorption of lipids; International review of science physiology; in Guyton, International review of science, physiology, ser. 1, vol. 4, chap. 10 (Butterworth, Lancaster 1974).
23 Sobotta, J.; Hammersen, F.: Histology: a color atlas of cytology, histology and microscopic anatomy (Urban & Schwarzenberg, Baltimore 1980).
24 Sutherland, S. D.: The intrinsic innervation of the gallbladder in *Macaca rhesus* and *Cavia porcellus.* J. Anat. *100:* 261–268 (1966).
25 Sutherland, S. D.: The neurons of the gallbladder and gut. J. Anat. *101:* 701–710 (1967).
26 Swell, L.; Gregory, D.; Vlahcevic, Z.: Current concepts of the pathogenesis of cholesterol gallstones. Med. Clins N. Am. *58:* 1449–1472 (1974).

7 The Pancreas

Structure of the Pancreas

In the adult human, the pancreas is a large (80–100 g, 12–15 cm long), soft, grayish-pink gland located retroperitoneally at approximately the level of the second lumbar vertebra. Its head and uncinate process lie within the concavity of the duodenum on the right, but its body and tail extend to the left as far as the spleen. The pancreas is covered and separated from adjacent structures by a fibrous capsule that is remarkably thin. Connective tissue septae, also very thin, extend into the gland from the capsule, dividing it into lobules. In vivo the lobulation within the gland is not as distinct as it is in section, where fissures in the septal planes (artifacts of fixation) may clearly demarcate the pattern. Because the septae are so delicate, it is questionable how much structural support they render. They do, however, constitute the tissue plane in which fine blood and lymph vessels and nerve plexuses (including parasympathetic ganglionic cells) are distributed. The condensations of connective tissue surrounding the main duct and its branches probably supply most of the internal support of the organ.

The parenchyma of the pancreas is composed of two quite separate types of glandular tissue which are, however, in intimate topographic association; the pancreas is both an exocrine and an endocrine gland. The main mass (84%) of the tissue (the portion with which we are concerned) is composed of exocrine cells arranged in acini. Embedded within the exocrine tissue are clusters of endocrine cells, the *pancreatic islets* (Langerhans), constituting only about 2% of the gland. The capsule and other connective tissue elements of the extracellular matrix account for 10% of the mass, with ductular tissue and blood vessels accounting for the remaining 4% [13].

The Pancreatic Acinus

The *acinus* is a flask- or tubular-shaped mass of pyramidal exocrine (acinar) cells arranged with their apices facing a central lumen (*centroacinar lumen*) (fig. 7/1).

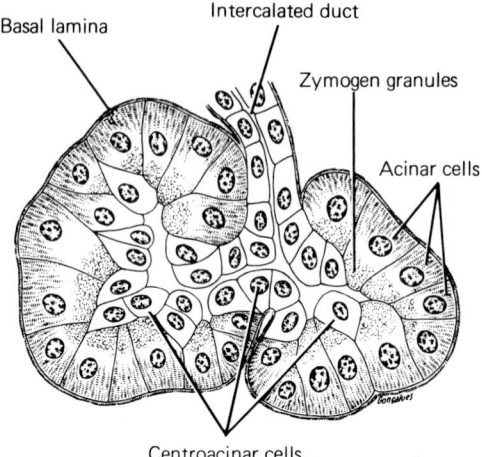

Fig. 7/1. Diagram of a pancreatic acinus. The pyramidal acinar (exocrine) cells have granular apices with rough endoplasmic reticulum located basally. An intercalated duct invaginates the acinus. Pale-staining cells are seen both in the luminal region of the acinus (centroacinar cells) and lining the initial portion of the duct (ductule cells). From *Junqueira* et al. [18].

The acinus has long been accepted as constituting the structural unit of the organ [21]. It is bound by a connective tissue matrix, including a basal lamina, which does not intervene between the lateral aspects of adjacent cells. A fine ductular tributary of the intralobular duct, called the intercalated duct, invaginates the secretory mass. The initial (intra-acinar) portion of the duct is lined by low cuboidal *(ductule)* cells, those directly related to the centroacinar lumen being designated as *centroacinar cells* (fig. 7/1).

Several recent studies have questioned whether the acinus also represents the functional as well as the structural unit of the gland. Casts of the terminal subdivisions of the ductular system [3] and electrical coupling experiments [17] have suggested a functional subunit much larger in size (approximately 500 acinar cells) than the 20- to 50-cell acinus [13].

On histological section, the pancreas is very similar in appearance to the parotid, the other large protein-secreting exocrine gland of the GI tract. The pancreas is characterized, however, by the relative absence of conspicuous intralobular ducts (and the absence of striated ducts in particular – see 'Microstructure of Salivary Glands', chapter 2). The intercalated ducts of the parotid do not penetrate the acinus nor have they centroacinar portions as in the pancreas [18]. The occurrence of pancreatic islets is, of course, unique to the pancreas.

The Pancreatic Acinar Cell. The acinar cell of the human pancreas is PAS-positive and is therefore a serous cell. It has all the characteristics described under 'The Structure of Protein-Secreting (Serous) Cells', chapter 1, and has, in fact, served as the paradigm of that cell type.

In terms of RNA content, the acinar cell is one of the richest cells of the body – a characteristic which coincides with its prodigious protein-synthesizing activity. It has been calculated that in the rat, under conditions of maximal stimulation, the pancreas can synthesize for secretion up to 1.5% of the gland's initial weight of protein in 1 h [18].

Zymogen granules are restricted to the apical cytoplasm of the acinar cells (fig. 7/1). The number of zymogen granules in the cells is variable, depending on the stage of development [19], supply of nutrients [35], and the digestive phase, i.e. the state of stimulation of the gland by neurohormonal agents [9]. A maximum level, accounting for 20% or more of the total cell volume, is attained in the fasted animal. The powerful enzymatic constituents of the pancreatic secretion contained within these membrane-bound bodies are in an inactive (proenzymatic) form, except for lipase, amylase, RNase and DNase. Their secretion as inactive proenzymes, as well as their sequestration into membrane-bound secretory granules, serves to protect the acinar cell, as well as the other elements of the gland, from autodigestion.

The Centroacinar Cell. If the acinus is sectioned in the proper plane, a few pale-staining cells may be seen in the luminal region of the acinus marking the origin of the duct system of the gland (fig. 7/1). These centroacinar cells are smaller than the acinar cell [11]; their relatively sparse cytoplasm lacks zymogen granules and they have only a small Golgi complex. They are quite irregular in shape, sometimes seeming to be squeezed between adjacent cells. The contiguous cells lining the fine, initial portion of the intercalated duct (the ductule cells), although more regular in shape, are in general structurally similar to the centroacinar cells. The centroacinar and ductule cells are believed to be responsible for secreting the bicarbonate-rich fluid of the pancreatic juice. Typical of cells actively involved in fluid and electrolyte transport (such as the parietal cells of the stomach and the striated duct cells of the salivary glands), the centroacinar cells are characterized by an abundance of mitochondria. On the other hand, the ductule cells do not exhibit the abundance of the mitochondria that the centroacinar cells do, and neither do they nor the centroacinar cells demonstrate some of the other structural adaptations (e.g. basal infoldings) associated with some known electron-transporting cells. However, neither do some other cells whose ability to transport electrolytes is undisputed. There is no single, identifiable cell organelle which is uniquely and invariably associated with salt and water transport [10]. However, several nonmorphological findings support the contention that the centroacinar and ductule cells are responsible for fluid and electrolyte secretion. These include the results of micropuncture experiments [37], the histochemical localization of carbonic anhydrase to these cells [5], and determination of the relative abundance of sialomucins (implicated as cation filters in other systems [28]) in their basolateral cell membranes [20]. The close association of these cells to the acinar cells, or more specifically, to the centroacinar lumen into which the acinar cells secrete, is probably a reflection of their role in solubilizing the content of the zymogen granules or during exocytosis (pancreatic secretion).

Ducts of the Pancreas

The *main duct of the pancreas* in humans (the embryological ventral pancreatic duct (Wirsung)) is lined by columnar epithelium in which goblet cells may be interspersed. Its lumen may be 2.5 mm wide. The duct is ensheathed in connective tissue and serves, more or less, as the structural 'backbone' of the gland. Side branches with low columnar epithelia flow into it at angles, so that the main duct and its tributaries, when dissected free from the parenchyma, have a 'herringbone' appearance. Because these side branches run in the septa between lobules, they are the *interlobular ducts.* The interlobular duct receive, in turn, the *intralobular ducts* which run within the lobules but the larger of which are nonetheless ensheathed by continuations of the connective tissue of the interlobular septa. The intralobular ducts have a low columnar to cuboidal epithelium and, as stated previously, are much less abundant and conspicuous than those of the parotid. Very fine ducts or ductules, issuing from the acini and lined with a flattened cuboidal epithelium, the *intercalated ducts,* drain into the intralobular ducts. As mentioned previously, of the ductule cells forming the initial (intra-acinar) portions of the intercalated ducts, those cells in contact with the centro-acinar lumen are designated as centroacinar cells. The ionic content of the alkaline fluid thought to be secreted by the centroacinar and other epithelial (ductule) cells of the intercalated duct, is probably modified somewhat as it courses through the remainder of the ductular system to the intestines.

7 The Pancreas

Nerves of the Pancreas

The pancreas is innervated both by parasympathetic (cholinergic) fibers from the vagi and by sympathetic fibers arising from the celiac, superior mesenteric and hepatic plexuses. Their function will be discussed below. Additionally, afferent pain fibers travelling with the sympathetic nerves reach the pancreas. Clinically, raised intraductal pressure in patients with constricted or occluded ducts is a common source of pancreatic pain, which is usually severe and poorly located.

Pancreatic Secretion

This is the most important single digestive gland in the whole system. Its enzymes are capable of digesting all the major categories of foodstuffs and its secretion of bicarbonate titrates the acid gastric chyme back up again to a pH at which enzymes can act optimally. Only the pancreas produces enough bicarbonate to do this. The biliary system we have already seen produces bicarbonate, but the volumes are smaller and the concentrations lower.

Gastric Phase

When food enters the stomach the gastric phase of pancreatic secretion starts. This ensures the presence of pancreatic juice in the duodenum when the stomach starts emptying. This is important for the initiation of the duodenal phase of secretion as we will see. The gastric phase has a dual mechanism. In the first instance distension of the body of the stomach starts the flow of an enzyme rich pancreatic juice. This depends on vagal innervation since bilateral thoracic vagal section abolishes it. The other phase is the hormonal. We have already seen that gastrin resembles CCK closely and indeed shares in some of its actions. One of these is the stimulation of pancreatic enzyme secretion. Gastrin is not as potent as CCK but, nevertheless, physiological blood levels of gastrin do stimulate pancreatic enzyme secretion.

Duodenal Phase

When the acid gastric chyme enters the duodenum rapid pancreatic secretion begins within a minute or two. This juice is rich in bicarbonate and in enzymes. The exocrine pancreas is really a double gland; the acinar part of it (fig. 7/2) (the cells of which in the unstimulated state contain zymogen

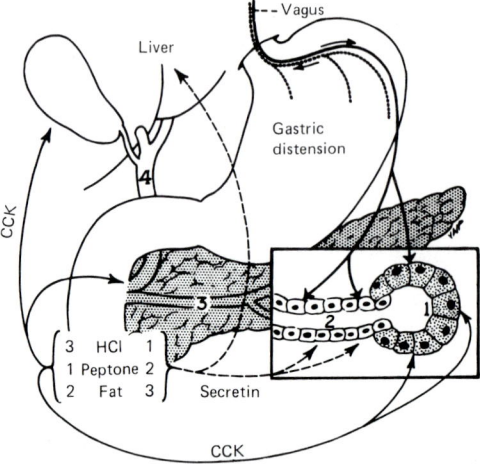

Fig. 7/2. Diagram illustrating the release and actions of the pancreatic hormones. HCl, peptone and fat, lower left, act on the intestinal mucosa. The numbers on the left represent the potency order in releasing pancreozymin and to the right in releasing secretin. CCK acts on the pancreatic acinar cells (1). It acts also to contract the gallbladder and relax the sphincter of Oddi, which is situated at the lower end of the bile duct (4) just before it joins the main pancreatic duct (3). In man secretin likely acts on the ductule (2) cells. It also acts on the liver to increase bile flow. From *Magee* [24].

granules) produces the enzymes. The collecting ducts and the centroacinar cells are the second gland. It is by these that water and sodium bicarbonate are added to the secretion. As usual the electrolytes are actively secreted and the water follows osmotically. Each of these has its own blood-borne stimulating hormone. CCK, which has already been mentioned as the gallbladder contracting hormone, stimulates enzyme secretion and secretin, which we have already seen promotes bicarbonate secretion by the bile cannuliculi, does the same to the ductule and the centroacinar cells of the pancreas. In some animals (rats, sheep and, perhaps, guinea pigs) secretin does stimulate enzyme secretion, but in man, dogs and cats it does not. In man and dogs, however, the stimulation of water and bicarbonate by secretin is augmented by the simultaneous administration of CCK. This may or may not be a physiological occurrence.

Secretin

Stimuli for Release. The most important stimulus for secretin release is an acid duodenum [15]. Acid will liberate secretin from the intestinal

7   The Pancreas

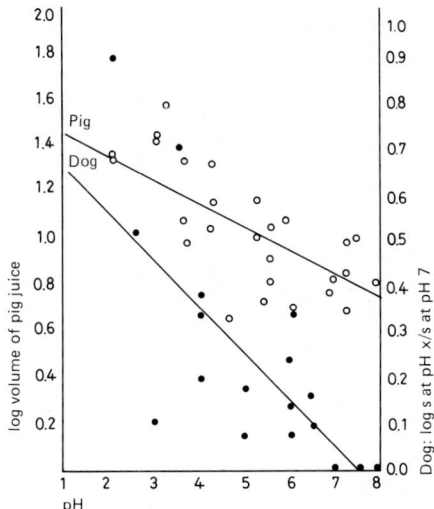

Fig. 7/3. The relationship between the pH of duodenal instillates, the log of the volume of pancreatic juice secreted in pigs, the log of ratio of secretion at pH 7 and the secretion at the other pHs tested in dogs. From *Hong* et al. [15].

mucosa down to about mid jejeunum, but progressively less effectively as one moves distally from the proximal duodenum. Modern extraction methods can recover more secretin per unit weight from duodenal mucosa than from elsewhere and the S cells now known to be the producers and storers of secretin are most abundant in the duodenum. Their numbers decline to mid-jejeunum, beyond which they are absent. S cells are pH sensitive. When acidic solutions are run through loops of upper small intestine the secretory rate of water and bicarbonate is linearly related to the [H$^+$] [15]. This implies that there is no threshold for secretin release, but a progression from no stimulation at neutrality to powerful stimulation at pH 2. The term 'acid load' has been used in discussing this subject and has caused confusion [28]. It simply means that even though the individual S cell responds to [H$^+$] a large volume (load) of acid introduced into the duodenum will produce more secretin and therefore, more pancreatic juice because it will be neutralized less rapidly and therefore, stimulate more S cells than a small volume (fig. 7/3).

Acid, although without doubt the most potent S cell stimulant, is not the only one. Partially digested fats and proteins in the upper small intestine will also cause the S cells to discharge their secretin granules into the blood

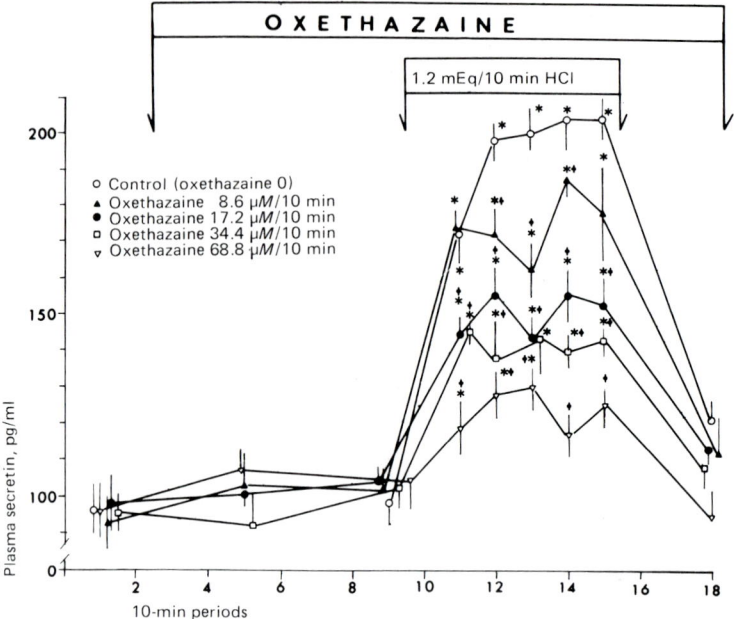

Fig. 7/4. Plasma secretin levels in response to intraduodenal acid before and after four concentrations of local anesthetics (oxethazaine) applied to the duodenal mucosa. Asterisks denote significant difference from control. From *Yamazaki* [46].

stream. Persons suffering from achlorhydria can and do secrete pancreatic juice.

Mechanism of Release. The mechanisms by which secretin is liberated from the mucosa and enters the portal blood is unclear. It is known that local anesthetization of the mucosa prevents liberation. After local anesthetics intraduodenal acid will no longer raise plasma levels of secretin and, of course, water and bicarbonate secretion from the gland will not increase (fig. 7/4, 7/5) [39, 46]. The consequences of vagal interruption, by atropine or by ganglionic blockade on S cell stimulation are harder to determine since it is difficult to separate modification of secretin liberation from modification of its action on the gland. It is a seemingly straightforward experiment to give secretin before and with vagotomy, atropine or ganglionic blockade, but in actual fact such experiments have been going on inconclusively for years [22]. The probable reason is that in the fasting animal, as all animals

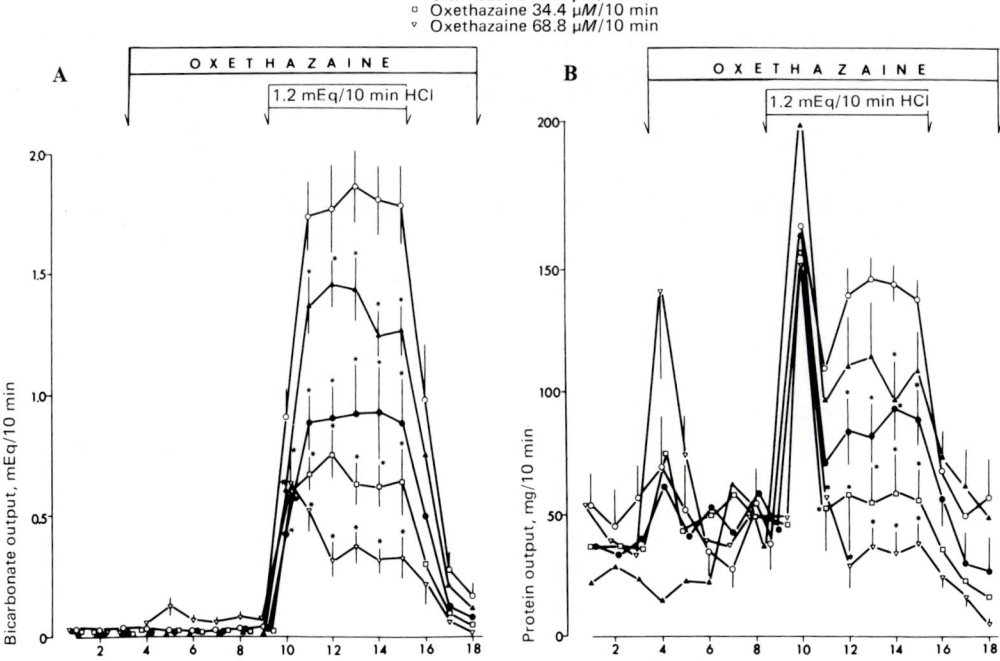

Fig. 7/5. Effect of four doses of oxethazaine given intraduodenally on the pancreatic response to intraduodenal HCl. **A** Bicarbonate secretion. **B** Protein secretion. From *Yamazaki* [46].

used in this sort of experiment invariably are, the pancreatic secretion of water and enzymes waxes and wanes in phase with the interdigestive motility complexes. This was not recognized until recently, so spontaneous changes, in short experiments, could not be separated from the experimental. This has been recognized now and experiments planned and interpreted accordingly have made it clear that atropine and vagal block do depress the response of the gland to secretin and probably do not affect the liberation of the hormone from the mucosa. In some animals (pigs and cats) vagal stimulation results in a copius secretion of an enzyme rich juice [26]. Anticholinergic agents will block the enzyme secretion thus stimulated, but not the water and bicarbonate. Secretin does not appear to be the mediator of this secretion as its plasma levels do not rise with stimulation; it is stated, however, that VIP levels do. This polypeptide is found in peripheral nerves and is thought by some to be a neurohumoral mediator. In dogs vagal stimulation does not increase the volume of pancreatic secretion.

Cholecystokinin (CCK)

Stimuli for Release. The acini of the gland are stimulated to produce and release enzymes by the hormone CCK as previously mentioned. This hormone is liberated from much the same region of the small bowel as secretin, but from different cells. Not as much is known about this hormone as of the others, since no satisfactory radioimmunoassay has yet been devised for it.

Like secretin it is liberated by acid chyme, but unlike secretin, acid is far from being the most potent liberator. In fact, there have been recent claims that acid does not liberate CCK from the duodenum at all. This is incorrect since all who have tried the experiment have obtained good gallbladder contraction and enzyme secretion when the duodenum is exposed to a pH below 3 (fig. 7/5). Secretin liberation, on the other hand, seems to have no threshold (fig. 7/2).

The most potent stimuli for CCK release are partially digested fat emulsions and partially digested proteins. Some single amino acids, e.g. tryptophane and phenylalanine, are potent releasers. Although the enzyme package secreted by the acini will digest starch and glycogen in addition to fats and proteins, neither starch, glycogen, nor any of their digestion products including monosaccharides will stimulate the duodenum to release CCK.

As in the case of secretin, dispute surrounds the relationship between the role, if any, of the cholinergic nerve supply to the gland in the action of CCK [23]. Ganglionic blocking agents greatly depress the action of CCK on the gland [16].

Some investigators find that atropine will depress the action of CCK on the intact gland but not its action on the denervated or transplated gland [39]. The interpretation is that the cholinergic innervation sensitizes the gland to CCK. This is abolished when the gland is depressed, whether by atropine or denervation. There must be some other nerves of importance since ganglionic blocking agents exert a much more profound depression of CCK-mediated stimulation than either vagotomy or atropine. It is unknown whether these other nerves are intrinsic or extrinsic since no one seems to have studied the action of ganglionic blockers on transplanted glands.

Mechanisms of Release. The liberation of CCK from the intestinal mucosa is, like secretin, prevented by local anesthetization (fig. 7/5) [39, 46]. It is also quite clear that atropine or vagotomy reduce the glands' responsiveness to a meal much more than can be accounted for by the consequent reduced response of the gland itself to CCK [14]. Vagal stimulation obviously increases enzyme secretion; this is blocked by atropine, but there is no direct evidence yet that vagal stimulation causes CCK liberation from the duodenum.

Since vagotomy or atropine reduce, but do not abolish meal stimulated secretion this must mean that either less CCK is liberated after cholinergic interruption or else that some pancreas sensitizing cholinergic connection between the intestinal lumen and the gland has been interrupted. The latter is an old idea; *Singer* is its contemporary champion [39]. He has found a prompt enzyme response to feeding in intact glands and a slow one from transplants, but no difference to intravenous CCK. Whatever the mechanism, it is clear that intact cholinergic innervation enhances the enzyme secretion by the gland to both meals and intravenous CCK. It is to be wondered at that the unequivocal almost complete suppression of secretion in response to either CCK or meals produced by ganglionic blockade is almost uninvestigated. This blocker seems to block both the action of CCK on the pancreas and its release from the duodenum because it prevents gallbladder contraction following meals, but does not block the action of CCK on the gallbladder.

Biochemistry of Secretin and Cholecystokinin

The nature of the hormones which stimulate the pancreas is now well known. Secretin is a straight chain polypeptide containing 27 amino acids. This is its only form either inside cells or in the blood stream. All 27 are required for its activity. It is quite unlike gastrin and CCK either chemically or biologically, but in both respects it is very much like the pancreatic islet hyperglycemic hormone glucagon and the gut hormones GIP and VIP (see chapter 12). Like these it is an agent which increases cyclic AMP within the cells sensitive to it whether they be in the biliary tract, pancreas, small intestine or stomach. Cyclic AMP seems to be an obligatory intermediate between it and the secretory elements within the cell.

Physiological experiments in 1903 made it evident that hormonal stimulation of pancreatic water and bicarbonate secretion was a logical necessity. The word hormone was coined to describe this as yet uncharacterized chemical messenger. The hormone was laboriously isolated within the last decade and its existence, cells of origin and physiological importance definitely established.

Radioimmunoassay of secretin proved to be difficult as it has no amino acid which can be iodinated. As a result, the physiological role of this the oldest hormone was, for a time, doubted. Increases in plasma levels could not be detected when acid was placed in the duodenum. Improved techniques have made the demonstration of increases following meals or acidification of the duodenum possible (fig. 7/4).

The nature of the hormone CCK has already been dealt with (chapter 6). Like gastrin, to which it is related structurally and biologically, it does not act via cyclic AMP. Neither CCK, vagal stimulation, nor cholinomimetics increase cyclic AMP in pancreatic acinar cells; all three, however, require calcium for their action. Stimulation of enzyme secretion seems to involve the freeing of bound Ca^{2+} within the cell since the gland in isolation can respond to stimuli for a time in calcium free media. Calcium in turn stimulates the discharge of zymogen granules. Some have suggested cyclic GMP as the replacement for AMP in the system.

Cephalic Phase

Numerous attempts have been made to obtain evidence for a cephalic phase of pancreatic secretion. The only evidence for a direct vagal influence on the pancreas in man and dogs is that which established the gastropancreatic reflex. In all of the other studies in which a cephalic phase has been claimed the positive results could well be due to acid running from stomach to duodenum or vagally stimulated gastrin release from the gastric antrum. Failure to recognize periodic secretion in fasting animals accounts also for the disagreement over the question of resting pancreatic secretion. Contemporaneously with motility peaks there is evident secretion of bicarbonate, water and enzymes, but between peaks there is virtually none. This could not be appreciated in experiments lasting less than 300 or 400 min, i.e. 3 or 4 spontaneous cycles (see chapter 9).

Composition of Pancreatic Secretion

The fluid of pancreatic juice contains protein over 90% of which is enzyme – not plasma – protein and the electrolytes found in plasma, but there is an important difference from saliva in that pancreatic juice, regardless of flow rate, is always isotonic with plasma. Its anionic composition does change with flow rate, but its cations do not. In the pancreas there is an acinar secretion which corresponds to the primary secretion of the parotid

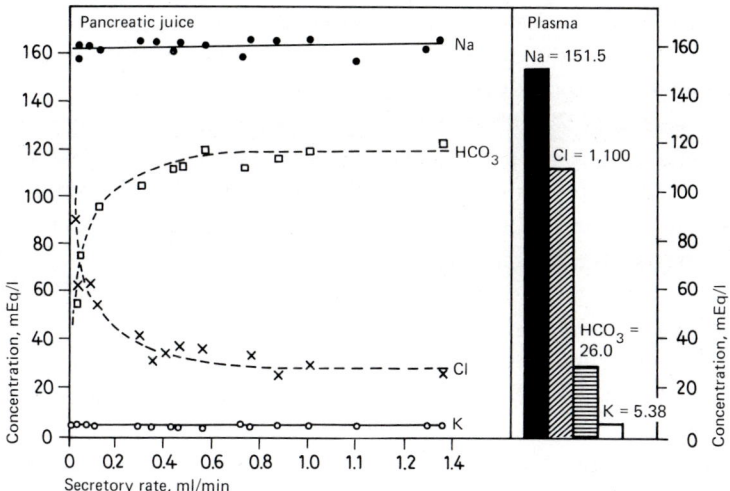

Fig. 7/6. Relationship between rate of secretion and electrolyte concentrations in the pancreatic juice of a dog compared with plasma concentrations. From *Bro-Rasmussen* et al. [4].

gland [4]. As the secretion passes from the acini during digestion the centroacinar and ductule cells, responding to the stimulus of circulating secretin, add an isotonic solution of sodium bicarbonate to it. At low rates of secretion the acinar secretion is a large fraction of the total and bicarbonate concentrations are low, but as the rate of secretion increases the contribution of the acinus to it becomes insignificant and bicarbonate concentrations reach a plateau (120 mmol/l). Chloride concentrations, which represent acinar secretion in the main, reach a nadir and of course Na^+ and K^+ concentrations remain constant throughout secretion (fig. 7/6). The pH of the blood leaving the gland falls noticeably as secretion increases. Even though bicarbonate concentration reaches a plateau before the maximal secretory volume is reached its total secretion per unit time increases with increasing juice volume.

That this modification of the acinar secretion is due to the active secretion of sodium bicarbonate has been confirmed by micropuncture and microperfusion studies, just as in the salivary glands [38]. There is circumstantial evidence also. The ductule cells and centroacinar cells are rich in carbonic anhydrase; blockade of this enzyme greatly reduces the glands response to secretin. In the pancreas as in the parotid gland it has been possible to destroy the centroacinar and ductule cells and leave the acini and vice versa. Destruction of ductule and centroacinar cells ends the ability of the pancreas to respond to secretin.

The source of the bicarbonate of pancreatic juice is largely CO_2, but plasma bicarbonate contributes also.

The electrolyte composition of the acinar secretion in mmol/l is as follows $[Na^+]$ 160, $[K^+]$ 8, $[HCO_3^-]$ 52, $[Cl^-]$ 100. This is not an ultrafiltrate of plasma, but an active secretion.

Inhibition of Pancreatic Secretion

Little attention has ever been paid to the sympathetic nerve supply of the pancreas, but what little there has been makes it clear that adrenergic nerves and neurohumours depress pancreatic secretion of both water and enzymes. Following bilateral division of the splanchnic nerves the daily secretion of the pancreas rises. Bilateral vagal section results in a fall in secretion; subsequent splanchnic division raises it in animals eating normally. The effect of splanchnic interruption is duplicated by β-adrenergic blocking agents.

The stimulating mechanisms so far described, for enzyme secretion at least, seems to be a positive feedback system – the more protein and fat digestion products, the more secretin and CCK; the more enzymes, the more digestion products – until such time as there is no more chyme in the intestine. With the acid stimulus, the feedback is negative since low pH means copious secretion of bicarbonate which in turn neutralizes acid. Perhaps adrenergic nerves keep things under restraint, but no one knows how. In certain animals the presence of pancreatic juice itself in the intestine depresses the gland. Some of this is simply due to the neutralization of gastric acid pouring in from the stomach [32]. This effect is seen only in rats, pigs and domestic fowl, all of which have high fasting gastric acid secretion; these experiments are invariably done in fasted animals. There is no doubt, however, that in these species the intraduodenal introduction of trypsin inhibitors does increase enzyme and water secretion. In dogs intraduodenal pancreatic juice or trypsin stimulates rather than depresses secretion. It has been claimed recently that in man intraduodenal trypsin inhibitor increases enzyme secretion. A solution to this problem is in sight for pigs and rats, but not for man and dog. In the rat there is reasonable evidence that trypsin depresses the sensitivity of the secretin and CCK producing cells to intraduodenal stimuli. Recently, it has been found that the presence of oleic acid in the ileum depresses the responsiveness of the pancreas to exogenous

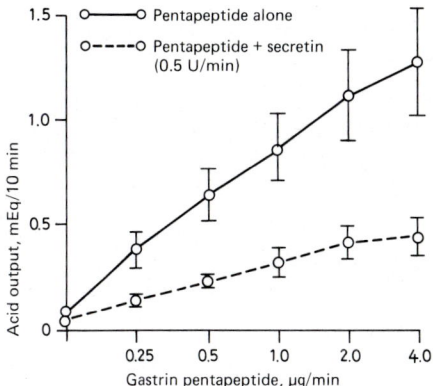

Fig. 7/7. Dose-response curves for acid output in response to gastrin pentapeptide alone and pentapeptide plus a fixed dose of secretin. Each point is the mean of 10 experiments in 5 dogs, the vertical bars represent SEM. From *Nakajima* et al. [32].

hormones. Since this has been seen in cross-circulation experiments a hormonal mechanism is presumed. The physiological importance is obscure.

Acid duodenal pH increases secretin and CCK production by the small intestine which in turn increases HCO_3^- secretion raising the duodenal pH to a level optimal for the action of the pancreatic enzymes. We have here, in essence, an automatic titrating apparatus. The mucosa of the duodenum is the pH sensor, the blood stream is the wire, secretin the electrical current and the pancreas the burette containing the alkaline solution. A unique feature of this automatic titration is that it can depress the delivery rate and acidity of the sample reaching it. Both secretin and CCK depress the motility of the stomach and hence its emptying rate and they both depress the action of gastrin on the parietal cells (fig. 7/7). CCK depresses pepsin also, but secretin stimulates it (fig. 7/8) [31]. This, of course, means that during the course of digestion the stomach is constantly under restraint from these two hormones. This has been known for years and a hypothetical hormone 'enterogastrone' was postulated to explain it. No currently known gastrointestinal hormone has yet been recognized as filling the role of 'enterogastrone' except CCK and secretin. Indeed in dogs if gastric chyme is diverted from the duodenum the daily secretion of gastric juice is greatly increased. Since acid is not a very potent CCK releaser it is only at low duodenal pH (below 3) (fig. 7/9) that characteristic depression of gastrin-

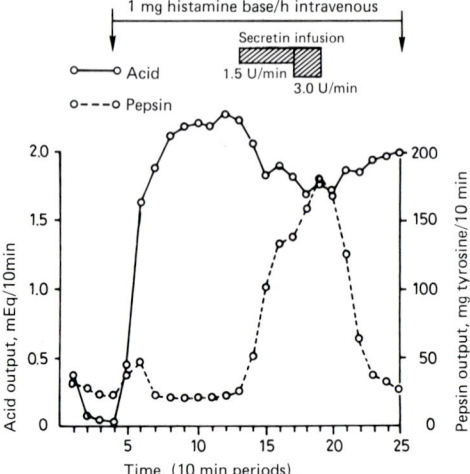

Fig. 7/8. Effect of high doses of secretin superimposed on background stimulation by 1 mg histamine/h on acid and pepsin secretion from the Heidenhain pouch. Each point is the mean of five experiments in 5 dogs. From *Nakajima* et al. [32].

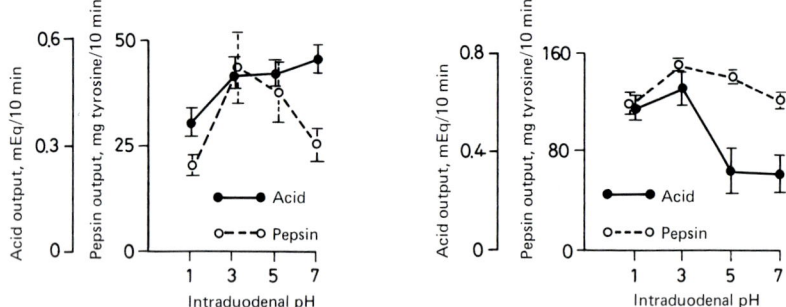

Fig. 7/9. Relationship between intraduodenal pH and Heidenhain pouch secretion of acid and pepsin with methacholine (2 µg/min i.v. rt.) or gastrin pentapeptide (2 µg/min i.v. lt.) as background stimulation. Duodenal pouch was irrigated with citrate buffer solutions of pH values indicated. Each point represents mean of 8 experiments in 4 dogs. Vertical lines indicate SEM. From *Nakajima* and *Magee* [31].

stimulated acid and pepsin is seen [30]. At higher pH, however, secretin is more important, but now although gastrin-stimulated acid is depressed slightly pepsin is augmented. The inhibitory effects can be seen in vagally denervated gastric pouches. As might be expected the other liberators of CCK and secretin also depress gastrin-stimulated secretion; fat in the

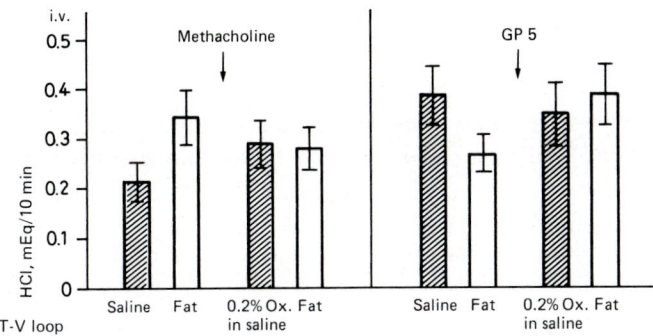

Fig. 7/10. The mean effect (SE) of fat emulsion in the T–V loop on acid secretion from the Heidenhain pouch stimulated either with methacholine or GP5. At the arrows the local anesthetic oxethazaine (Ox.) was introduced into the loop and the experiment repeated (n = 10). From *Odori and Magee* [34].

duodenum is the most obvious example. Fat greatly depresses gastric motility and gastrin-stimulated pepsin and acid secretion as does CCK (fig. 7/10) [33]. A physiological role for secretin in the negative feedback mechanism is universally accepted, but the role of CCK is not. Those who do not accept it explain the inhibition from the acid duodenum solely on the basis of secretin and have no explanation at all for fat inhibition. Fat in the intestine, however, depresses gastrin-stimulated motility, acid and pepsin, but like CCK and, again, unlike secretin it augments cholinergically stimulated motility, acid and pepsin (fig. 7/9, 7/10). CCK, as one might imagine from its structural similarity, acts competitively against gastrin, but synergistically with cholinergic stimuli. Secretin is a noncompetitive antagonist to all acid stimuli; perhaps because it increases mucosal cyclic AMP. Virtually every compound that does this depresses gastric acid secretion.

A complication in this story was that commercially available preparations of CCK were found to contain a gastric secretory inhibitor as an impurity, GIP (gastric inhibitory polypeptide). This polypeptide, released as are CCK and secretin from the mucosa of the small bowel, resembles secretin and glucagon closely; its action on the stomach resembles that of secretin more than of CCK or fat in the intestine. GIP is not well liberated from its cells of origin by fat, but is by carbohydrates and glucose in the small bowel, which produce only slight duodenal depression of gastric secretion. This reduces the likelihood that GIP is an important physiological gastric secretory depressor ('enterogastrone'). Some will state that, while CCK does compete with gastrin on the stomach, this action is pharmacological. We believe it to be physiological and to be in part responsible for acid inhibition of duodenal origin and wholly responsible for fat inhibition. A deficiency in this feed back mechanism has been advanced as one of the more reasonable hypotheses to explain the occurrence of duodenal ulcer and the hyperchlorhydria which often accompanies it.

Intraluminal Digestion and Pancreatic Enzymes [7]

Digestion in the intestine can be divided into intraluminal and epithelial. The pancreas is the almost exclusive supplier of enzymes which effect intraluminal digestion of all the major classes of foodstuffs: lipid, carbohydrate, protein and nucleoproteins. These are reduced by pancreatic enzymes to small fragments of which only the fats are absorbed directly; the others require further digestion at the epithelial cell surface. Although no vertebrate is without the full range of pancreatic enzymes the proportions differ from species to species and with age. In cattle, for example, in which the bodies of ruminal bacteria bulk large, ribonuclease is much more abundant than in the pancreatic juice of man and dog; in these nucleoproteins are not such an abundant component of the diet. In suckling infants, the diet of which contains no starch or glycogen, the pancreatic juice contains virtually no amylase or lipase. At weaning both increase.

Of the protein in pancreatic juice, which is usually between 1 and 10 g/dl, 90% is enzymic. Of the nonenzymic proteins there are trypsin inhibitor, a cofactor for lipase, a trace of plasma proteins and lactoferrin, which we have noted above in salivary secretion and which is an ubiquitous constituent of almost all external secretions. The human pancreas is estimated to produce about 1 liter of juice daily which means that the gland can manufacture, store and secrete up to 100 g of protein in 24 h. Thirty to 50 g is about the average.

All of this enzyme protein is stored in the zymogen cell granules of the acinar cells and every cell, even every granule contains the whole spectrum of enzyme activity. Whether or not the proportions are the same from granule to granule is not known. Some use purported granule differences to explain the observation that the enzyme proportions in the secreted juice from a single gland are not always identical. It is often assumed that granule formation is an absolute requirement for secretion, but after 30 or so minutes of strong stimulation the granules disappear and protein, water and electrolyte secretion continues with only slight abatement for many hours longer.

All of the pancreatic enzymes are hydrolasases.

Amylase

Amylase accounts for about 10% of the juice enzyme protein. It is fully active within the zymogen granules and as secreted. It is the enzyme which hydrolyses starch and glycogen. Most dietary carbohydrate is in this form; a little is eaten as sugars which do not require amylolytic hydrolysis.

Amylase itself is a protein (MW 50,000). Of all the pancreatic enzymes it is the most robust. The others, including the proteases, are very subject to proteolytic destruction. Each molecule of amylase contains an atom of calcium, which does not seem to be necessary for its activity, but is for its stability. By refined separation methods pancreatic and salivary amylase can be distinguished from each other structurally, but biochemically they are identical. The pH optimum for both is 6, below pH 4.5 they are inactivated. They require the presence of halogen ions for full activity. The Cl^- of pancreatic juice and saliva fill this requirement, but other halogens can substitute for it. Both enzymes are α-amylases and hydrolyse their substrate in the same manner producing identical end products. Cellulose is an isomer of starch, but its linkages are β. It is, therefore, completely resistant to mammalian amylase.

Starches are long chains of glucose molecules. The common dietary starches and glycogen are made up of long glucose chains, but they have branches in addition. The linkages between glucose molecules at branching points are 1–6 and the others 1–4. Amylases cannot hydrolyse the 1–6 linkages nor can they hydrolyse the terminal glucoses because they are *endoamylases*. This means that the end points of amylase activity will be (fig. 7/11) two glucoses linked 1–4 (maltose) and 3 glucoses linked 1–4 (maltotriose). (Amylase prefers to hydrolyse chains of 5 glucoses.) Any amylase end product larger than a maltotriose must have a 1–6 linkage. These are found with 4 glucoses molecules (two 1–4 and one 1–6 linkage) and 5 glucoses (one 1–6 and three 1–4 linkage). These are called *limit dextrins*. The 1–6 linkage is not only resistant itself to amylase, but also hinders hydrolysis of any internal 1–4 linkage next to it, hence the 5 glucose limit dextrin. One would expect very little free glucose to result from amylase hydrolysis and this indeed is the case. None of the above end

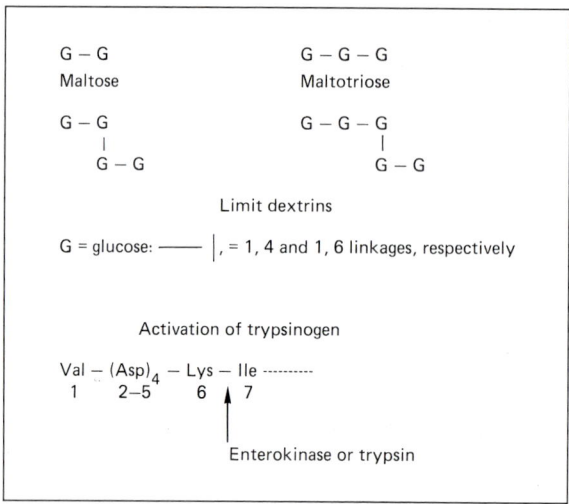

Fig. 7/11. End products of amylase digestion and the activation of trypsinogen.

products can be absorbed as such, since only monosaccharides cross the intestinal mucosa substantially. More hydrolysis is required elsewhere.

Proteases

All of the proteases in pancreatic juice are activated on entry into the duodenum by limited proteolysis of trypsinogen by the proteolytic enzyme *enterokinase*. Trypsin thus formed then activates the others. Enterokinase is a very large glycoprotein (MW 300,000) which is found on and produced by the brush border of the duodenal surface epithelial cells. The specificity of enterokinase is remarkable, in that it acts only at the lysine, isoleucine 6–7 linkage close to the N terminal end of the trypsin molecule and on no other linkage (fig. 7/11). The size of the peptide split off (6 amino acids) is small and unique. This is the *trypsin activation peptide* mentioned later as a possible gastric secretory inhibitor. Trypsin once activated is capable of activating further trypsinogen at the same site, but trypsin, unlike enterokinase, can further cleave the trypsin amino acid chain. It does so between amino acids 131 and 132 creating two chains or further between 176 and 177 creating 3 chains. Fragments in these two and three chain trypsins are usually bound together by disulphide bridges and have full proteolytic activity. Further proteolysis results in inert protein. This happens readily when, in solution, the pH exceeds 4.5 and more readily still in the absence of calcium. Without enterokinase trypsinogen activation is very slow. Once enterokinase has started activation going, then the concentration of trypsin present becomes enough to sustain it. Persons with deficiencies in enterokinase production suffer obvious impairment of protein digestion.

Trypsin hydrolyses only basic bonds on the N side of lysine or arginine. Trypsinogens are about 19% of the protein in pancreatic juice. Now that there are active trypsins in the intestine these proceed to activate the other proteases which include *chymotrypsinogens.* There are two of these in man. This activation again is limited proteolysis. Trypsin breaks the bond between Arg 15 and Ile 16. The molecules then rearrange themselves by bending and become proteolytically active. Chymotrypsins also act autocatalytically on themselves to produce a number of still active fragments which are not of much interest to physiologists. Chymotrypsins are not as specific as trypsins. They prefer the bond on the N side of aromatic amino acid residues but also of leucine, methionine and asparagine.

Carboxypeptidases are the only *exopeptidases* in pancreatic juice, that is, they act at the ends of peptide chains. The carboxypeptidases act at the carbonyl end and the aminopeptidases, absent from pancreatic juice, hydrolyse peptides sequentially from the N terminal end.

Inactive zymogens of carboxypeptidases (there are two of these, procarboxypeptidases A and B) exist as three subunits joined together. The activation of procarboxypeptidase is poorly understood. It is effected by trypsin, but in vitro this is very slow, in vivo it is not. The inactive fragments removed on activation are very large, about 1/3 the weight of the whole zymogen. This enzyme is curious in that it contains one zinc ion per molecule. The zinc is essential for full enzymatic activity.

Carboxypeptidase B differs from A in that it has no affinity for basic residues in the C terminal position but carboxypeptidase A is not at all fussy.

Pancreatic juice contains two proelastases. Once again trypsin is the activator by limited proteolysis of an 11 amino acid peptide. *Elastase* is so named for its ability to hydrolyse elastin, the fibrous protein of elastic tissue, but it also hydrolyses bonds susceptible to both trypsin and chymotrypsin.

Nucleases are all phosphodiesterases, that is they hydrolyse the bonds uniting the mononucleotides. Ribonucleases have an affinity for adenosine nucleotides in position 3. This enzyme has been extensively studied, because it is stable and has a low molecular weight (16,000). In man, this enzyme is represented by less than 0.1% of the juice protein. Deoxyribonuclease if present in human pancreatic juice is insignificant. Lipase is dealt with in chapter 10. It, like amylase, is stored in and secreted from the pancreas in active form.

Phospholipases are the only nonproteases which undergo activation in the gut. Trypsin hydrolyses a 7-amino acid peptide from the N-terminal end and thus confers activity. Phospholipases hydrolyse the middle chain from 3-chain phospholipids forming lysophospholipids (see chapter 10).

The importance of trypsin in bringing about activation of proteases and phospholipases will be evident now and so will be the digestive impairment consequential to enterokinase failure. The converse, activation of trypsinogen within the pancreas, is catastrophic in that the gland digests itself and its active secretion leaks into the abdominal cavity hydrolysing the constituents of almost every tissue with which it comes in contact. This is particularly evident in the case of fat which manifests the action of lipase as white deposits of calcium soap. It is not known how or why trypsinogen is occasionally activated within the gland except that rupture of the ducts or cellular damage releases tissue activators.

The injection of a variety of substances into the ducts under pressure has proved to be a fairly reliable method of producing experimental acute pancreatitis. Bile is the favored agent. For years it was a surgical dogma that bile salts activate trypsinogen, but this as we have seen above is not the case. The idea was that in man, because of the common bile-pancreatic channel in the ampulla, regurgitation of bile into the pancreas was possible and could lead quickly to acute pancreatitis. Bile does regurgitate quite often without consequence. Acute pancreatitis has been seen in people without the usual common channel and finally experiments in which all the bile has been directed through the pancreas for months have not resulted in pancreatitis in dogs or goats. Bile does seem to reduce the rate at which proleolytic activity declines in standing pancreatic juice, but it does not activate trypsinogen; only enzymes can do this.

Should trypsinogen become activated in pancreatic juice there is a defense mechanism in the shape of trypsin inhibitors. These immediately block the enzymic activity of trypsin both as an enzyme activator and as a digestive enzyme. The concentration of this material in the juice varies directly with that of the enzyme protein. Trypsin inhibitors are small polypeptides (6,000), stable and specific for trypsin with which they combine 1:1. They are present in large amounts in man (about 0.5% of the total protein). Pancreatic trypsin inhibitor does not cause permanent inhibition. It combines with trypsin at the activation site, but free trypsin can disrupt this again; therefore, in the gut the complex breaks up and in the pancreas, if the activation of trypsinogens is massive, the inhibitor will be useless.

Circulating Enzymes

Most digestive enzymes can be found in small amounts in blood and, consequently, in urine. It seems as if a minute amount from the secreting glands leaks backwards. Blood enzymes from the digestive tract that have diagnostic importance are amylase, lipase, trypsinogens, chymotrypsinogens and pepsinogen.

Both pancreatic and salivary amylase can be found in plasma. The two isozymes can be column separated: pancreatic amylase disappears following pancreatectomy. The blood concentrations of these enzymes are constant in healthy people, with salivary amylase predominating. Some of the blood amylase (about 24%) appears in the urine but here pancreatic amylase predominates. Amylase clearance seems to vary with the glomerular filtration rate; to eliminate this variation for diagnostic purposes amylase clearance is usually expressed as a fraction of creatinine clearance.

In healthy people, blood lipase, which is all of pancreatic origin, is virtually undetectable. The existence of lipase in urine is debatable.

Trypsinogen and chymotrypsinogen can be detected in plasma by radioimmunoassay. Assay based on enzymic activity is not possible as plasma contains potent trypsin inhibitors.

Serum lipase and serum and urinary amylase are used clinically to establish a diagnosis of pancreatic disease. However, fluctuations are often transitory and the large variety of conditions unconnected with the pancreas which can alter values greatly reduces the diagnostic value of these estimations. Since amylase is the easiest to do, it is the most often used.

Adaptation to Diet

The ratio of enzyme activities to one another in pancreatic juice varies with the diet in both rats and dogs and, it is presumed, in man. High carbohydrate diets fed for weeks elevate the levels of amylase and reduce the proteases and lipase [24]. High protein diets do the converse (fig. 7/12). Carbohydrates raise the proportion of juice amylase whether given as starch or glucose either by mouth or intravenously. The effect of protein is seen clearly with intact protein, but in dogs certain amino acids have the propensity to increase protease activity within a day or two [25]. Lipase activity seems to follow the proteases. There have been claims that high fat diets promote elevated lipase [12], but the evidence is not as substantial as that for

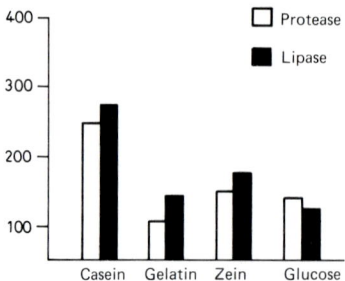

Fig. 7/12. The effect of various dietary proteins on the enzyme content of the rat pancreas. The proteins listed were the only ones fed the animals. The glucose group received 5% casein. Each diet was fed for three weeks. Gelatin and Zein are both deficient in essential amino acids. Protease units are mgm of tryosine liberated from hemoglobin substrate per 10 min. Lipase units are ml of N/20 NaOH required to titrate the fatty acid liberated from olive oil emulsion in 10 min. From *Magee and Anderson* [25].

amylase and protease activity. There seems to be a connection between blood glucose and amylase secretion which has gained credibility by the finding that in experimental rats diabetes (alloxan) lowers the ratio of juice amylase to chymotrypsin and insulin injection does the converse, indeed insulin accelerates the rate at which amylase is synthesized in this species. The mechanisms by which this is brought about are obscure, individual duodenal stimulatory hormones for individual enzymes have been claimed, e.g. chymedin for chymotrypsin but CCK action plus overriding regulation by the pancreatic islets (Langerhans) seems to be a sufficient explanation at the moment. GIP mentioned earlier, which is released by carbohydrates from the gut mucosa and is a potent insulin releaser, is a logical hormonal candidate for the leading role in amylase adaption to high carbohydrate diets. A connection between the absorption of glucose and the secretion of insulin is an obvious expectation. It has been noted above that there is such a connection through GIP. It has been known for years that oral glucose was much more effective as a releaser of insulin from the pancreatic islets than was the same amount given intravenously (fig. 7/13) [36]. The gastrointestinal hormones known at that time all had some ability to release insulin, but carbohydrates or sugars are not potent releasers of these; acid, fat and protein digests which are good gastrin, CCK and secretin releases are not spectacular insulin releasers. Radioimmunoassay of GIP soon made it evident that its plasma concentrations were augmented by carbohydrate and glucose. It does indeed cause insulin release. This hormone, which was once

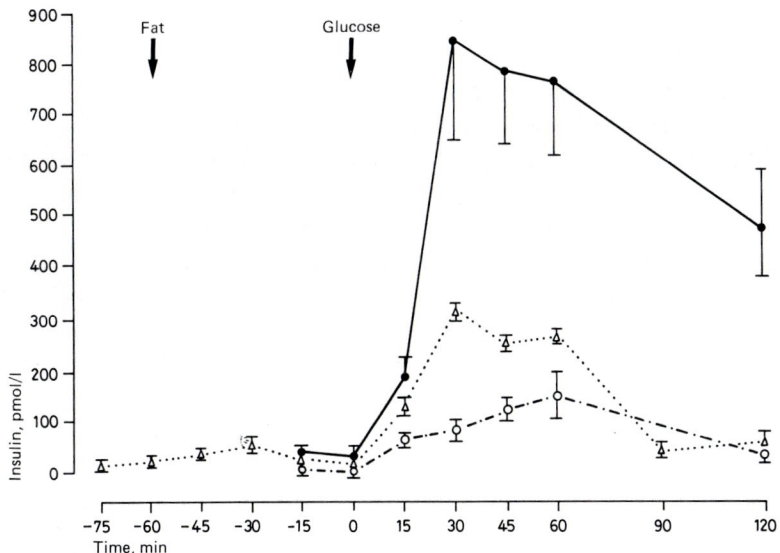

Fig. 7/13. Blood levels of insulin following oral fat and oral glucose fed at the arrow. In the oral fat experiment the glucose was given intravenously at this point. △ = Oral fat followed by intravenous glucose; ● = oral glucose; ○ = intravenous glucose. From *Sarson and Bloom* [37].

suggested as the mediator of gastric secretory inhibition from the duodenum, hence its name, is now held to be the mediator of the entero-insulotropic effect rather than the enterogastric and the name glucose-dependent insulin-releasing peptide has been suggested for it.

Such a mechanism would clearly be a poor arrangement if it occurred in people who were hypoglycemic when they ate carbohydrate, but fortunately GIP can only act on the pancreas when there is hyperglycemia suggesting either that insulin itself exerts a negative feedback on the GIP cells or that GIP sensitizes the β-cells to the well-known stimulatory action of high plasma glucose.

The Pancreas in Protein Deficiency Disease

The turnover of protein in the pancreas is high, about 50 g/p day in man. It is hardly a surprise that pancreatic function is seriously affected in protein deficiency disease. This is a common disorder in many parts of the world in which food is inadequate and first class protein and fat expensive [6]. Caloric intake may be adequate, but protein grossly inadequate. Atrophy of the pancreas particularly of the almost agranular acinar cells, fatty infiltration of the liver and in advanced cases cirrhosis are seen. The exocrine function of the pancreas fails early in the disease and the gland may ultimately become fibrotic [1]. Before this irreversible stage is reached addition of protein to the diet produces a prompt return to normal exocrine secretion. In studies conducted in post war Budapest it was evident that protein by mouth was much more effective treatment than intravenous protein [44]. This indicates that this disease is not solely a deficiency in the amino acid building blocks for pancreatic enzymes (these can be built at the expense of body protein) but also a deficiency in the duodenal release of CCK which requires protein and fat. This intriguing and important deduction seems not to have been followed even by these who made the original observation.

Plasma proteins are uniformly low in this disorder, but following treatment improved pancreatic enzyme secretion precedes improvement in plasma protein levels.

The consequences of impairment of the exocrine function of the pancreas are serious and obvious: rapid weight loss and undigested fat, carbohydrate and protein in the feces. Excess fat in the feces, *steatorrhea,* is the symptom that causes the most complaints. The stools are loose, bulky and foul smelling. A further discussion of steatorrhaea is found in chapter 11. A universal consequence of impaired pancreatic secretion is gastric hypersecretion. The explanation usually advanced for this is failure of the duodenal secretin and CCK feedback mechanisms which are activated by pancreatic digests of fats and proteins. This is not altogether satisfactory since the low duodenal pH itself ought to be sufficient. The absence of the trypsin activation peptide, claimed by some to depress gastric secretion, may be another explanation.

In experimental animals the secretory capacity of the pancreas must be reduced by 90% to produce evident maldigestion of food, but even after complete ligation of the pancreatic duct, dogs still absorb 50% of their protein and 75% of their fat. Similar figures are reported for man. Vitamin B_{12} is often poorly absorbed in pancreatic deficiency because pancreatic proteases split the B_{12} R factor complex, but inorganic iron is better absorbed than usual as in the resulting acid duodenal pH it is less ionized.

7 The Pancreas

A long-term consequence of pancreatectomy or ligation of the pancreatic duct is fatty infiltration of the liver. This occurs even in animals adequately maintained with insulin. The possible cause for this was hotly debated for years [2], but is now in oblivion unresolved. The rival ideas were that deficient protein digestion brought about a deficiency of lipotropic factors, notably, methionine. The supporters of this idea held that inclusion of trypsin in the diet could prevent the condition. The other faction held that raw pancreas by mouth cured the condition, not because of its content of proteolytic enzymes, but because it contained a specific factor christened *lipocaic*. This factor has not been purified or isolated and there the controversy rests.

Digestion and Absorption in the Infant

No one needs to be told that the infant animal even when fed entirely on its mother's milk produces feces. It will be pointed out later that in the adult, feces are composed of indigestible and unabsorbable food residues and the bodies of dead bacteria plus a little water. The bowel of the human infant is sterile for the first four weeks after birth and milk, in the adult, leaves no indigestible residue. It is clear, therefore, that the infant intestine must waste a good deal of its dietary intake [22].

The normal infant has a fatty stool to the extent that would be called steatorrhea in the adult. The artificially fed infant loses up to 30% of its ingested fat in the feces. This is because the infant pancreas does not produce much lipase, indeed lipase concentrations in the duodenum fall following meals rather than rising, as happens in the adult. This simply means that the small pancreatic lipase secretion is overwhelmed by the increased volume of duodenal contents. By 2 months' postnatum, pancreatic lipase has doubled. Another important factor in reducing fat digestion in the infant intestine is the paucity of bile salt. The infant bile acid pool/kg is less than half that of the adult and bile acid concentration in the intestine, like the lipase, falls following meals.

In the human infant the fat from maternal milk is much better absorbed than that from other sources because of the presence within it of lipoprotein lipase and of a bile acid-stimulated lipase, both of mammary origin. In human milk, in addition, the triglycerides usually have the long chain, poorly soluble fatty acid (palmitic as a rule) in position two. This is the lipase-resistant position. The result is that the poorly soluble acids remain as monoglycerides, which are bipolar and therefore much more soluble in water.

In contrast to lipase, protease secretion by the newly born infant is virtually at adult levels. The normal infant digests and absorbs protein almost as efficiently as the adult. The complete spectrum of mucosal (brush border) peptidase activities is present at birth. Close to 40% of the infant's dietary protein is utilized for growth.

The infant pancreas produces no α-amylase. This appears slowly after weaning. This is no great drawback to the suckling infant since milk contains neither starch nor glycogen. The normal intestinal mucosa, however, contains all of the disaccharidases necessary to digest the sugars of milk and also those necessary to digest sugars such as sucrose and maltose which may be in artificial foods. At weaning all pancreatic enzyme production starts to increase, especially the amylase. This is a good example of dietary adaptation, referred to earlier. Absorption of sugars by the infant is therefore good.

References

1 Barbezat, G.; Hansen, J.: The exocrine pancreas and protein caloric malnutrition. Pediatrics, Springfield *42:* 77–92 (1968).
2 Bennett, L.: What causes fatty liver after pancreatectomy. An unresolved and forgotten controversy. Perspect. Biol. Med. *26:* 595–613 (1983).
3 Brockman, D.: Anastomosing tubular arrangement of dog exocrine pancreas. Cell Tissue Res. *189:* 497–500 (1978).
4 Bro-Rassmussen, F.; Killmann, S.; Thaysen, G.: The composition of pancreatic juice as compared to sweat, parotid saliva and tears. Acta physiol. scand. *37:* 97–113 (1956).
5 Churg, A.; Richter, W.: Histochemical distribution of carbonic anhydrase after ligation of the pancreatic duct. Am. J. Path. *68:* 23–30 (1972).
6 Davis, J.: Essential pathology of Kwashiorkor. Lancet *i:* 317–320 (1948).
7 Desnuelle, P.; Figarella, C.: Biochemistry; in Howat, Sarles, The exocrine pancreas, (Saunders, Philadelphia 1979).
8 Desnuelle, P.; Reboud, J.; Ben Abdeljil, A.: The exocrine pancreas: normal and abnormal functions; in Renk, A.; Cameron, Ciba Foundation Symp., London 1962).
9 Doyle, C.; Jamieson, J.: The development of secretogogue response in rat pancreatic acinar cells. Devel. Biol. *65:* 11–27 (1978).
10 Duthie, H.; Wormsley, K.: Scientific basis of gastroenterology (Churchill-Livingstone, Edinburgh 1979).
11 Ekholm, R.; Zelander, T.; Edlund, Y.: The ultrastructural organization of the rat exocrine pancreas. II. Centroacinar cells. J. Ultrastruct. Res. *7:* 73–83 (1962).

12 Gidez, L.: Effect of dietary fat on the pancreatic lipase level in the rat. J. Lipid Res. *14:* 169–177 (1973).
13 Gorelick, F.; Jamieson, J.: Structure-function relationships of the pancreas; in Johnson, Physiology of the gastrointestinal tract; pp. 773–794 (Raven Press, New York 1981).
14 Hayama, T.; Magee, D.; White, T.: Influence of autonomic nerves on the daily secretion of pancreatic juice in dogs. Ann. Surg. *158:* 290–294 (1963).
15 Hong, S.; Nakamura, M.; Magee, D.: The relationship between duodenal pH and pancreatic secretion in dogs and pigs. Ann. Surg. *166:* 778–782 (1967).
16 Hong, S.; Magee, D.: Pharmacological studies on the regulation of pancreatic secretion in pigs. Ann. Surg. *172:* 41 (1970).
17 Iwatsuki, N.; Petersen, O.: Electrical coupling and uncoupling of endocrine acinar cells. J. Cell Biol. *79:* 533–545 (1978).
18 Junqueirea, L.; Carneiro, J.; Contopoulos, A.: Basic histology (Lange Medical Publications, Los Altos 1975).
19 Kallman, F.; Grobstein, C.: Fine structure of differentiating mouse pancreatic exocrine cells in trans filter culture. J. Cell Biol. *20:* 399–413 (1964).
20 Katsuyama, T.; Spicer, S.: The surface characteristics of the plasma membrane of the exocrine pancreas. Am. J. Anat. *148:* 535–554 (1977).
21 Langerhans, P.: Beiträge zur Mikroskopischen Anatomie der Bauchspeicheldrüse; in Aug. Diss., Berlin 1869. Cited in Opie, E.: Diseases of the pancreas, p. 67 (Lippincott, 1903 Philadelphia).
22 Lebenthal, E.; Lev, R.; Lee, P.: Perinatal development of the exocrine pancreas; in Lebenthal, A textbook of gastroenterology and nutrition in infancy; No.1 (Raven Press, New York 1981).
23 Magee, D.: A consideration of the cholinergic nerve supply to the pancreas. Mount Sinai J. Med. *49:* 2 (1982).
24 Magee, D.: Secretions of the digestive tract; in Ruch, Patton, Physiology and biophysics; chap. 49 (Saunders, Philadelphia 1970).
25 Magee, D.; Anderson, E.: Changes in pancreatic enzymes brought about by alterations in the dietary protein. Am. J. Physiol. *181:* 79–82 (1955).
26 Magee, D.; Hong, S.: Daily output of pancreatic juice and some dietary factors which modify it. Am. J. Physiol. *197:* 27–30 (1959).
27 Magee, D.; White, T.: Influence of vagal stimulation on secretion of pancreatic juice in pigs. Ann. Surg. *161:* 605–607 (1965).
28 Martin, B.; Philpott, C.: The biochemical nature of the cell periphery of the salt gland secretory cells of fresh and salt water adapted mallard ducks, for 1 hour. Cell Tissue Res. *150:* 193–211 (1974).
29 Meyer, I.; Way, L.; Grossman, M.: Pancreatic response to acidification of various lengths of proximal intestine in the dog. Am. J. Physiol. *219:* 971–977 (1970).
30 Meyer, J.: Control of pancreatic exocrine secretion; in Johnson, Physiology of the gastrointestinal tract (Raven Press, New York 1981).
31 Nakajima, S.; Magee, D.: The influence of duodenal acidification on acid and pepsin secretion of the stomach. Am. J. Physiol. *218:* 545–549 (1970).
32 Nakajima, S.; Nakamura, M.; Magee, D.: Effect of secretin on gastric acid and pepsin secretion. Am. J. Physiol. *216:* 87–91 (1969).
33 Noda, A.; Magee, D.; Sarles, H.: Role of gastric secretion in post-diverted pancreatic hypersecretion in conscious rats. J. Physiol. *326:* 453–459 (1982).

34 Odori, Y.; Magee, D.: Cholecystokinin-pancreozymin as a physiological mediator of gastric acid inhibition. Pflügers Arch. *318:* 287–293 (1970).
35 Palade, G.: In Hayashi, Subcellular particles (The Ronald Press, New York 1959).
36 Polak, J.; Bloom, S.: The hormones of the gastrointestinal tract; in Duthie, Wormsley, Scientific basis of gastroenterology (Churchill-Livingstone, Edinburgh 1979).
37 Sarson, D.; Bloom, G.: GIP and the entero-insular axis, chap. 38; in Bloom, Polak, Gut hormones; 2nd ed. (Churchill-Livingstone, Edinburgh 1980).
38 Schulz, I.; Yamagata, A.; Weske, M.: Micropuncture studies on the pancreas of the rabbit. Pflügers Arch. *308:* 277–290 (1969).
39 Singer, M.; Solomon, T.; Grossman, M.: Effect of atropine on secretion from intact and transplanted pancreas in dog. Am. J. Physiol. *238:* G18–27 (1980).
40 Singer, M.; Solomon, T.; Wood, J.; Grossman, M.: Latency of pancreatic enzyme response to intraduodenal stimulants. Am. J. Physiol. *238:* G23–31 (1980).
41 Slayback, J.; Swena, E.; Thomas, J.; Smith, L.: The pancreatic secretory response to topical anesthetic block of the small bowel. Surgery *61:* 591–596 (1967).
42 Sullivan, J.; Burch, R.; Magee, D.: Enzymatic activity and divalent cation content of pancreatic juice. Am. J. Physiol. *226:* 1420 (1976).
43 Thomas, E.: The external secretion of the pancreas (Thomas, Springfield 1950).
44 Veghelyi, D.: Pancreatic function in nutritional edema. Lancet *i:* 497 (1948).
45 Wormsley, K.: Pancreatic secretion; physiological control; in Duthie, Wormsley, Scientific basis of gastroenterology (Churchill-Livingstone, Edinburgh 1981).
46 Yamazaki, Y.: Inhibitory mechanism of oxethazaine on release of endogenous secretin and pancreatic response in dogs. Dig. Dis. Sci. *27:* 821–830 (1982).

8 Intestinal Motility

Our main concern in this subject is the movement of the fluid chyme within the intestine and how this contributes to digestion and absorption. As is so often the case, we know little about this, our primary concern, but a good deal about more peripheral things which we are sure contribute and alter fluid movement. In the intact man a barium meal can be traced radiographically. Within from 80 to 190 min of leaving the stomach barium reaches the terminal ileum and within 4 h, all of an average sized meal will be there. The transit of the entire intestine, from mouth to anus, has been estimated as 31–118 h (mean 60.3 h) with markers. Therefore, most of the delay between mouth and anus occurs in the colon and cecum. Anything which increases the activity of the intestinal muscle (such as distension or mucosal irritation) will shorten the transit time. If, for example, lactose is not hydrolysed it will not be adsorbed and the distension caused by the resulting osmotically retained water will cause increased motility which may well result in diarrhea. This is the mechanism by which all the saline or osmotic laxatives act. Castor oil can also produce diarrhea, but it does it by irritating the mucosa, which increases both secretion and motility; so that a larger than normal volume of fluid is hurried down into the colon, which reacts in its turn to these same stimuli.

From simple observation it is known that intestinal contents move downwards, but how they are propelled and mixed on the way down is surmise.

Intestinal Motility Patterns in Intact Animals

Until about 50 years ago physiologists without electronic recording equipment were dependent largely on observation and as a result they usually observed well. *Alvarez* [1, 2] was such an observer. He has classified intestinal movements as follows: rhythmic segmentation, pendular movement, the peristaltic rush, and reversed waves. The commonest types of

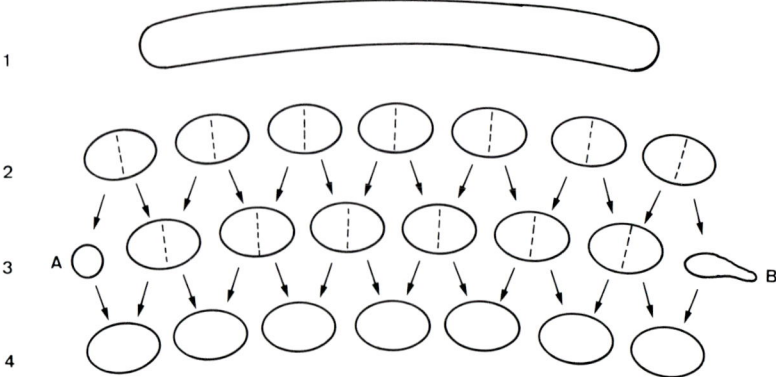

Fig. 8/1. Diagram of rhythmic segmentation as it effects a barium or food bolus. Lines 1, 2, 3 and 4 indicate the sequence. The arrows show the rearrangement of contracted muscle at the succeeding stages. The direction of propulsion was from left to right (A→B). Reproduced, with permission, from *Cannon* [6].

contraction are known as *segmentation*. These are simultaneous contraction involving a centimeter or two of gut, spaced at 4 or 5 cm from each other, over 15–20 cm of small bowel. Contractions pass off after 3–4 s to be replaced by simultaneous contractions in the previously quiescent segments (fig. 8/1). The maximal frequency at which these can occur is the frequency of the BER (see chapter 4). These are often known as *standing contractions* as they move neither cranially or caudally.

Alvarez has taken X-ray motion pictures of segmentation, with barium sulphate as the contrast. On lining these up, he found that rhythmic segmentation contractions do not recur at precisely the same spot but, on the contrary, move slowly down the intestine.

Pendular movement, according to *Alvarez,* is often seen in the exposed intestines of anesthetized rabbits, but is also seen in X-ray examination of conscious man as well. A larger segment (2–10 cm) contracts than in segmentation movement. 'The gut seems to be pulled over its contents first in one direction and then in the other.' The description of this is not clear. In some instances it sounds like swaying movement and in others as movements up and down the long axis. Most modern observers believe it to be a form of rhythmic segmentation.

Alvarez was greatly taken with the *peristaltic rushes.* He produced them by introducing fluids into the stomach or duodenum in a fasting animal. In

8 Intestinal Motility

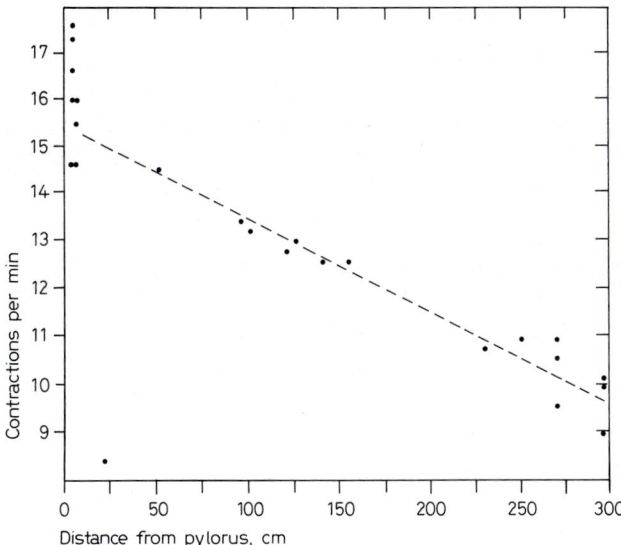

Fig. 8/2. Rate of rhythmic contractions in the rabbit small intestine as a function of distance from the pylorus. Reproduced, with permission, from *Alvarez* [2].

man he sometimes saw them following food. The latter may be what we call the gastrocolic reflex today. Many of his rushes may have been examples of periodic fasting motility, discussed below, or overdistension.

Periodic Activity

What has already been said of the BER applies to the small bowel, with the difference that the intestinal BER declines in frequency caudally. Throughout the duodenum the frequency in man is about 12/min and in the dog 18. This declines to about 8 in the terminal ileum in man and to about 10 in the dog (fig. 8/2). When the gut is empty, the intestine is quiescent, except that every 110 min or so rhythmic activity increases to a maximum just as was seen in the stomach. This powerful activity, in which every slow wave fires, in the succeeding 110 min passes all the way to the ileocecal valve, by which time a new wave will have started in the duodenum. This is not a travelling peristaltic wave such as was seen in the esophagus, but is a group of segmentation contractions which involves successive areas of the gut and thus moves downward. This interdigestive activity has been divided into four phases which seem rather artifical. An adequate description is simply quiescence followed by increasing activity, both frequency and amplitude, to a peak (maximum) and then a decline again to quiescence. The whole active phase, although variable, lasts about 30 min in man and dog and the peak, 10 min. When a meal is eaten and food enters the intestine, the *feeding pattern,* which is a continuous but slower and much less powerful activity without the quiescent periods, supervenes. This happens when food or fluids are placed in the intestine, but it occurs also when food is confined to the stomach. This periodic activity was first noted at the turn of the century, but was quickly forgotten. Failure to recognize its existence and its uniqueness in form, timing and genesis, undoubtedly has caused confusion. It seems to be the most likely explanation for the peristaltic rush, which also is unique, and resembles the fasting spontaneous motility closely.

The genesis of the spontaneous motility is unique in that it is dependent on extraluminal factors, nerves (intrinsic and/or extrinsic), and hormones. Since postcibal motility is intraluminal in purpose and origin, it is scarcely surprising that the forms are very different.

Propagated Peristaltic Waves

The role of propagated peristaltic waves is uncertain. *Baylis and Starling* [3] found that distension produced by a balloon first caused contraction of the longitudinal muscle followed by relaxation and then contraction of the circular muscle. Immediately below the balloon, both layers of muscle were relaxed. This *peristaltic reflex,* as it is called, will drive a balloon downwards, much as happens in the esophagus. This is the physiological response to a distending foreign body, but may not be, in its entirety at least, the response to the normal fluid content of the intestine. Peristaltic waves when they occur in vivo, which seems to be rarely, move only 4–5 cm caudally. It is not known what decides that some waves should be peristaltic and others not. It is evident, however, that the small bowel exhibits caudal directional polarity. That this is intrinsic to the gut itself can be dramatically demonstrated if a length of small bowel is replaced upside down following excision. The polarity will not reverse to fit the new situation. Partial or complete obstruction (if the reversed segment is more than 25–30 cm long) will result. A much shorter segment would produce obstruction were the predominant intestinal movements aborally directed ring contractions, but if the segment is long enough even rhyhtmic segmentation (standing) contractions which do not completely occlude the lumen will cause obstruction.

Intestinal contractions displace the fluid contents in both directions, but they also hinder displacement; therefore, if contraction is more rapid orally than aborally [2], resistance to flow in the aboral direction will be the lesser. If one irrigates the lumen of the intestine, flow is much more rapid through the quiescent gut, which simply behaves as a pipe, than through the active intestine. Flow is slowest [11] or absent through active gut which is perfused in the wrong direction. In dogs a partial obstruction will result if a piece of terminal ileum is translocated to the duodenum with the correct orientation. In the course of time, the muscle and villi will hypertrophy, but the intrinsic rhythm will always remain too slow for the great volumes of fluid handled by the duodenum.

Reverse Peristalsis

Despite this obvious intrinsic aboral directional polarity it is evident clinically that intestinal contents do regurgitate back into the stomach and, during vomiting, into the mouth. A barium enema in a fasting patient will often ascend high up into the small bowel and in cases of prolonged colonic obstruction fecal vomiting is sometimes seen. Early workers, using both balloons and radiological methods have claimed to have detected reversed peristalsis in both the stomach and small intestine. Recent studies of the electrical activity of the small intestine in cats following administration of centrally acting emetics have shown a marked decrease in the amplitudes of the BER waves and also orally migrating spike activity of the type that indicates active intestinal muscle [22]. Vomiting occurred when the activity reached the duodenum. Several investigators using balloons introduced into Thiry-Vella loops or into small bowel fistulae have noted and measured orally directed peristalses. Some others have failed to find this; however, antiperistalsis seems to have better evidence in favor than against. As might be surmised this central interruption of the normal course of events in the intestine (vomiting) is interrupted by bilateral vagotomy or depression of acetylcholine receptors in the gut (atropine). Neither of these procedures will prevent vomiting since the propulsive force in vomiting is a great increase in intraabdominal pressure caused by contraction of the abdominal muscle and the diaphragm with relaxation of the gastric and esophageal sphincters.

Intestinal Pacemaker

As in the stomach, the built-in rhythmicity seems in most species to be a property of the longitudinal muscle. Unfortunately, the guinea pig has been used almost exclusively to study the electrical phenomena in intestinal muscle and this animal does not show longitudinal muscle slow waves. Slow waves can be recorded from isolated strips of intestine from most other animals [7]. The further from the duodenum, the slower they are. However, if these are recorded before the intestine is cut into segments it will be found that at every point caudal to the first one, the rate in situ is higher than after cutting, although it does decline with distance from the duodenum even after cutting (fig. 8/3). The rate throughout the duodenum is constant, unless it is transected; this will lower the rate distal to the cut. What this means is that there are pacemakers throughout; but the upper dominate the lower. The dominance of one pacemaker in situ seems to extend downwards only for 4–5 cm except in the duodenum. Thus the decline in intrinsic rate is as a series of steps. One pacemaker dominates the entire duodenum, making it by far the broadest of the steps. If the upper ones are cut off the lower one with the highest intrinsic rate will dominate (see fig. 8/3).

Slow waves are myogenic and they spread through the circular muscle electronically, that is without nerves (see chapter 1). Tetrodotoxin, a poison which eliminates innervation, alters neither the rate nor spread of slow waves. The nature of the electrical couplings between cells of the same layer, or across layers which obviously exist, is unknown. Texts dealing with this subjects obscure it for many by stating that the gut behaves like a series of coupled relaxation oscillators, without explaining what a relaxation oscillator is and does. This is simply a fancy word for a pacemaker for which gut physiologists, but not cardiac physiologists, have an affection. In a pacemaker, a change gradually builds up and on reaching a certain point, dissipates. Water dripping from a tap is an example in that the water is flowing constantly, but drops only when enough has accumulated to overcome surface tension. The neon light is an electrical example. When the potential difference (PD) between the two ends of the tube has grown high enough, the gas ionizes and current passes, but in doing so the PD is dissipated and the current flow across the tube ceases until the PD again builds up. The word relaxation is confusing since it has nothing to do with muscle relaxation. These oscillators or pacemakers are said to be coupled, because, as noted above, the upper fast ones dominate and drive the lower ones faster than their intrinsic rate, just as in the heart, except that in the heart the dominance of the SA node is absolute and that of the gut pacemaker is distance limited.

What causes slow waves to reach threshold, fire and set-off a contraction? The stimuli seem to be extrinsic to the muscle and its nerves, but even

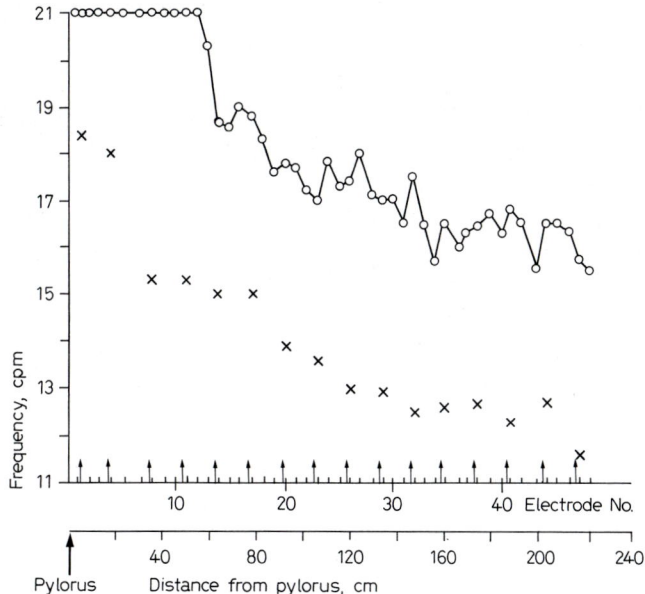

Fig. 8/3. Intrinsic frequencies recorded from 48 electrodes along the length of the small intestine in a reserpinized dog. × indicates frequencies following transections made at the arrows in the same dog. Reproduced, with permission, from *Sarna* et al. [20].

in the empty small intestine the feeding pattern which follows gastric distension makes it evident that the gut is primed for action, the long quiescent periods of the interdigestive period having been replaced by continuous low level mechanical activity. Isolated longitudinal muscle will often fire spontaneously, presumably in response to stretch and other unphysiological factors in its new environment. The physiological stimuli for activity are luminal in origin.

Peristaltic Reflex

The peristaltic reflex described by *Bayliss* and *Starling* [3] has, as noted above, been dismissed as the reaction to an abnormal object, since the normal intestinal content is fluid. Within the last few years, however, it has become possible to impale single cells in both Auerbach's and submucous plexuses of guinea pig small intestine and to determine that in response to transmural stimulation or to distension both inhibitory and excitatory

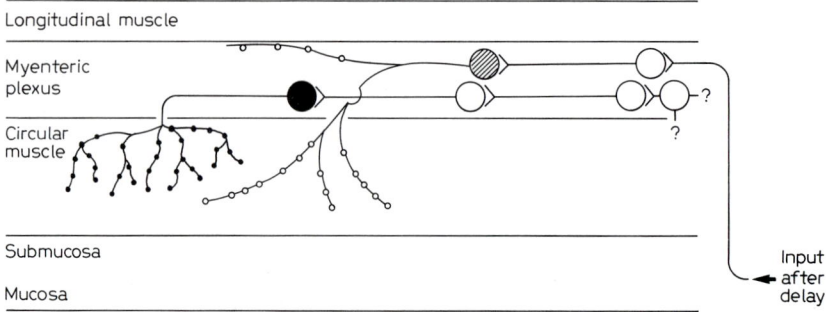

Fig. 8/4. Schematic representation of the two descending nerve pathways. Lower pathway = the descending inhibitory pathway. An AH cell (–?–?) is in some way activated by distension; its efferent process impinges on a chain of cholinergic interneurons (open circles), which in turn activate an inhibitory neuron (large filled circle). Inhibitory fibers, running in an anal direction, terminate in the circular muscle; inhibitory transmitter is assumed to be released from varicostities (small filled circles). The upper pathway represents the descending excitatory pathway. Input to this pathway occurs only after a delay. Both interneurons (large open circles) and excitatory neurons (hatched circles) are activated. Excitatory transmitter (ACh) is then assumed to be released from varicosities (small open circles) within the circular muscle layer at the surface of the longitudinal muscle layer. Reproduced, with permission, from *Hirst* et al. [13].

impulses are indeed generated (fig. 8/4) [13, 14]. The latter induced depolarization of the circular muscle and spiking, which could be abolished by atropine. The former induced hyperpolarization (inhibition) of the circular muscle, which was atropine insensitive. The inhibitory phenomenon was always transient, even if distension was maintained. Excitatory impulses, on the other hand, originated from both cells which rapidly adapted and ceased firing and from others which were capable of continuous discharge.

It is clear that excitatory cholinergic innervation does reach the longitudinal muscle and causes it to contract when the BERs reach their peaks in animals other than guinea pigs, but can excitatory nerve stimulation cause the circular muscle to contract or depolarize independently of the longitudinal muscle? Perhaps it does in guinea pigs, but what is known of the BER in other animals and the absolute dependence of the rate of contraction on it, leads to the conclusion that excitatory nerves must first excite the longitudinal muscle and then excitation spreads to the circular. Inhibition might well be directly to circular muscle. The work seems to confirm the initial stages of the peristaltic reflex of *Baylis and Starling*. Both excitation and inhibition in these experiments were descending so that here also, in the arrangement of the myenteric plexus, we have intrinsic directional polarity.

Luminal distension will not produce descending excitation if the submucosal plexuses are removed or blocked with local anesthetics, but inhibition can still take place. These two must, therefore, be of different origin.

Luminal Distension

The course of events, when the lumen of the guinea pig small bowel is distended, is a contraction at, and 1–2 mm above, the point of distension. Since there are almost no orally directed nerves, this must be myogenic in origin and electrotonically spread. The longitudinal muscle is the next to contract; it is here, in other species, that slow waves have their origin, so this presumably is where spreading depolarization starts. Three to four centimeters below a distending balloon, within 1 s of distension, the muscle cells become hyperpolarized, i.e. inhibited, 2–3 s after this they are depolarized and contracted. The cause of this latency is unknown. The ganglion cells themselves manifest the electrical consequences of stimulation within 0.5 s of distension. This means that a balloon could pass all the way through the intestine following distension, which of course does not happen very often; to explain why not one must speculate. A given contraction wave does not usually move downwards more than 4–6 cm. Perhaps it cannot proceed beyond the BER plateau in which it was initiated or, on arrival at the next slower one, it may find itself out of phase; or since feeding activity seems to occur at random, it may become involved in preexisting activity in the lower segment. Does the above mean that the gut is quiescent when empty, except during periodic fasting activity? Evidently it is not, because the small bowel of the fed animal develops the feeding pattern of activity while the meal is still confined to the stomach. Unfortunately, a gap still remains between results obtained from isolated intestine and observation in intact animals.

The peristaltic reflex is dependent on the intrinsic nerves of the gut. Without them and some stimulus to activate them, the slow waves will not easily reach threshold. The stimulating mediator for this seems to be cholinergic since it is blocked by atropine, but the inhibitory one is something else, since inhibition persists after atropine. ATP has been suggested [5] and many refer to these inhibitory fibers as purinergic, although the evidence for ATP is far from conclusive. It seems hardly credible that the body's major energy source should be diverted to become an inhibitory neurohumor.

Effect of Pharmacologic Agents

Pharmacologically, the relationship between the inhibitory and excitatory innervation of the gut is confusing. The negative results with tetrodotoxin and local anesthetics make it clear that the slow waves are myogenic and that the muscle can contract unaided. Indeed, rhythmic contraction in embryogenic smooth muscle, before it has developed innervation, is well known. Denervated intestinal smooth muscle will contract if stretched. What then is the role of the intrinsic innervation if rhythmicity is intrinsic to the muscle? It seems that the denervated muscle is much less sensitive to stretch, and that, as mentioned already, the peristaltic reflex and propagation, even over the usual short distances, also require intrinsic innervation. But, and this is confusing, tetrodotoxin, which is a nonspecific blocker of nerve terminals, when added to segments of intestine in vitro causes every slow wave to fire so that contraction at the maximal rate occurs [4]. Others, who have measured inhibitory nerve impulses and hyperpolarization in muscle cells in isolated guinea pig gut preparations, state that both tetrodotoxin and local anesthetics do block inhibition. Thus, it appears that luminal distension acts to overcome an intrinsic tonic inhibition. Aganglionic segments of in situ bowel are toxically contracted. This occurs spontaneously in man in a disease called megacolon (Hirschsprung's disease), so called because the colon proximal to the aganglionic segment becomes grealy dilated. The nerves of isolated intestine can be stimulated electrically by passing current across the wall. This causes contraction, but after atropine, inhibition. Morphine and compounds related to it cause spastic contraction of the intestine and are successfully used for the symptomatic treatment of diarrhea; these compounds depress the release of acetycholine.

In striated muscle and in nerves the potential difference across the cell membrane is -90 mV; in smooth muscle it is a little smaller, $60-70$ mV. The spontaneous slow waves have an amplitude of 13 mV. When the muscle contracts the membrane seldom depolarizes to zero, indeed the muscle will contract in response to acetylocholine stimulation even if the transmembrane potential is abolished by suspending the muscle in a high potassium bathing solution [8]. This will not happen if calcium is excluded. Therefore, it appears that in smooth muscle the calcium mobilization stage in the chain of events leading to contraction can be reached directly without the membrane depolarization change involving Na^+, K^+ movement, which is essential in striated muscle and nerve. It is thought that the action potential in smooth muscle is representative of the influx of Ca^{2+}.

8 Intestinal Motility

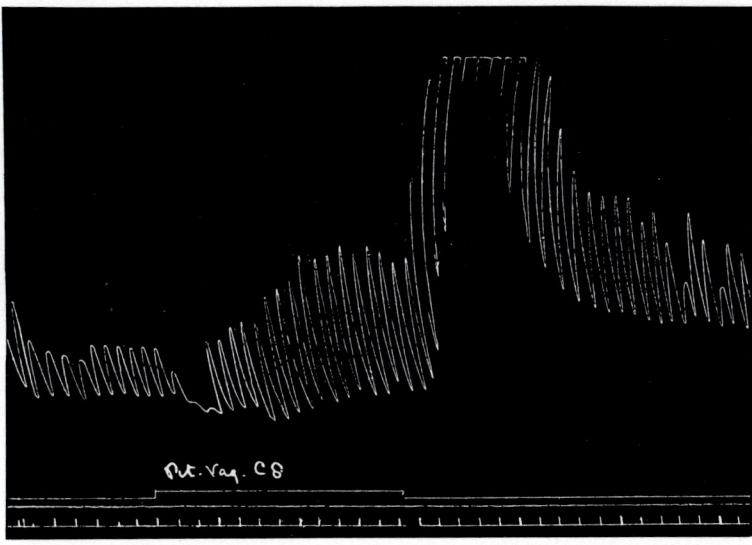

Fig. 8/5. Balloon recording the effect of electrical vagal stimulation, at point indicated, on the small intestinal activity of an anesthetized dog. Reproduced, with permission, from *Bayliss and Starling* [3].

Extrinsic Nerves

Unlike the stomach the small bowel seems to be uninfluenced by bilateral division of the vagus nerves [17]. Stimulation of the peripheral ends of either cut vagus increases intestinal motility. Some have described contraction followed by relaxation (fig. 8/5). We have seen rhythmic segmentation or peristalses which travel 3–4 cm for the duration of a 10 or 20 min continuous stimulation. A single shock produces no activity, continuous stimulations at between 5 and 25/s produces rhythmic activity, often with an initial delay of about 30 s, which continues for a minute or so after the stimulus has ceased. After atropine many have described vagal inhibition. Others, ourselves included, have found atropine to be an ineffective blocker of vagally stimulated activity. This accords well with the clinical experience that anticholinergic agents are of doubtful value in treating painful spastic conditions of the gut. Sympathetic interruption in anesthetised animals, either by nerve section or by blocking agents, generally increases activity and stimulation of sympathetic nerves depresses

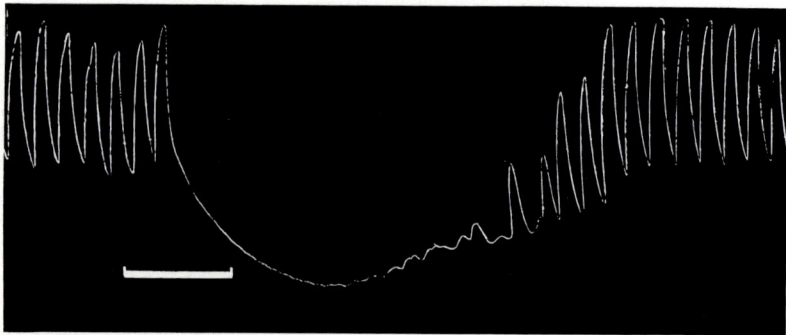

Fig. 8/6. Balloon recording of the effect of bilateral splanchnic excitation on small intestinal activity in an anesthetized dog. Reproduced, with permission, from *Bayliss and Starling* [3].

motility (fig. 8/6). Indeed, sympathetic pathways carry important gut inhibitory reflexes to be mentioned below. Sympathomimetic mediators (norepinephrines) diminish the relase of acetylcholine which follows stimulation of excitatory nerves in vitro. Very few sympathetic nerve terminals reach the muscle cells themselves, but they are abundant in the ganglia of Auerbach's plexus where their action is thought to be presynaptic depression. Stimulation of perivascular sympathetics does depress activity produced either by balloon distension or transmural electrical stimulation of isolated preparations. Sympathetic denervation of the bowel ends the conduction of pain impulses. From the gastro-esophageal junction to the middle sigmoid colon pain impulses are carried exclusively in sympathetic nerves.

The small intestine like the stomach is insensitive to most noxious stimuli. Only distension and spasm clearly produce painful sensations. Chronic inflammation apart from the associated spasm does produce pain in the stomach and may do so in the small intestine. The pain of a duodenal ulcer, for example, is not entirely due to spasm. Intestinal pain may be acute but it always is poorly localized and referred to the mid line, usually to an area of skin having the same segmental innervation as the sympathetic supply to the viscus involved. Pain from the lower ileum, cecum and appendix is referred to the umbilicus. The skin here is supplied by T 10. The sympathetic supply to this segment of gut comes from this thoracic segment. The midline reference of pain is, no doubt, due to the fact that the gut itself is a much folded midline organ.

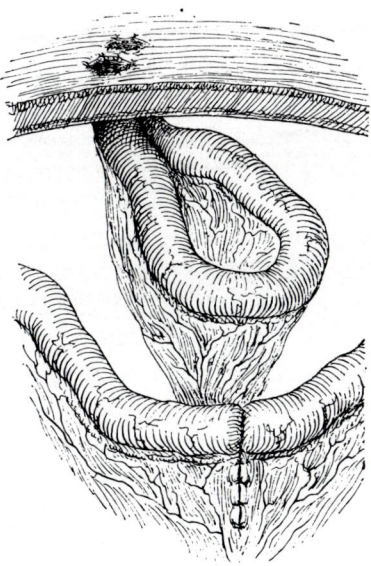

Fig. 8/7. Diagram of a Thiry-Vella loop of small intestine. Note the mesenteric pedicle. [Reproduced, with permission, from Markowitz, J.: Experimental surgery; 3rd ed. (Williams & Wilkins, Baltimore 1954).]

Intestinal Paralysis

Ileus (paralysis of the intestine) is a dangerous condition which sometimes follows surgery. When it occurs intestinal secretion continues, but absorption is greatly depressed. *Bayliss and Starling* [3] noted that rough handling of a loop of intestine almost immediately inhibited spontaneous activity in the remainder of the gut. Others subsequently have noted that unphysiological distension with a balloon will paralyse the intestine, even in a Thiry-Vella loop (fig. 8/7) out of continuity with the bowel. An animal with a balloon-distended Thiry-Vella loop will die from the consequences of paralytic ileus unless the balloon is deflated or the nerves supplying the loop are cut. Peritoneal irritation and, at times, surgical intervention which does not involve the gut at all can result in ileus. Once started this disorder is a positive feedback system in that paralysis causes distension and distension increases the paralysis. Treatment to be effective must reduce the distension and replace the fluid intravenously which, of necessity, has been removed from the small intestine.

The intestinal aspects of this condition are an example of the *intestino-intestinal reflex* which results from sympathetic inhibition of bowel, small and large [25]. It is assumed that the afferents from intestinal stretch receptors also travel in sympathetic nerves. Early workers showed that bilateral splanchnicectomy abolished the reflex. The role of the sympathetics is unquestioned today, but the pathways are debated. The majority claims that the spinal cord is unnecessary and that the reflex can be elicited as long as the celiac and other prevertebral ganglia are intact. This seems to be so in the cat and guinea pig, but not in some other species including man. In those animals in which the reflex can be elicited after the spinal cord connections have been severed, the necessary distension pressure is much higher than before decentralization of the ganglia. The name for this reflex is misleading because it can be elicited from the stomach and colon as well as from the small intestine. Moreover, all of these organs are subject to the ileus which results regardless of the viscus distended. There is, for example, a gastrocolic inhibitory reflex, although the term is usually used only for the excitatory response which is discussed below under colonic motility.

Many have described an increase in intestinal activity following feeding. *Gregory* [11, 12] found that activity increased in Thiry-Vella loops (fig. 8/7) when dogs bearing them were fed. This disappeared with loop denervation. Sham feeding has been claimed to increase activity. In man ileal movements increase following breakfast. It has been stated that the longer the fast, the quieter the intestine. Today these results and conclusions must be reinterpreted in view of the rediscovery of fasting periodic activity [19]. It is remarkable that the studies quoted above, especially those in man, did not disclose periodic activity. Undoubtedly what *Gregory,* and those who did the human experiments, saw was the change over from the fasting to the feeding pattern of activity. This is discussed further below.

Central Effects

Disturbances in emotional states can cause gastrointestinal abnormalities. The diarrhea of fright is well known. *Cannon* [6] observed that worry or fear often caused a decrease in gastric motility. The dyspepsia of anxiety has been the experience of many of us. Despite this obvious involvement of the CNS precise information is lacking, except for the expected results of stimulating the central connections of nuclei of the vagi and the sympathetics. Stimulation of the anterior hypothalamus and the thalamus has produced both increased and decreased gut motility, but only with intact vagi. Stimulation of the posterior hypothalamus gave evidence of general sympathetic discharge involving the heart, circulatory system and the gut. From the medulla, the dorsal motor nucleus of the vagus and the tractus solitarius, vagally dependent excitatory and inhibitory responses have been obtained. Direct stimulation of the spinal cord and of the telencephalon have produced discordant results. Even the olfactory area, which one would suspect would excite the gut, has yielded discordant results.

The most positive results are those which have followed studies of the hypothalamic *thyrotropic releasing hormone* (TRH) [15]. TRH is a tripeptide. When injected intraventricularly in rabbits it produces intestinal hypermotility and diarrhea. This could be abolished only by vagotomy and by transection of the sacral cord, the latter in lieu of section of the pelvic splanchnic nerves. To effect chemical blockade, both atropine and 5-hydroxytryptamine (5-HT) antagonists were needed.

A pharmacological agent with a suggested central effect upon the gut is *metoclopromide* [21]. This increases the amplitude of contractions of the entire gut, except the colon. It also increases the closing pressure of the lower esophageal sphincter. It has a central antidopaminergic effect and its action on the gut is prevented by atropine and vagotomy. This suggests a central action; however, it does act in vitro. At the present time, a more likely explanation for its action is that it increases the peripheral release of acetylocholine. Thus the gut, quiescent because it can produce no acetylcholine, is not activated by metoclopromide. It has little effect on the colon where cholinergic nerve terminals are sparse. This implies that its central antidopaminergic effect is irrelevant to its intestinal effect, but is very relevant to its antemetic action, which is potent and specific.

Regulators of GI Activity

Many hormonal and neurohumoral factors have potential to be physiological regulators of gut activity [24]. Of the hormones released during digestion, gastrin and CCK both increase intestinal motility when given intravenously. Secretin, on the other hand, is reputed to cause inhibition. It certainly does so to the stomach. It is difficult to tell, however, whether these actions and the blood concentrations needed to cause them are physiological or not. Of the other substances extractable from gut mucosa of obscure physiological function so far, substance P, prostaglandins and 5-HT will all contract the isolated intestine. Only 5-HT is of obvious practical importance, since from time to time the argentophile cells which produce it

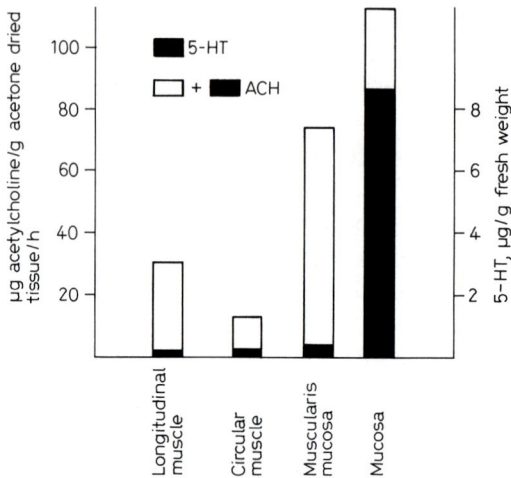

Fig. 8/8. Synthesis of acetylcholine by various layers of dog intestine and distribution of 5-HT. Data taken from *Feldberg and Lin* [10] and from *Feldberg and Toh* [9].

become neoplastic and it enters the portal blood. Its pharmacological effects become obvious if it escapes the usual and normal hepatic degradation. 5-HT has the interesting property of sensitizing receptor endings; for this reason it has been proposed as a sensitizer or facilitator of the peristaltic reflex.

Acetylcholine is mentioned often above. It causes the BERs to reach threshold and the muscle to contract. The rate of its manufacture in the gut, not surprisingly, is directly related to the motility gradient (fig. 8/8).

Norepinephrine is released from sympathetic terminals in the neighborhood of the neurons of the gut plexuses. It hyperpolarizes these and thus prevents release of acetylcholine from the postsynaptic terminals. A direct effect on the muscle also is claimed. Hyperpolarization of the membrane following both norepinephrine and splanchnic stimulation has been described. These inhibitory muscle receptors are β-receptors. If these are blocked norepinephrine can cause contraction by acting on the present but much scarcer α-receptors. Inhibition is the predominant effect of adrenergic nerve stimulation or sympathetic neurohumours.

5-HT increases gut motility. The mucosal enterochromaffin cells of the gut contain 5-HT and moreover, the endogenous production of 5-HT by the small intestine like acetylcholine exactly parallels the motility gradient (see fig. 8/8) [9, 10]. There are 5-HT containing neurones in the enteric plexuses.

The enzymes for making and activating it are there, as are those for inactivating it once it has done its job. Stimulation of activity in intestine and stomach, in vitro, is accompanied by a release of 5-HT. TRH releases 5-HT from the gut in intact rabbits [15]. Chronic inhibition of 5-HT formation in animals causes gastrointestinal distension. As mentioned above there are inhibitory influences on the small bowel, notably the descending inhibition which proceeds the contraction of the peristaltic reflex. This is resistant to both adrenergic and cholinergic blockers, but not to nerve blockers like local anesthetics or tetrodotoxin. ATP has been suggested to fill the role of inhibitory mediator [5]. This notion is now in considerable doubt. Because ATP is ubiquitous, it is hard to prove that its release is a specific result of nonadrenergic inhibitor nerve stimulation and, moreover, ATP, although it can relax smooth muscle, has many other actions and chemical modifiers of those actions not shared by nonadrenergic inhibitory nerve stimulation.

Peptidergic Transmitters

Vasomotor Inhibitory Peptide. This peptide has been suggested as an alternative to ATP as an inhibitory neurohormone. It is found in vesicles in nerve terminals in the CNS and throughout the gut, including the enteric plexuses, the vagal endings in the gastric antrum and the lower esophageal sphincter. At the latter site VIP is a strong contender for neurohumoral status. Immunoantagonists to VIP are claimed to prevent physiological relaxation of the lower esophageal sphincter. Recently it has been stated that vagal stimulation, which in the pig increases the volume of pancreatic secretion, is accompanied by an increase in the plasma concentration of VIP [18]. One's confidence in VIP as a discrete neurohormone is somewhat disturbed by its remarkably wide distribution in the mucosa and submucosa of the GI tract.

Somatostatin. This hypothalamic factor inhibits the release of growth hormone. It is a peptide of 14 amino acids which was subsequently found in the myenteric plexuses of the gut and also in the D cells of the pancreas and gut mucosa. This peptide, given intravenously or in vitro, inhibits intestine motility. It depresses both fasting and interdigestive contractions. Since its action is resistant to both anticholinergic and antiadrenergic drugs somatostatin is a candidate for the inhibitory nerve mediator.

Substance P. A peptide of 11 amino acids, substance P was first noted in extracts of small bowel in 1931. It proved to be a vasodepressor. Much later, it was found to be present in the hypothalamus, spinal cord and vagus nerves. It is widely distributed throughout the gut muscle, but most especially in the duodenum. Immunofluorescence examination has revealed that most substance P is in the myenteric plexus in both cell bodies and nerve terminals. It is also in the enterochromaffin cells which contain 5-HT. It induces strong contraction of gut smooth muscle in vivo and in vitro. It also stimulates salivary and lacrimal secretion and lowers blood pressure. The concentrations required for these actions far exceed normal plasma concentrations. If, therefore, it has any physiological action on gut motility it most likely is as a local transmitter.

Neurotensin. A 13-amino acid peptide isolated from the hypothalamus, neurotensin was later extracted from the intestines of cattle, man and rat. In dogs and man intravenous injection depresses gastric and intestinal motility. In rats some have claimed stimulation of duodenal activity. This peptide has therefore alternately been suggested as a noncholinergic stimulant and a nonadrenergic inhibitor.

Encephalins and Endorphins. Endogenous opioids (that is, they act like morphine), encephalins and endorphins were originally isolated from the brain, but later were found to be present in the gut, especially in the nerves of the upper levels of the small intestine, where it is believed they inhibit rhythmic activity in the manner of morphine, i.e. by decreasing the release of acetylcholine. These effects are antagonized by the morphine antagonist naloxone.

Bombesin. This peptide was originally isolated from frog skin and later found in the tissue of the stomach (except the pylorus) and in intestines, again mainly in the small nerve fibres. Its most notable effect is that, on intravenous injection, it brings about release of gastrin, but it also inhibits upper intestinal motility including the periodic interdigestive activity.

Gastric Inhibitory Peptide (GIP). In keeping with its suggested role as an enterogastrone, GIP inhibits gastric and intestinal motility on intravenous administration.

Physiologically, the established hormones gastrin, CCK and secretin exert an indisputable effect on gastrointestinal motility, but since these

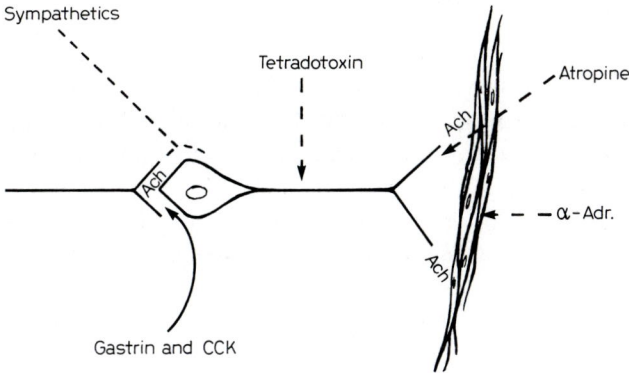

Fig. 8/9. Diagram of the action of CCK and gastrin to stimulate gut motility. Ganglionic blocking agents do not block the action of CCK and gastrin but tetrodotoxin does. Their ganglionic action cannot, therefore, be nicotinic. Modified, with permission, from *Vizi* et al. [23].

hormones are not found in myenteric nerve endings and evidently must act via the blood stream they have been the subject of much less work and speculation than the more uncertain polypeptides and other claimed transmitters.

Gastrin. Gastrin and CCK both increase the activity of the stomach, small intestine and colon in vivo and in vitro. It is argued, vigorously by some, that the increase in the tone of the lower esophageal sphincter is a pharmacological phenomenon not seen with physiological plasma gastrin levels. The most interesting aspect of gastrin action on intestinal muscle is that it seems to be indirect [23]. It is abolished by atropine and by tetrodotoxin, which blocks peripheral nerves, but not by ganglionic blocking agents (fig. 8/9). The evidence that gastrin acts by releasing acetycholine is excellent. It cannot stimulate the intestine in vitro if pharmacological tools which prevent acetylcholine release are administered. Its pharmacological characteristics on smooth muscle resemble those which characterize its action on gastric acid secretion (see above), but curiously orthodox opinion of this action does not countenance the possibility of an indirect cholinergic mechanism for stimulation of acid or pepsin secretion.

Cholecystokinin. A stimulator of small and large bowel activity, and of course the gallbladder, cholecystokinin inhibits gastric activity in vivo in both man and dog. The action on small intestinal smooth muscle is exactly like that of gastrin. This remarkable duality of action remains a puzzle. The action on the stomach in vivo may be an expression of competition with the stimulating action of fasting plasma gastrin. Gastrin and CCK have not been described as competing on any other smooth muscle. In experiments before 1970 the 10% commercial CCK, which contained GIP (a possible explanation for its gastric inhibitory action) was used, but in recent studies, pure CCK octapeptide has been employed and has verified the earlier results without exception.

Secretin. With its close relative glucagon, secretin inhibits gastrointestinal motility, including the lower esophageal sphincter. This has been seen in vivo and in vitro in man and animals against basal, gastrin or cholinergically stimulated activity.

Motilin. Like CCK and gastrin, motilin indubitably increases gastrointestinal motility. It is found throughout the upper small intestine in enterochromaffin cells like those which store 5-hydroxytryptamine. It is not usually considered when one talks of peptidergic transmission because like CCK and gastrin it is not found in peripheral nerves and because a fairly definite function has been ascribed to it. It initiates typical interdigestive motility patterns in the stomach, intestine and gallbladder following intravenous injection and indeed its basal blood levels wax and wane in time with the onset and disappearance of the *interdigestive activity* [16] (see chapter 9). Atropine and ganglionic blockade depress the action of this substance.

Pancreatic Polypeptide. A contaminant of certain insulin preparations, pancreatic polypeptide is found almost exclusively in the pancreas. When given intravenously in inhibits basal, CCK or secretin-stimulated pancreatic secretion. It relaxes the gallbladder but stimulates gastric motility and emptying and speeds intestinal transit. It seems, except for its action on the small intestine, to be a CCK antagonist. This polypeptide is present in blood, where it rises following meals and very obviously following atropine-sensitive vagal stimulation or intravenous cholinomimetics. No role has yet been suggested for it in the physiology of digestion, but its blood levels have become important in the diagnosis of pancreatic disease. Levels are low in the hypofunctional pancreas or in chronic pancreatitis and high in neoplasms and diabetes.

It will clearly require some Herculean intellectual and organizational effort before the disparate work of the polypeptide specialists, the microelectrical people and the pharmacologists can offer a coherent explanation for the physiological activity of the gut so beautifully observed and described by the early observers. Even conceptually, there has been little advance on what the early people did. *Alvarez* [2], for example, described automaticity, polarity, the activity gradient and *Bayliss and Starling* [3] elucidated the peristaltic reflex. The present situation is perhaps the classical one of the wood and the trees. Apparatus is so refined now and we are so good at isolating peptides and little pieces of cells that we have lost sight of the original problem.

References

1 Alvarez, W.: An introduction to gastroenterology, chap. 1–12 (Hoeber, New York 1940).
2 Alvarez, W.: Functional variations in contraction of different parts of the small intestine. Am. J. Physiol. *35:* 177–193 (1914).
3 Bayliss, W.; Starling, E.: Movements and innervation of the small intestine. J. Physiol. *24:* 99–143 (1899).
4 Biber, B.; Fara, J.: Intestinal motility increased by tetrodotoxin, lidocaine and procaine. Experientia *29:* 551–552 (1973).
5 Burnstock, G.: Purinergic nerves. Pharmac. Rev. *24:* 509–581 (1972).
6 Cannon, W.: The movement of the intestine studied by means of the roentgen rays. Am. J. Physiol. *6:* 251–277 (1902).
7 Daniel, E.; Sarna, S.: The generation and conduction of activity in smooth muscle. Annu. Rev. Pharmacol. Toxicol. *18:* 145–166 (1978).
8 Evans, D.; Schild, H.; Thesleff, S.: Effect of drugs on depolarized plain muscle. J. Physiol. *143:* 676–685 (1958).
9 Feldberg, W.; Toh, C.: Distribution of 5-hydroxytryptamine in the wall of the digestive tract. J. Physiol. *119:* 352–362 (1953).
10 Feldberg, W.; Lin, R.: Synthesis of acetylcholine in the wall of the digestive tract. J. Physiol. *11:* 96–118 (1950).
11 Gregory, R.: Some factors influencing the passage of fluid through intestinal loops in the dog. J. Physiol. *111:* 119–137 (1950).
12 Gregory, R.: The nervous pathways of intestinal reflexes associated with nausea and vomiting. J. Physiol. *106:* 95–103 (1947).
13 Hirst, G.; Holman, M.; McKirdy, H.: Two descending nerve pathways activated by distension of guinea pig small intestine. J. Physiol. *244:* 113–127 (1975).
14 Holman, M.: The intrinsic innervation and peristaltic reflex of the small intestine; in Bulbring, Smooth muscle, chap. 14 (University of Texas Press, Austin 1981).
15 Horita, A.: Peptides in the central regulation of the parasympathetic nervous system. Proc. West. Pharmacol. Soc. *25:* 217–221 (1982).

16 Itoh, Z.; Honda, K.; Hewatashi, S; Takeuchi, I; Aizawa, R.; Takayanagi, R.; Couch, E.: Motilin induced mechanical activity in the canine alimentary tract. Scand. J. Gastroent. *111:* suppl., pp. 39, 93–110 (1976).
17 Kewenter, J.: The vagal control of duodenal and ileal motility. Acta physiol. scand. *65:* suppl. 251, pp. 1–68 (1965).
18 Larsson, L.; Fahrenkrug, J.; Holst, J.; Schaffalitzky de Muckadell, O.: Innervation of the pancreas by vasoactive intestinal polypeptide (VIP) immunoreactive nerves. Life Sci. *22:* 773–780 (1978).
19 Magee, D.; Naruse, S.: Neural control of periodic secretion of the pancreas and the stomach in fasting dogs. J. Physiol. *344:* 153–160 (1983).
20 Sarna, S.; Daniel, E.; Kingman, Y.: Simulation of slow wave electrical activity of small intestine. Am. J. Physiol. *221:* 160–173 (1971).
21 Schultze-Delrieu, K.: Metaclopramide. New Engl. J. Med. *305:* 28–32 (1981).
22 Stewart, J.; Weisbrodt, N.; Burks, T.: Intestinal myoelectrical activity after activation of central emetic mechanism. Am. J. Physiol. *23:* E131–137 (1978).
23 Vizi, S.; Bertaccini, G.; Impicciatore, M.; Knoll, J.: Evidence that acetyl choline released by gastrin and related polypeptides contributes to their effect on gastrointestinal motility. Gastroenterology *64:* 268–277 (1973).
24 Walsh, J.: Gastrointestinal hormones and peptides; in Johnson, Physiology of the gastrointestinal tract, chap. 3 (Raven Press, New York 1981).
25 Youmans, W.: Handbook of physiology, sect. C, vol. 4, chap. 28. Innervation of the gastrointestinal tract (American Physiological Society, Washington 1968).

9 Interdigestive Activity

So far we have dealt mainly with the secretory and motor responses of the gut to food and hormonal or pharmacological stimuli. It might be presumed, and indeed it was until 10 years ago, that in the absence of stimuli the gut and its glands are inactive. This idea was current because the work of *Boldyreff* [1] had been completely overlooked and forgotten. It is remarkable that until recently no physiologist had noticed the marked and regular fluctuations in motility and secretion which *Boldyreff* had seen and reported.

Periodic Motility of the Gut

In fasting dogs *Boldyreff* found that at intervals of approximately 100 min the stomach became spontaneously active, the activity rising to a peak and then falling, to start up again 100 min later. This active phase lasted about 30 min and contractions occurred at the maximal gastric rate (3/min). This wave of activity travelled to the duodenum with exactly the same type of waxing and waning, except that the contraction rate in the duodenum was about 18/min (the duodenal maximum). This activity wave travels down the gut to the terminal ileum. The contraction rate falls as it goes, but it is always the characteristic maximum for each segment of intestine it involves. The complete journey from gastro-esophageal junction to ileocecal sphincter takes about 100 min so that by the time one MMC *(migrating motor complex)* is ending its journey a new one is starting in the stomach. This means that the gut can have two complexes occurring simultaneously, but no more, and that during fasting there is always activity somewhere along the intestine.

Periodic Gallbladder Contraction

In phase with this intestinal periodicity is periodic gallbladder contraction. This is one of the earliest events. It occurs 10–20 min before gastric and 30 min before duodenal activity. This activity empties into the duodenum about 50% of the bile which has accumulated in the gallbladder in the previous 100 min. This implies that the sphincter of Oddi relaxes in phase also. It is evident from earlier work that bile secretion waxes and wanes in phase with gastric and gallbladder activity. We have mentioned earlier that logically the gallbladder must empty spontaneously, otherwise a fast of a day or two would produce obstructive jaundice, and indeed in the course of gastric analysis in fasting subjects, bile is often seen in the first sample withdrawn.

Periodic Secretion

The stomach increases the volume of its secretion in phase with increased motility (fig. 9/1). Pepsin is unequivocally increased, but in dogs acid is variable. In man acid seems to be increased at activity peaks [7]. In the dog, if there is a basal acid secretion, it is never during the inter-peak phase [6].

Between peaks, which are in synchrony with those in the duodenum, the secretion from the dog pancreas is virtually zero. At peaks it rises from 0.2 to 1.5 ml/min and its protein output, in the 10-min peak, rises to 55–60% of the maximal stimulated secretory rate.

Pattern Alteration

In most fed animals all phases of this interdigestive spontaneous activity are superseded by continuous feeding patterns of motility and secretion, but with the clear exception of ruminants and continuous eaters. Pigs on intermittent feeding show the usual haphazard feeding pattern after meals, but on ad libitum feeding food consumption does not interrupt the fasting periodicity. The feeding pattern starts at the cephalic phase of feeding and ends when the small bowel is empty. The feeding pattern consists of low level continuous activity throughout the small intestine. The force of one contraction is only about 1/4 that of the interdigestive peaks and the rate of

9   Interdigestive Activity

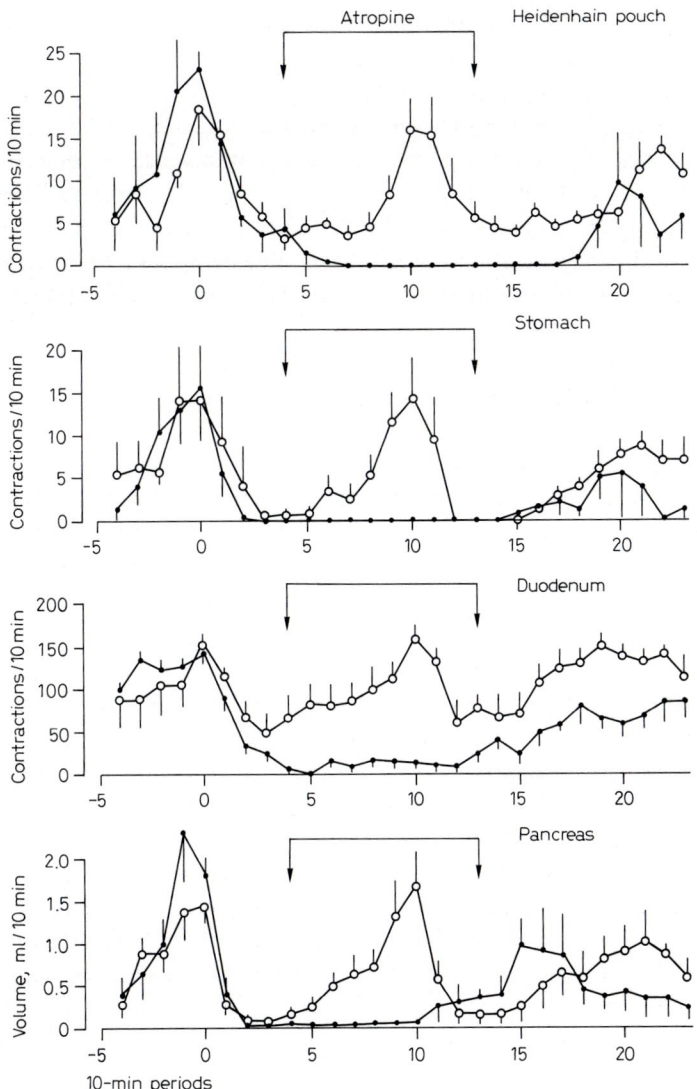

Fig. 9/1. Simultaneous periodic interdigestive contractions in the stomach, duodenum and a Heidenhain (vagally denervated) gastric pouch and pancreatic secretion showing the effect of atropine (closed circles). Open circles = control data. Zero time was taken as the first duodenal peak. Reproduced, with permission, from *Magee and Naruse* [4].

contraction is slower since not every slow wave fires, as they do at the interdigestive peaks. The feeding pattern of gastrointestinal motility, or at least interruption of the interdigestive activity, can be initiated by balloon distension of the stomach, by placing water in the stomach or by intravenous injection of gastrin. Secretin and gastric distension increase the secretion of water and bicarbonate by the pancreas; but during continuous infusion of submaximal doses or gastric distension, periodic 100-min fluctuations in volume and protein secretion are still seen. This means that at peaks the gland is more sensitive to these stimuli and between them less sensitive.

Hypothetical Mechanisms of Interdigestive Activity

The mechanism behind interdigestive activity is problematic and probably mixed. Acute vagal block abolishes gallbladder and decreases gastric contraction. It also depresses periodic duodenal motility and pancreatic secretion. In the duodenum and pancreas vagal block does not reduce the height of the peak, but only its duration. Atropine and ganglionic blocking agents abolish all peaks, both secretory and motor (fig. 9/1). This makes the mechanism seem to be nervous and the vagi apparently important. The vagi cannot, however, be all important since periodic secretory and motor activity is seen in Heidenhain pouches, and vagal block diminishes both. Evidently, intrinsic innervation is important, but hormones cannot be excluded since atropine and ganglionic blocking agents depress the release of gastrin, CCK, and motilin.

Motilin is a prime candidate for the role of *hormonal mediator*, since it does cause caudally migrating motor activity of gut and gallbladder contraction. Its plasma concentrations wax and wane with the periodicity in the upper gut (fig. 9/2). Feeding depresses plasma motilin levels, and not only that, but after feeding, intravenous motilin no longer stimulates motility. Motilin cannot be the whole answer, since by the time spontaneous activity has reached the lower gut, plasma motilin will have passed its peak. It may serve, therefore, only as an initiator of activity.

In the small intestine the *intrinsic plexuses* seem to be all important, because denervated transplanted loops of jejunum retain their periodic activity, which may be out of phase with that of the rest of the intestine, and the waxing of plasma motilin levels. After ganglionic blockade or atropine, motilin will not influence the small intestine, gallbladder, or pancreas at all, but it will still initiate peaks in the stomach, either innervated or vagally

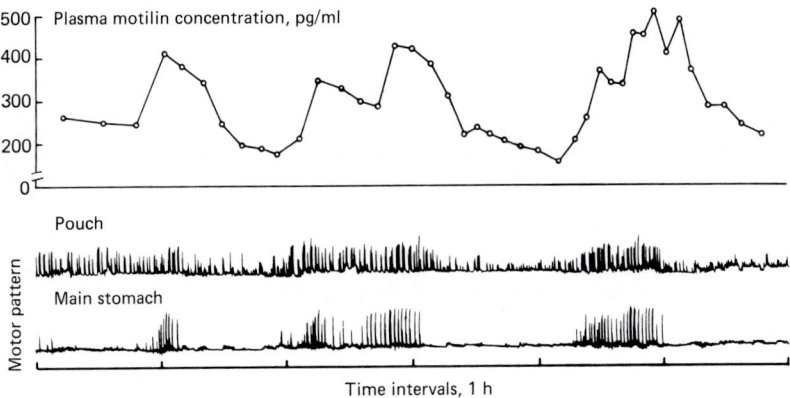

Fig. 9/2. Changes in plasma motilin concentration and contractile activity in Heidenhain pouch and main stomach in a conscious dog. Reproduced, with permission, from *Itoh* et al. [2].

denervated. It thus seems that motilin cannot initiate activity by itself on gallbladder, small intestine and pancreas, but can only function through the extrinsic innervation in the former and the intrinsic innervation in the latter.

There is considerable doubt even in the case of the stomach that motilin is a direct stimulant of the muscle because continuous infusions of motilin do not produce continuous activity; rather, motilin seems to advance the time for the spontaneous peak but subsequently rhythmic peaking activity continues as before. The size or intensity of the peaks is not related to the dose used; it is all or nothing, but the interval between the start of motilin infusion and peak activity varies indirectly with the dose employed.

It seems to us that the earliest sign of interdigestive activity, gallbladder contraction which precedes the motilin peak, *depends* on the vagi; the stomach activity, which coincides with motilin peaks, seems to be *triggered* by the vagi, while the small bowel seems to depend on its own intrinsic innervation, but is kept in phase by the activity of the stomach above.

Gastrin is not a candidate for the initiator as its plasma concentrations do not fluctuate in-phase.

Periodic activity seems to be the normal basal state of the gut. In prolonged starvation it continues to the point of death. The feeding pattern seems to be something that is superimposed upon it from time to time in intermittently digesting animals because low doses of CCK, secretin or gastrin only mask but do not abolish the periodicity of their respective secretions. Pharmacological doses do, however, produce secretory plateaus.

The situation in the ruminant is interesting. These animals are continuous digesters. The rumen under normal circumstances is always full and dribbles digesta continuously into the gut. One might expect, in this case, a constant feeding pattern, but the converse is the case. The gut below the rumen exhibits uninterrupted periodicity. The young ruminant, before weaning, is not functionally a ruminant at all. Its stomach behaves exactly like that of a carnivore because milk bypasses the rumen; like a carnivore feeding causes a typical feeding pattern to replace the interdigestive periodicity.

Interdigestive Activity and Secretory Studies

Failure to recognize periodicity is the explanation for the long and unresolved controversy about basal gastric and pancreatic secretions. Now we have the answer: at peaks there is secretion and between them there is almost none. The rediscovery also necessitates a reassessment and reevaluation of earlier studies on intestinal and gastric motility. An attempt has been made herein to do this where possible.

Now that it is known that this fluctuant activity exists, it is easy to spot in the results and tracings in old research papers where it was missed or if noticed unexplained. [5].

References

1. Boldyreff, W. (1911): Quoted in Itoh and Sekeguchi [2].
2. Itoh, Z.; Aizawa, I.; Sekiguchi, I.: The interdigestive migrating complex and its significance in man. Clin. Gastroent. *11:* 497–521 (1982).
3. Magee, D. F.; Naruse, S.: Neural control of periodic secretion of the pancreas and the stomach in fasting dogs. J. Physiol. *344:* 153–160 (1983).
4. Magee, D. F.; Naruse, S.: Periodic secretion and motility in fasting dogs: stomach, gallbladder, pancreas and duodenum (in press).
5. Schoefield, B.: The inhibition of pepsin output in separated gastric pouches in dogs following feeding and its correlation with motility changes. Gastroenterology *37:* 169–181 (1959).
6. Szurszewski, J. H.: A migratory electrical complex of the canine small intestine. Am. J. Physiol. *217:* 1757–1763 (1969).
7. Vantrappen, G. R.; Peeters, T. L.; Jansens, J.: The secretory component of the interdigestive migrating motor complex in man. Scand. J. Gastroent. *14:* 663–667 (1979).

10 Absorption

The Small Intestine

General Structure and Function

Running from stomach (pyloric valve) to colon, the small intestine of the human consists of three portions: duodenum, jejunum and ileum, respectively. The first 10 inches (25 cm) or so is the *duodenum*. The duodenum is mostly retroperitoneal and hence is relatively fixed in position as it pursues its horseshoe-shaped course around the head of the pancreas. The *duodenal-jejunal junction* is marked externally by the *suspensory muscle (ligament) of the duodenum (Treitz)*. Otherwise, the three segments are not distinctly demarcated from each other, the transitions being gradual, and although certain macro- and microscopic features characterize each segment, they share the same basic organization. The *jejunum*, constituting the next two-fifths of the small intestine, primarily occupies the upper left abdominal cavity while the *ileum,* the last three-fifths of the small intestine, is situated mainly in the lower abdominopelvic cavity. The jejunum and ileum are intraperitoneal – suspended from the dorsal body wall by the mesenteries – and hence are freely moveable.

The structure of the small intestine is similar in most mammals. It is a thin-walled tubular viscus (fig. 10/1A) which conforms closely to the 'General Structure of the Gastrointestinal Canal' (see corresponding paragraph, chapter 1). Structural specializations, as elsewhere, are reflections of functional demands. The functional demands placed upon the small intestine are threefold: (1) completion of the digestion of food delivered to it with enzymatic juices secreted by its own glands and its accessory glands, the liver and pancreas; (2) selective absorption of water and the final products of digestion (nutrients) into its blood and lymph vessels, and (3) continuing

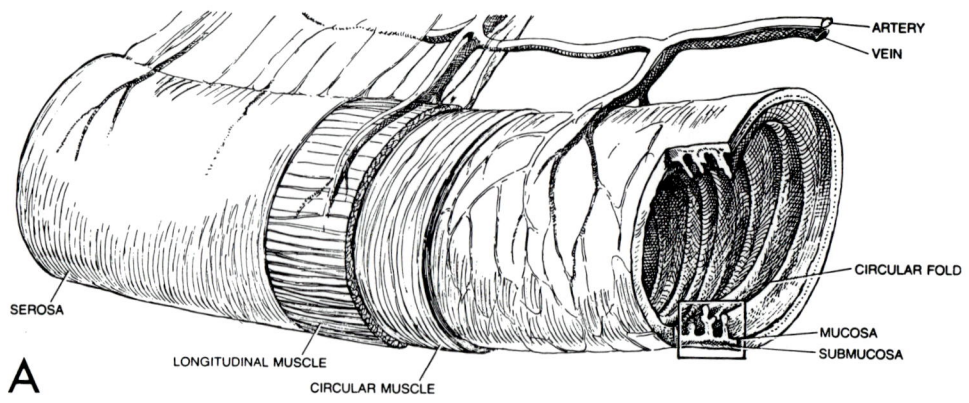
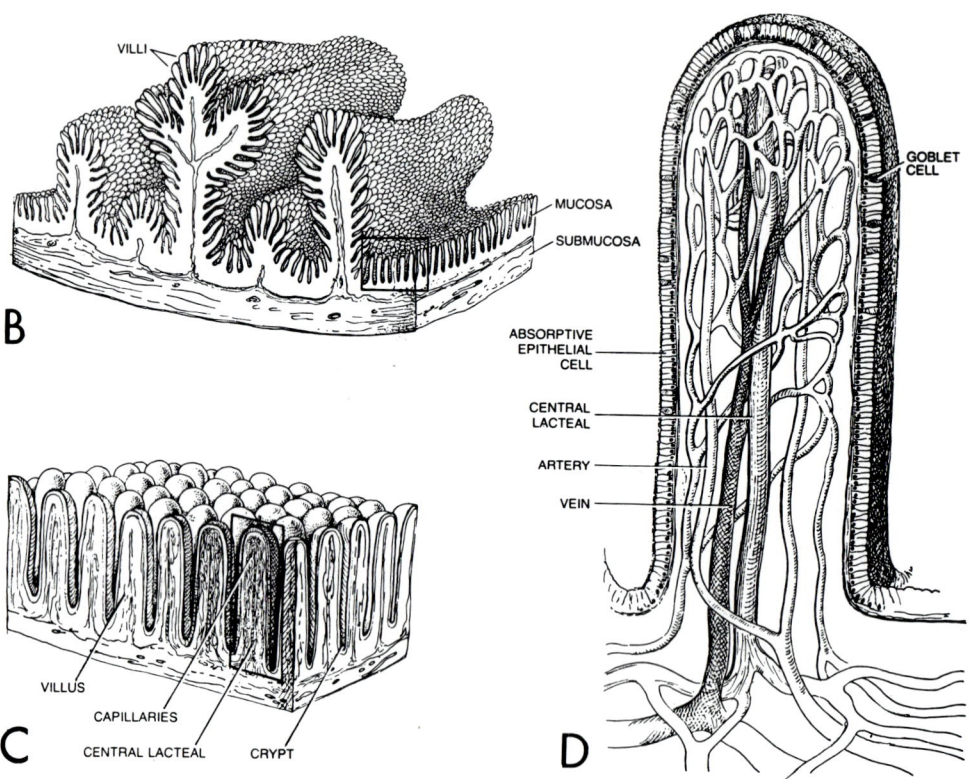

Fig. 10/1. Structure of the small intestine. In order of increasing detail, the morphology of (**A**) the intestinal wall; (**B**) the circular folds; (**C**) the mucous membrane, and (**D**) the intestinal villus is demonstrated. Depressions (intestinal crypts) exist between villi, but are not depicted in **C** and **D** (see fig. 10/3). Reproduced, with permission, from *Moog* [44].

the passage of the unabsorbed materials and gas along the tract. In addition, the small intestine produces some hormones.

The third function listed is shared with all other parts of the tract, and reguires little specialization here. The first two functions require specializations mainly of the lining coat, or mucosa. The function perhaps most unique to this portion of the canal is absorption. To perform this function efficiently, a vast mucosal surface composed of absorptive epithelial cells is required. The greater the surface area, the greater the absorption possible, but it must of course fit within the confines of the abdominal cavity.

Augmentation of the Luminal Absorptive Surface
Macroscopic Augmentation

Enlargement of the luminal surface is accomplished by several progressively finer yet increasingly effective (and hence physiologically more important) means. On the gross or macroscopic level, this is accomplished in two ways. First, this portion of the tract is greatly lengthened. Although the pyloric sphincter and ileocolic junction are within 6 inches (15 cm) or so of each other 'as the crow flies,' the intestinal tube linking them is some 20 feet (6 m) in length in the adult human. Of course, to fit the confines of the abdominopelvic cavity, a grerat deal of coiling and convolution of the tube is required. A tube 20 feet in length having a diameter between $1\frac{1}{2}$ and $1\frac{1}{4}$ inches (the small intestine tends to become narrower throughout its course) has a surface area of roughly 7 square feet.

Secondly, on the internal aspect of the tube, the mucosa and submucosal layers are thrown into crescentic folds (*plicae circulares,* or 'valves' of Kerckring) which project into the intestinal lumen at right angles (fig. 10/1A, B). Unlike the rugae of the stomach, these folds are not obliterated with distention. Most extend $\frac{1}{2}$ to $\frac{2}{3}$ the girth of the lumen, but some completely encircle it: others bifurcate, perhaps joining adjacent folds: still others spiral about the lumen, occasionally taking 2 to 3 full turns. The folds begin about 2 inches (5 cm) beyond the pylorus and reach maximum size and closest approximation just distal to the entrance of the common bile and pancreatic ducts into the 2nd (descending) part of the duodenum. Here the larger ones may be 8–10 mm in height, 3–4 mm in thickness, and up to 5 cm in length, but most are smaller. The folds remain large and numerous throughout the proximal half of the jejunum, but from here to mid-ileum they diminish considerably, the terminal ileum being nearly devoid of them. Hence, the comparable thinness of this portion of the intestine as compared with duodenum and jejunum. These folds may increase the mucosal surface area by as much as a factor of 3, bringing the luminal surface area roughly to 20 square feet by these macroscopic means. As well as increasing the absorptive area, the plicae circulares are said to have an effect in retarding the passage of food, or at least in preventing it from passing too rapidly for effective digestion and absorption.

Microscopic Augmentation

Villi. A third and more effective augmentation of the mucosal surface area is achieved at the nearly microscopic level (just visible to the naked eye, but best observed with magnification) in the form of enormous numbers of *villi* (fig. 10/1B–D, 10/2). Villi are minute projections or mucosal evaginations having a length of 0.5–1.5 mm, depending upon the degree of distention of the intestinal wall and the state of contraction of smooth muscle fibers in their own interior. They cover the entire mucosal surface at a density 10–40/mm^2, giving it a characteristic velvety appearance in the fresh condition. There are species differences in the length and shape of villi along the small intestine. In the human, the villi are large and numerous in the duodenum and jejunum, but are smaller and fewer in the ileum. In the proxial duodenum they generally are broad, ridge-like structures, changing to tall, leaf-like or spatulate villi in the distal duodenum/proximal jejunum. Thereafter, they gradually transform into shorter, finger-like extensions in the distal jejunum and ileum [42, 64, 65]. However, these shapes do vary with the individual. At first, it would appear that the size of villi varies with the intensity of absorption. However, experimental work indicates that the large villus size in the duodenum is maintained by some factors present in the secretion arising locally as well as in that coming from stomach and pancreas. When the duodenum is connected to the terminal ileum in such a way that secretions are shared, ileal villi become taller and duodenal villi smaller than normal [1, 23, 37]. The structure of the villi will be explained shortly. The villi increase the luminal surface area about eightfold, bringing our modification of the surface to approximately 160 square feet (15 m^2).

Mircovilli. Each villus is covered by an epithelium one cell thick, composed mostly of absorptive cells. Their most distinctive feature, the striated border, is the most diminutive but the most effective modification of the absorptive surface area. The structure of the highly refractile luminal surface (border or cuticle) of these cells is just beyond the limit of resolution of the light microscope. *Granger and Baker* [19] were the first to show, by

10 Absorption

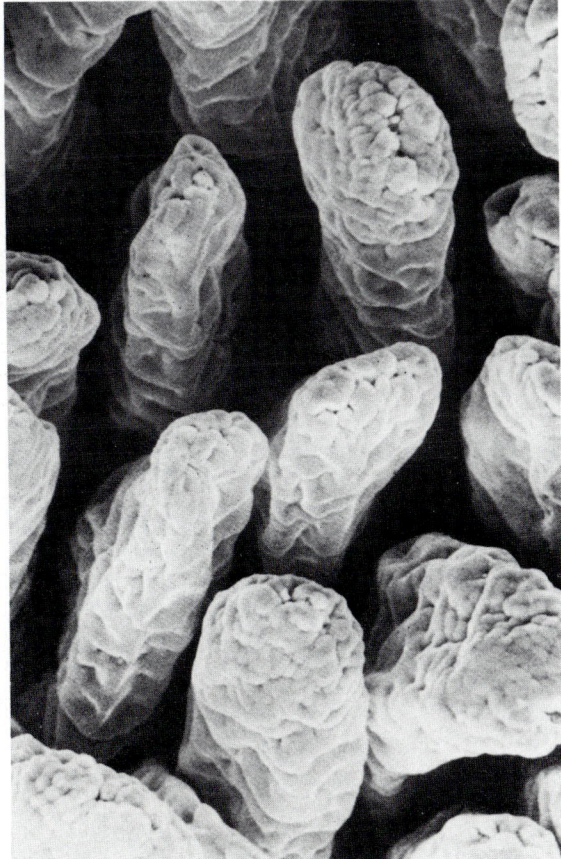

Fig. 10/2. Scanning electron micrograph of the duodenal portion of the small intestine of a rat. × 270. Reproduced, with permission, from *Moog* [44].

means of electron microscopy, that the luminal cell border is a 'brush border' comprised of a third set of projections, the *microvilli* (fig. 10/3A, B). The microvilli are very minute, parallel cylindrical projections about 1 µm long and 0.1 µm broad, packed at a density of about 200000/mm^2 in the human jejunum. Their presence increases the luminal surface or absorptive area of the small intestine by a factor of about 20, so that our final figure for the absorptive area is some 3200 square feet (300 m^2)!

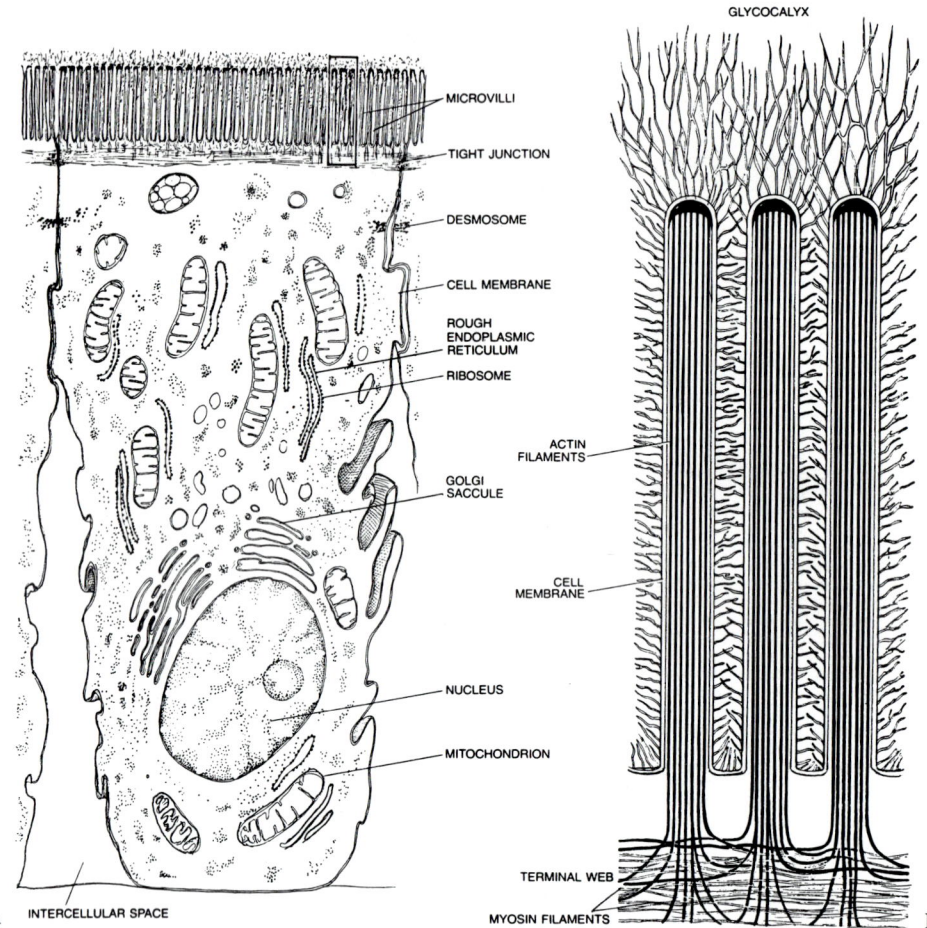

Fig.10/3. The columnar absorptive cell ('enterocyte'). **A** The cell as seen in longitudinal section. **B** Detail of **A** demonstrating the ultrastructure of the absorptive cell. Reproduced, with permission, from *Moog* [44].

The small intestines of smaller mammals (such as the commonly studied laboratory rodents) lack the plicae circulares, having only the villous and microvillous modifications.

Diseases that affect mucosal function often perturb normal villus architecture, reducing substantially the absorptive surface of the small intestine. Atrophy of the villi and microvilli of the jejunum, wherein they become severely blunted and distorted – and may even disappear, is the characteristic anatomic finding of the malabsorption syndrome known as nontropical sprue (gluten enteropathy). Consequently, the absorption of metabolites is greatly hindered in these individuals. On the other hand, villi increase in height both during lactation [25, 39], and following partial resection of the small intestine [24, 66] increasing the absorptive surface over a given length of the intestine [60].

Having established the massive size of the surface area available for absorption, further understanding of how the functional demands of continued digestion and absorption are met requires further examination of the structure of the mucosa of the small intestine and the cells that compose it.

The Mucous Membrane of the Small Intestine

The tunica mucosa is thick and highly vascular in the upper part of the small intestine, but thinner and less vascular in the lower part. It has the usual three layers of muscularis mucosa externally, lamina propria and then epithelium internally (fig. 10/4). The muscularis mucosa, which extends with the underlying submucosa into the circular folds, is typical (see 'General Structure of the Gastrointestinal Canal', chapter 1). The lamina propria also shares the features common throughout the tract, but plasma cells and lymphocytic cells are especially numerous, the latter often being amassed into solitary or aggregated lymph follicles (Peyer's patches), some extending through the muscularis mucosa into the submucosa. The amount of lymphatic tissue progressively increases from proximal to distal small intestine. The core of each villus is formed of extensions of the lamina propria, as will be explained in detail further on. Internal to the lamina propria is a simple, columnar epithelium, supported by a basal lamina and covering the free surface of the mucosa.

It was explained previously that the free surface of the mucosa is modified (evaginated) to form villi in order to increase the absorptive surface area. Not explained earlier is the fact that between the villi mucosal invaginations, known as *intestinal glands* or *crypts* (of Lieberkuhn), dip into the lamina propria (fig. 10/4B) further increasing the epithelial surface but not necessarily increasing the area available for nutrient absorption. It is unclear what contribution, if any, the crypts make to intestinal absorption. The crypts are simple, tubular glands or pits, 320–345 µm in depth, which extend down into the lamina propria nearly to the muscularis mucosa

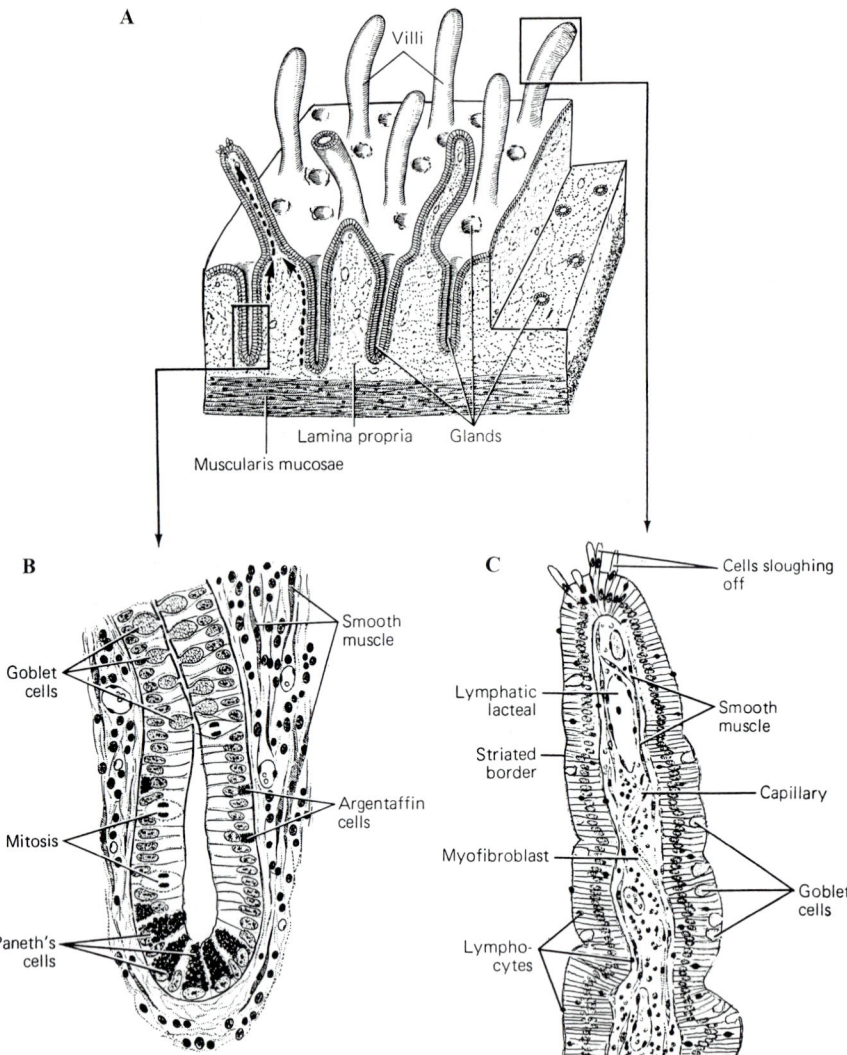

Fig. 10/4. Microanatomy of the mucous membrane. **A** Macroscopic diagram of the mucosa of the small intestine. On the left side of the figure, dotted arrows indicate the pathway of enterocyte migration leading ultimately to desquamation at the villus tip. The ostia (and, on section, the depths) of the intestinal crypts (glands of Lieberkuhn) are demonstrated between the bases of the villi. **B** Detail of **A** demonstrating the cellular structure of the intestinal crypt. **C** Microscopic detail of **A** demonstrating the cellular structure of the intestinal villus. Reproduced, with permission, from *Junqueira and Carneiro* [31], figure was redrawn and reproduced, with permission, from *Ham* [23].

(fig. 10/4B). Their openings on the luminal surface are difficult to distinguish in the living animals because their orifices are tightly closed – a fact which also brings their role in absorption into question.

Cellular Migration and Epithelial Renewal

The single layer of epithelial cells that lines the crypts and covers the villi forms a continuous sheet that is constantly renewed. In 1958 *Leblond and Messier* [38] demonstrated that most crypt epithelial cells migrate upward to become villous epithelial cells, there continuing their migration to the villous tip. Leblond injected thymidine labelled with radioactive tritium (tritiated thymidine) into mice, thymidine being the only one of the four nucleotides in DNA that is not also a component of RNA. It is incorporated into the DNA molecule only when the DNA strands are duplicated just prior to cell division. Cellular uptake of tritiated thymidine, demonstrable through radioautography, is then a clear sign of DNA replication (and hence cellular division). Within hours of injection, tritium-labelled cells were found among the intestinal epithelial cells, but only in the crypts. After 24 h, cells containing radioactive thymidine were found midway up the villous shaft; after 48 h, labelled cells were detected at the villous tip. Once the villous tip was achieved, the cells were sloughed into the lumen, carrying with them any contained digestive enzymes [38, 43]. Most of the enzymes of the intestinally secreted digestive fluid *(succus entericuis)* are not, therefore, secreted in the usual merocrine manner but are extruded holocrine fashion into the intestinal lumen.

Approximately 17 billion cells per day are discarded along the length of the human small intestine. The overall rate of production must, of course, equal the rate of cell loss by migration or death. The amount of time required for the cycle of proliferation and migration of intestinal epithelial cells is approximately the same in all mammals, the intestinal epithelium being completely renewed about every third day [44]. The replacement time for the parietal and zymogenic cells of the gastric glands and intestinal crypts is much greater, extending from 3 weeks to many years [31].

With the recent marked increase in the therapeutic use of irradiation and antimitotic compounds (i.e. chemotherapy) to destroy rapidly dividing cancer cells, epithelial renewal has significant clinical implications. Such therapy also affects the normal, rapidly renewing epithelial cells of the GI tract. Normal cell sloughing continues in spite of the death of proliferating cells, and consequently the mucosa atrophies. Since there is no replacement of the mature functional epithelial cells, malabsorption and diarrhea are the sequelae of irradiation and chemotherapy. Fortunately, sufficient numbers of quiescent stem cells usually remain to reestablish the mucosa when the treatments cease [47].

Cellular Differentiation

The epithelium lining the crypts, being relatively undifferentiated, is histologically different from the villous epithelium. The principal cell type in the crypt is the *undifferentiated columnar cell*. These cells have slightly basophilic cytoplasm and, since cellular division occurs frequently, mitotic figures are commonly seen. Their progeny differentiate into four cell types, three of which migrate out of the crypt and onto the villus. The principal cell type on the villus (accounting for approximately 90% of the cells present) is the brush-bordered columnar absorptive cell (fig. 10/3, 10/4B), with moderately basophilic cytoplasm. Mitotic figures are not seen among the cells of the villous epithelium. In the crypts and on the villi, interspersed among the undifferentiated and absorptive columnar cells respectively, are mucous-secreting *goblet cells* (accounting for approximately 9.7% of the cell population on the villus) and *enteroendocrine (argentaffin) cells* (approximately 0.3% of the villous epithelium) (see fig. 10/3A). Found only at the bases of the crypts are the zymogenic *Paneth cells,* which have the cytological characteristics of protein-secreting (enzyme-producing) cells (see 'Glandular Epithelial Tissue', chapter 1). Migratory lymphocytes are commonly found between the cells of the villous epithelium (fig. 10/4C) [47].

The Structure of the Intestinal Villi

Contained within the lamina propria core of each villus is a capillary bed with a supplying arteriole and draining vein, a central lymphatic vessel or *lacteal,* some nonstriated muscle and some nerve fibers (see fig. 10/1D, 10/4C).

According to the size of the villus, one or more arterioles arise deep to the villus from the arteries of the submucosa, penetrate the muscularis mucosa and then ascend into the villus. There a rich plexus of capillaries is formed which ramifies through the lamina propria of the villus, closely applied to the basement membrane of the absorptive epithelium. Arterioles arising separately from the submucosal arteries also penetrate the muscularis mucosa to supply the capillary networks which surround the intestinal crypts. The vascularity of the villi is, however, considerably greater than that of the tissue surrounding the crypts. Further, the capillaries of the villus are fenestrated on the surfaces facing the lumen, making them permeable to macromolecules; the capillaries of the crypts are not. (Further evidence that the crypts do not play a significant role in nutrient absorption.) Both the abundance and the fenestration of the villus capillaries are undoubtedly significant in ensuring the rapid uptake of nutrients which have passed through the villus epithelium [9].

The lacteal, usually single but occasionally double in a large or broad villus, has its blind, dilated beginning near the villus summit and courses along the axis of the villus in its center to the base. The lacteal walls consist only of a single layer of thin, nonfenestrated endothelial cells surrounded by a discontinuous basal lamina and supporting reticular fibers. Although tight junctions do seal most of the margins of adjacent lymphatic endothelial cells, gaps occur of a size sufficient to allow passage of macromolecules and even chylomicrons [7, 15, 50]. Whether most macromolecules and lipoprotein particles (including chylomicrons) cross the lymphatic endothelial wall through the gaps between adjacent cells or via the vesicular transport system in the endothelial cytoplasm remains unsettled [60].

Delicate slips of smooth muscle extend from the muscularis mucosae into the villous core. They are arranged in bundles which surround the lacteal and which run parallel to the long axis of the villi, extending from base to apex and attaching to the basement membrane peripherally and the lacteal centrally. Thus, the villi sway and contract (shorten) intermittently when the muscularis mucosae contracts. These asynchronous movements occur at the rate of several strokes per minute. The rate increases when bathed in solutions known to be absorbed, and hence it is assumed to do so during digestion. In fasting animals, the rate decreases considerably [31]. It is proposed that these movements and contractions have the effect of 'milking' the lacteals, forcing the lymph and absorbed nutrients (fats) into the more basal lymphatic vessels and toward the mesenteric lymphatics en route to the thoracic duct and the venous system. A hormone liberated from the intestinal mucosa by chyme has been postulated as an activator of the villi. Only one group, that of *Kokas and Ludany* [33] has ever been interested in it and, therefore, 'villikinn', the name they gave it, rests in the realm of the 'might be' hormones.

The unmyelinated nerves of the lamina propria (including the villous cores) are associated with blood vessels and the smooth muscle fibers. The entire villus, then, is covered on its external surface with the single-cell thick epithelium of absorptive cells interspersed with goblet cells (fig. 10/1D, 10/4C).

Columnar Absorptive Cells (Enterocytes)

As stated previously, the single layer of epithelial cells that lines the crypts and covers the villi is a continuous sheet which is constantly being renewed. The predominant cell type is the columnar cell – undifferentiated and multiplying in the crypts, and differentiated and migrating as 'absorptive cells' on the villi.

Undifferentiated Columnar Cells. In the base of the crypts where they are formed, the undifferentiated columnar cells are small cells interspersed ('squeezed') among an approximately equal number of zymogenic (Paneth) cells. Here the columnar cells are poor in cytoplasm (cytoplasm: nucleus ratio approximately 2). Their lateral membranes are smooth. They contain only a few mitochondria, little rough endoplasmic reticulum and a small Golgi apparatus, but they are packed with ribosomes. These cytoplasmic features are characteristic of their undifferentiated state. Of course, this being the 'cell-production site' of the epithelium, mitotic figures are commonplace.

Mature Columnar Absorptive Cells. As the cells move up the crypt walls, differentiation (maturation) occurs, and mitotic figures no longer appear. Size increases progressively as the villus base is approached. Here, the full-developed cells are tall and narrow (25 µm high and 8 µm wide) with abundant cytoplasm (cytoplasm: nucleus ratio approximately 3.5) (fig. 10/3A). The clear, oval nuclei are located in the basal half of the cytoplasm. Free ribosomes have become scarce and mitochondria abundant. Smooth and rough endoplasmic reticulum, scattered throughout the cytoplasm, and Golgi saccules, supranuclear in position, have become well developed. The apical cytoplasm contains numerous membrane-bound lysosomes, and hence has a granular appearance. These cytoplasmic features indicate intense oxidative and synthetic activity.

Just below the level of the bases of the microvilli, each cell is tightly bound to the adjacent cells by junctional complexes involving an exceptionally close approximation of cell membranes (fig. 10/3A). These tight junctions exclude the possibility of macromolecular nutrients entering the villi – and hence the vasculature of its core – by intercellular channels, as was once held. Immediately beneath the tight junctions are looser intercellular connections (desmosomes) which augment the tight junctions in binding the cells of the epithelium into a continuous sheet. The lateral plasma membranes have become highly convoluted, interdigitating to variable degrees with adjacent cells apically but separated from them in the basal half of the epithelium by intervals constituting intercellular canaliculi. The intercellular canaliculi may be greatly dilated when ions and water or lipid are being actively absorbed. The plasma membrane bounding the intercellular canaliculi is devoid of digestive enzymes but is rich in Na, K-dependent ATPase, an enzyme intercalated into the membrane which pumps sodium (and hence passively draws water) into the lateral intercellular space. Thus, the enzyme activity of the lateral cell membrane of the absorptive cell results in the movement of ions and water between those cells (the 'paracellular pathway'), providing a rapid and efficient means for the retrieval of water from the intestinal lumen. Just how water and small ions pass through the tight junction is not known. There is some regional variation in the junctional structure (e.g. those of the crypt are less well developed) implying some regions may be 'leakier' than others.

The single most characteristic feature of the fully differentiated absorptive columnar cell is, of course, the vertically 'striated' apical or 'brush-border' (cuticle) of minute, rod-like microvilli (approximately 3,000 per cell) in parallel array, greatly increasing the cell's luminal or absorptive surface as discussed previously (fig. 10/3A). The outer surface (surface coat or 'glycocalyx') of the microvillus membrane is filamentous and rich in carbohydrates (glycoprotein) (fig. 10/3B). At 10–11 nm thickness, it is thicker than the coats of most cell membranes. The reason for this unique architectural feature is uncertain, but it may reflect its distinctive biochemical composition.

An array of enzymes and transport systems is incorporated into the membrane (fig. 10/5). Included among the enzymes are disaccharidases and dipeptidases, enterokinase, ATPase and alkaline phosphatase. These are all glycoproteins; each molecule of such enzymes consists of a protein with carbohydrate side chains attached to it. The current prevailing theory concerning the relationship of the digestive enzymes to the cell (microvillus) membrane has the protein component of the enzymes inserted into the lipid cell membrane to varying depths, with the carbohydrate side chains extending into the lumen, forming the filaments of the glycocalyx. In addition to the enzymes, transport proteins (e.g. those responsible for the cotransport of Na+ and D-glucose and Na+ and amino acids) and nonenzymatic proteins that function as receptors (e.g. those that selectively bind vitamin B_{12} and Ca^{2+}) are associated with the microvillus membrane. The chemical composition of the microvillus membrane appears to vary significantly in absorptive cells occurring at different levels of the small bowel. Enzymes such as

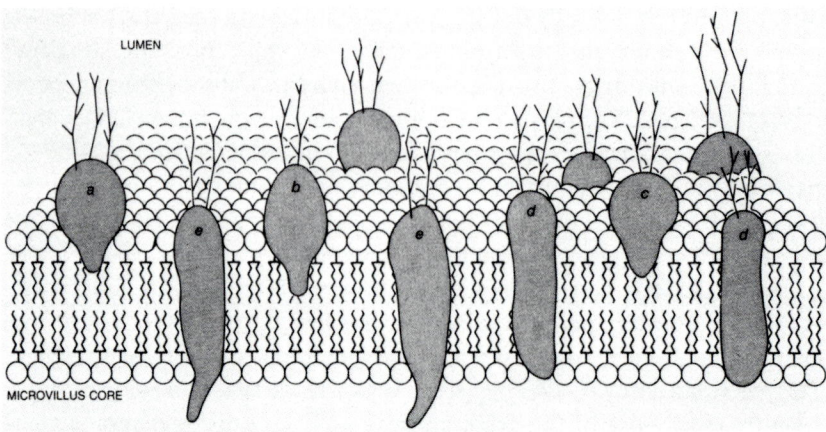

Fig. 10/5. Schematic illustration of the presumptive structure of the membrane of the microvillus based on the fluid mosaic model of the cell membrane. Integrated into the matrix of the bilaminar lipid unit membrane are the protein components (solid gray structures) of the glycoprotein digestive enzymes; the carbohydrate components are depicted as branching chains which protrude into the lumen, thus constituting the glycocalyx of the microvillus membrane. Disaccharidases (**a, b, c**) protrude from the membrane, which they only partially penetrate. Alkaline phosphatase (**d**) extend the depth of the unit membrane. Aminopeptidases (**e**) traverse the membrane and protrude into the microvillus core. Reproduced, with permission, from *Moog* [44].

alkaline phosphatase, lactase, trehalase, and sucrase-isomaltase demonstrate activity, and are, therefore assumed to be mainly present in the proximal small intestine while the receptor for intrinsic factor mediated vitamin B_{12} absorption is selectively located in the distal small intestine [13, 16, 26, 41].

Experiments utilizing autoradiography have shown that each absorptive cell continuously synthesizes components of the cell membrane/glycocalyx (including most of the enzymes incorporated therein) and transports them to the microvillus surface in the form of small vesicles. These activities are in keeping with the abundance and type of cytoplasmic organelles previously described (mitochondria, ER and Golgi saccules, etc.). Thus, the so-called absorptive cell also has important secretory functions.

Enzymes thus synthesized within the absorptive cell – the process is explained in detail under 'The Structure of Protein-Secreting (Serous) Cells', chapter 1 – and transported to the microvillus surface may be further modified at that site by pancreatic enzymes which, having been secreted into the intestinal lumen, become adsorbed onto the glycocalyx. Likewise, the intestinally produced enzymes may act on pancreatic enzymes at the microvillus surface. For example, intestinal enterokinase cleaves (and thus activates) pancreatic trypsinogen; the product of this interaction, trypsin, in turn cleaves and activates other pancreatic hydrolases.

Thus, two of the major functions of the small intestine occur in large part at this site. Because of the high concentration of active intestinal and pancreatic enzymes and their substrates on the outer surface of the microvilli, much of the digestive process is completed here. The breakdown products are thus conveniently and efficiently delivered to the transport proteins in the microvillous membrane for absorption. The microvillous membrane is, then, the actual site of admittance of nutrients into the body. Whether or not the glycocalyx of the absorptive cell functions in protecting the apical surface remains to be established.

Within the cytoplasmic core of the microvillus there is a longitudinal bundle of helical microfilaments composed of a protein similar in molecular weight and immunological characteristics to actin (fig. 10/3B). No myosin is found within the microvillus. Under phase microscopy, an in vitro organ culture of late gestational age fetal chick small intestine has demonstrated rapid coordinated microvillus motion [51]. Coordinated microvillus contractions have also been reported in other epithelial tissues [59], and thus it has been proposed that microvillus motility may represent a stirring mechanism which keeps the microvillus surface in continual contact with a fresh stream of luminal content [51]. However, it remains unclear to what extent the microvilli move in vivo.

Goblet Cells

Interspersed between the absorptive cells of the lining epithelium (fig. 10/4C) goblet cells are found with increasing frequency along the length of the small bowel, being least abundant in the duodenum and most numerous in the distal ileum. The apical two-thirds of these cells are often filled to the point of distention by an accumulation of membrane-bound mucin secretory granules, while their bases remain slender, containing elongated nuclei; hence, the characteristic goblet shape for which they are named (fig. 10/6). The mucous granules are poorly preserved in routine preparations so that the cells often appear empty; however, with special stains, the carbohydrate-rich, acidic glycoprotein secretory product is seen to be intensely basophilic, metachromatic and PAS-positive. Once secreted from these unicellular glands, the mucous forms a luminal lining lying on top of the glycocalyx of the microvilli.

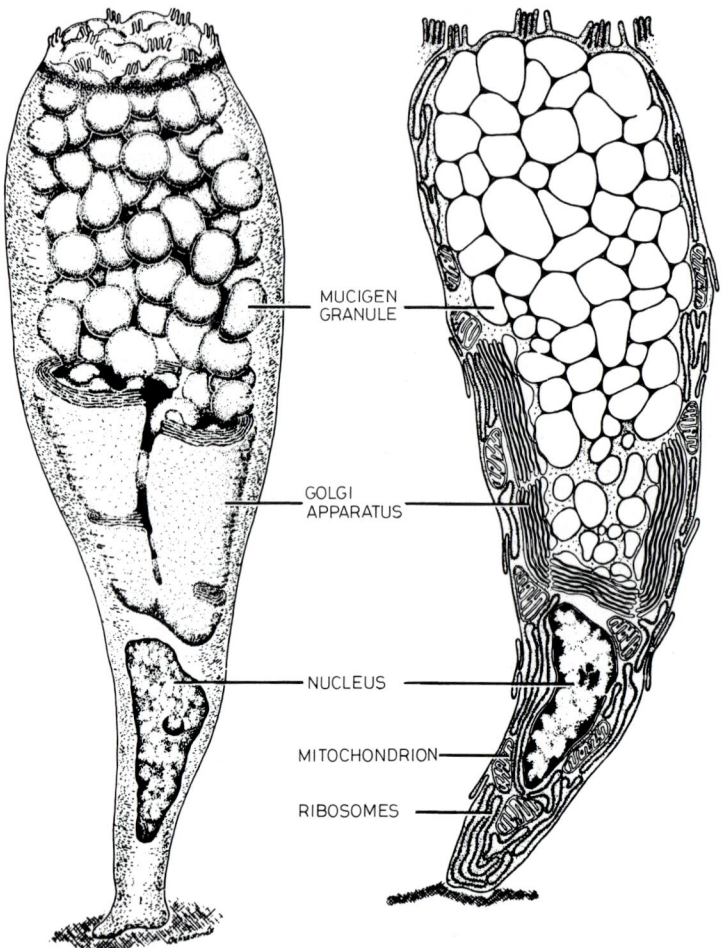

Fig. 10/6. The goblet cell. The structure of the cell is presented schematically in a three-dimensional reconstruction (left) and as it appears upon longitudinal section via electron microscopy (right). Reproduced, with permission, from *Neutra and Padkula* [47].

The function of mucous in the small intestine has not been established with certainty. It is widely assumed that the mucous secreted by the goblet cells lubricates and forms a barrier which protects the mucosal epithelium from potentially noxious intraluminal substances. Indirect support for this assumption comes from the observation that goblet cell secretion can be induced with a variety of noxious agents, such as *Escherichia coli* and cholera enterotoxins [45], immune complexes [35] and bile [57] in the postirradiated rat.

Goblet cells will be discussed further with 'The Large Intestine' (chapter 11).

Paneth Cells

Occurring in small groups found only at the depths of the intestinal crypts (fig. 10/4B), the Paneth cells are pyramid-shaped cells with conspicuous eosinophilic secretory granules located apically in an otherwise intensely basophilic cytoplasm. They are abundant in the small intestines of most mammals, including man and other primates, ruminants and several species of rodents, but are absent in cats, dogs, pigs and raccoons. The species variation remains to be understood. In those species in which they are present, the Paneth cells, unlike most cells of the intestinal epithelium, represent a relatively stable population, having a low rate of turnover (18–22 days in mice [8], 30 days in some other mammals [31]). Like goblet cells, they increase in number from duodenum to ileum.

The fine structure of Paneth cells is similar to that of other zymogenic cells with protein-rich secretions, such as the gastric-chief cells and pancreatic acinar cells (see 'The Structure of Protein-Secreting (Serous) Cells', chapter 1). That they are active secretory cells has been shown by autoradiography.

On the basis of the microchemical analysis of slices of mucosa at the level of the crypt bases [40] it was suggested that these cells might secrete a peptidase; however, this has not been substantiated in later studies.

Lysozyme (a bacteriolytic enzyme which partially degrades bacterial cell walls), immunoglobulin (thought to be synthesized elsewhere and taken up by the Paneth cells), and degenerating protozoa and bacteria contained in lysosomes have all been demonstrated in Paneth cells [17]. These findings have led to speculation that Paneth cells may play a role in regulating the microbiological flora of the small intestine, especially in the immediate region of the intestinal crypts [48, 49]. However, an antibacterial function has yet to be convincingly demonstrated. Others have speculated without substantive data that Paneth cells may, by virtue of their secretion, provide nutrients needed by other crypt cells [11]. Because Paneth cells are reported to concentrate radioactive zinc – a trace metal which is an essential component of a number of enzymes – it has also been suggested that Paneth cells may secrete enzymes important in intraluminal digestion of foodstuffs [63], but no digestive enzyme has been specifically localized to the Paneth cells.

In point of fact, the role of the Paneth cell in the function of the small intestine remains unknown even though their presence was described over a century ago [54, 60].

Intestinal Secretion

There have been no modern studies on the regulation of intestinal secretion with the exception of the Brunner's gland region. The empty intestine secretes slowly. *Boldyreff* [4] in 1904 described a periodic secretion which amounted to 50 ml/h for the entire small intestine of the dog. *Boldyreff*, as mentioned earlier, was the discoverer of fasting periodic secretion; since every other instance of this type of secretion coincides with periodic motor activity, it is quite possible that the observed intestinal secretion is the result of the mechanical activity of the gut and not a primary activity at all. Several have reported that atropine reduced this secretion, but atropine also abolishes periodic interdigestive motor activity. The secretion of fully innervated Thiry-Vella loops of intestine (see fig. 5/6) is unchanged by feeding. This is strong evidence against any hormonal or nervous control emanating from the fed gut. *Landor* et al. [36] made pouches of the entire duodenum in dogs from which bile and pancreatic juice had been diverted and obtained an amazing secretion of 1.5 l/day. After a few days this declined by more than half [21]. Since lipase and amylase were found in this it is suspected that pancreatic juice was not entirely diverted. However, neither prolonged fasting, feeding, nor vagal section or stimulation influenced the rate of secretion. Thus, there is no clear evidence for vagal control of secretion, but there is evidence that complete denervation of Thiry-Vella loops causes a great increase in motility and secretion [22]. Denervation is effected by severing the pedicle of such a loop after an alternative blood supply to it has been established. After a week and so of hypersecretion the loop returns to its predenervation state. This observation has a long and venerable past since *Bernard* [3] first noticed it in 1859. The possible explanation is removal of sympathetic inhibition or denervation hypersensitivity. It has been argued that the sympathetics are motor to the muscularis mucosa which may become hyperresponsive to circulating catecholamines following sympathectomy. The removal of inhibition is seen in the greatly increased propulsive activity, in the claimed increase in acetylcholine production and in the mucosal hyperemia. The intestinal juice after denervation is often blood stained.

Denervation hypersensitivity, which usually lasts about 10 days in other structures, could explain the return to the normal state whereas loss of inhibition cannot. This phenomenon is not seen in either the stomach or colon because, as the acetylcholine enthusiasts claim, acetylcholine production in them is slight.

Nassett et al. [46] have claimed that the completely denervated loop, once the initial hypersecretion phase has passed off, will increase its secretory rate in response to feeding. These workers made extracts from small intestine, stomach and colon. All of these on intravenous administration increased the fluid and enzyme secretion from their preparation. The suspected hormone was named 'enterocrinine'. Mechanical stimulation, insertion of balloons or catheters, will cause secretion from intestinal mucosa, but the secretion is still very small.

The traditional view is that the succus entericus is a highly important digestive secretion, but knowledge about it, its stimuli for secretion, regulating mechanisms, even its very existence during normal digestion is uncertain. A vast quantity of fluid does enter the small intestine during digestion, but how much of this is simple diffusion through a leaking epithelium? That secretion can occur is obvious from pathological states, e. g. cholera and the watery diarrhea syndrome.

It is clear that an increase in mucosal cyclic AMP is necessary for small bowel secretion. It is not unreasonable to expect that in the ordinary course of digestion cyclic AMP might build up in crypt cells under the influence of physiological amounts of secretin, VIP or glucagon. These hormones all increase it, but cholera toxin does so spectacularly.

Brunner's Glands [10, 21]

Brunner's glands are elaborately branched tubuloalveolar glands which characterize the first part of the duodenum, extending only a centimeter or two distal to the gastroduodenal junction in man. They lie within the submucosa, their small ducts penetrating the muscularis mucosae to open into the intestinal crypts. Brunner's glands may contain both serous and mucous cells. Much species variation has been reported regarding the ultrastructural details of the constituent cells. There is much more literature concerning them than of any other intestinal glands because study of them is more rewarding; in many experimental animals they are much more extensive than in man. They increase their secretion in response to food and to vagal stimulation in anesthetized preparations. Spontaneous secretion is usually less than 2 ml/h and it may increase 3- or 4-fold from isolated

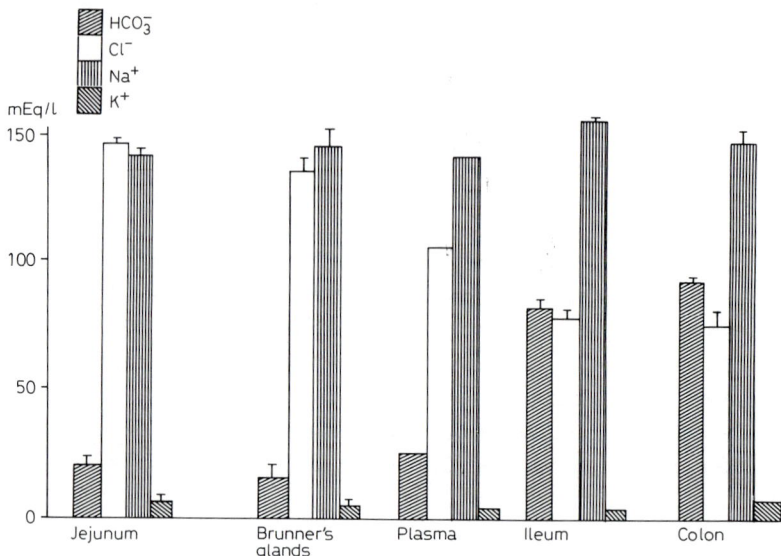

Fig. 10/7. The electrolytic composition of intestinal fluid (man) and Brunner's gland secretion (dog) compared with plasma.

pouches of the gland area following meals. It can continue for hours. Fat seems to be a good stimulant, but 0.1 N HCl is not. Intravenous infusion of intestinal extracts promptly increases secretion; most effective of these were the early crude extracts known to contain secretin. The more purified secretin available by 1950 did not stimulate secretion and so a new hormone, christened 'duocrinin', was postulated. A difficulty was and is that feeding stimulated pouches, but not flaps, indicating that mechanical stimulation from increased motility might be the effective secretory stimulus. Mechanical and chemical stimuli, just as in the case of jejunal glands, increase secretion.

Brunner's gland secretion is highly mucinous juice and in some animals its [HCO_3^-] is much higher than in other intestinal secretions. It is clearly alkaline in reaction (fig. 10/7) in the rabbit where [HCO_3^-] is similar to the pancreatic juice maximum. Brunner's gland secretion from man has never been obtained or studied.

Brunner's glands do not produce enough juice to play any substantial part in the neutralization of gastric juice. The secretion may be locally protective in the manner of the nonparietal secretion of the stomach. There is evidence that in dogs this region is less sensitive to the ulcerogenic effect of acid-pepsin solutions than is the more distal intestinal mucosa.

Absorption

Factors Contributing to Absorption and Absorbability
Thus far the source of energy for everything which passes from gut lumen to blood has been either the Na^+ pump or concentration gradients. Water movement has followed soluble ions or molecules according to osmotic requirements. Mechanical activity has received scant attention in the 1930s and more recently none at all. That the mechanical activity of the gut does contribute importantly to absorption is clear. The clinical consequence of ileus (paralysis of the intestine) is the rapid accumulation of luminal fluid to the extent that the patient finds himself in serious danger of dehydration. When normal motility returns so does absorption and the normal balance between secretion and absorption is restored.

Most drugs and medicines are not absorbed actively and do not ionize completely in solution because they are either weak acids or bases [29, 30, 34, 52]. Some organic acids and bases which are ordinarily encountered in the diet like acetates, lactates, or citrates, fall into this category. Other things being equal, ionized particles are not nearly as readily absorbed as the unionized. The other things are, first, molecular weight: small particles can and do penetrate water channels. If they are small enough it should make little difference whether they are ionized or not as long as they are water soluble, but the compounds we are dealing with here are large with molecular weights in excess of 180. Secondly, solubility: substances insoluble in water are not absorbed. Substances soluble in both water and lipid are much better absorbed than those with less lipid solubility. This is often expressed as the *lipid-water partition coefficient*. Nonionized chemicals are much more lipid soluble than are ionized forms and, as a result, diffuse across cell membranes at a rate which is almost independent of molecular size. Less lipid soluble ionized forms are greatly limited by molecular size since their diffusion is limited by the pore size of the water channels.

The factors which determine whether or not one of these weaker electrolytes is ionized are its pK (the pH at which an agent is half dissociated) and the pH of the aqueous solution in which it finds itself [27, 28].

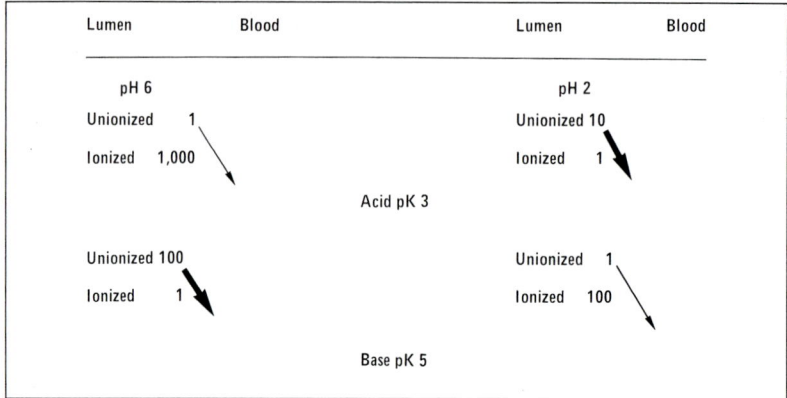

Fig. 10/8. The effect of ionization on the diffusion of weak acids and bases from intestinal lumen to blood.

An acidic drug like aspirin with a pK of between 3 and 4 will be substantially ionized at pHs above this and unionized below. It should, therefore, be better absorbed from the stomach than from the jejunum, whereas basic drugs with high pKs will be completely ionized in the gastric lumen and less so in the small bowel from which they will in consequence be better absorbed (fig. 10/8).

Such physicochemical considerations hold, but it is no surprise that the observed relationship between pK and amount absorbed falls short of the calculated absorption. Calculations based on lipid solubility do not predict in anything more than a general mathematically imprecise way the observed absorption. A possible explanation for this is the certainty that the pH of mucosal surfaces is not always the same as that of the luminal content, so that a substance which is unionized within the lumen of the stomach, for example, may become ionized in crossing the mucosal barrier to the cell surface. This is probably the case with aspirin. Therefore, one can say that although in vivo experiments bear out the pH lipid-water partition idea, quantitative agreement is usually poor. Movement of weak electrolytes from blood to gut lumen or into the secretion of digestive glands obeys the same pK lipid-water partition rules.

Water and Electrolytes

The intestine absorbs both exogenous fluids and its own secretion; it is therefore necessary to consider both absorption and secretion together. The

Table 10/I. Daily flux of gut water and sodium. Reproduced, with permission, from *Schedl* [53]

	Gut water	
	ml/day	Na, mM/day
Oral intake	2,000	50–100
Saliva	1,000	50
Gastric juice	2,000	100
Pancreatic juice	2,000	200
Bile	1,000	150
Intestine secretion	6,200	840
Sum	14,200	1,440
Transit ileum to colon	1,500	200
Fecal	100–150	5

secretions of the intestine and digestive glands are so great that, were they not absorbed, death from dehydration would result in 24 h (table 10/I). It is customary to speak of fluxes and net movement of water and electrolytes. In the GI tract considered as a whole the net movement is into the bloodstream; if it were not, diarrhea would result. In the duodenum the net movement, on the other hand, is into the gut during digestion, while in the ileum and colon it is very much the other way. The fluxes, particularly in the upper small intestine, may be ten times the net movement, thus even though more fluid and electrolytes enter the duodenum than leave it, nevertheless, more fluid is absorbed from the duodenum and upper jejunum than from any other part of the intestine. A modern way to express this phenomenon is to say that the duodenum and ileum have very leaky epithelia while the ileum, the colon and stomach are much less leaky, i.e. the tight junctions between cells are tighter. The tightest of all are in the stomach which is not an important absorbing surface. Solutions out of chemical equilibrium with plasma can remain so for long periods of time in gastric pouches. In the duodenum, on the other hand, diffusion is free in response to osmotic or concentration gradients and duodenal contents rapidly reach equilibrium with plasma. By the time the gastric contents have passed the short human duodenum they are in osmotic equilibrium with plasma; the concentrations, however, are not identical because the duodenal pH is a little lower (6.0–6.5), which means that the [HCO_3^-] is a little lower and the Cl^- a little higher than in plasma (fig. 10/7), and there may be some osmotically active substances in the gut which cannot be absorbed at all. As seen in the earlier section this pH is meticulously controlled by pancreatic secretion and ensures optimal conditions for the action of pancreatic enzymes.

Not all equally water-soluble ions diffuse freely to and fro across even the duodenal epithelium, some have active transport mechanisms, e.g. Na^+, Fe^{2+}, Ca^{2+}, HCO_3^- and Cl^-; many seem to diffuse, e. g. Li^+ I^-, Br^-, K^+, CNS^-; and some seem to be excluded to various extents. Many divalent and polyvalent ions are almost completely excluded, e. g., Al^{3+}, Mg^{2+}, Ba^{2+}, SO_4^{2-}, but there are exceptions, the water-soluble salts of mercury and cadmium, for example. A possible explanation for this is the size of ion. Small di- or polyvalent ions are densely charged and hold water molecules with the result that the hydrated ions may be very large; Mg^{2+}, SO_4^{2-} may hold 30 molecules of water. Another explanation is that the junctions between cells, i. e. the pores, are themselves charged; it would appear negatively. This would explain why densely charged anions might have difficulty in diffusing passively through the surface epithelial pores. The sulfate ion, for example, does seem to be actively excluded since it is much more readily absorbed from a mucosa poisoned with fluoride or cyanide than from a normal one. This intestinal selectivity of the intestine concerned physiologists in the 1920s and 1930s but since then it has been virtually ignored and attention has been focused almost entirely on the movement of Na^+ and Cl^- [18]. In 1888 *Hofmeister* ranked elements according to their ability to disrupt colloidal solutions. This, in fact, was directly proportional to the propensity for their ions to become hydrated in solution [2]. It did seem to offer an explanation, but the exceptions, Hg^{2+} for example, led to the conclusion that physical or chemical characteristics do not offer any consistent explanation for selectivity in the passive uptake of water soluble electrolytes. This discussion of course assumes water solubility; a completely water insoluble salt will not be absorbed at all. Barium sulfate, for example, is safe to use as an X-ray contrast medium in the GI tract because it is almost completely water insoluble. Barium chloride which is water soluble would not do, because even though the Ba^{2+} is only poorly absorbed, it is highly toxic. The selective ability of the intestine should tell us a lot about it, but unfortunately it is necessary in many instances simply to remember what is and what is not well absorbed.

If a poorly absorbed ion like SO_4^{2-} is taken by mouth as a water soluble salt it will remain in the intestine holding two cations with it, most likely Na^+. These ions will hold water in the intestine to the extent that they are in osmotic equilibrium with plasma and in this fashion they will pass into the large bowel. The unusual distension caused by this osmotically held water will decrease intestinal transit time.

In normal individuals, by the time the mid-jejunum has been reached, virtually all the fat, protein, and carbohydrate will have been absorbed. As has been noted already this absorption is accompanied by large volumes of water. Peptides, amino acids and sugars are all osmotically active and Na^+ absorption is a necessary part of their active transport. These are all absorbed as isomotic and electrically neutral solutions, because Cl^- accompanies the Na^+. In the human upper small intestine this type of Na^+Cl^- and water absorption seems to be almost exclusive since here Na^+Cl^- is not well absorbed from isotonic NaCl solutions. The lower small intestine, the ileum, does absorb both Na^+ and Cl^- even from pure hypotonic solutions. Since the luminal surface is negative to the serosal surface this means active absorption of Na^+ to the extent that it must oppose an electro-chemical gradient to get out of the cells. Obviously the surface epithelial cells are unidirectional, i.e. the sodium pump extrudes sodium across the bases and sides of the cells only, which is where the specific $Na^+ K^+$ ATPase is situated. This does not mean that there is less ATPase in the duodenum, but more likely it means that because of the leakiness of the intercellular tight junctions there, Na^+, Cl^- and water can flow back into the lumen.

It was noted above that the absorption of amino acids and sugars was facilitated by the presence of Na^+ and vice versa. This is not seen to the same extent in the human ileum possibly because the separate absorption of Na^+ is more important and evident in the ileum as back leakage of ions and water does not occur to the same extent here. In addition, Na^+ assisted nonelectrolytes (amino acids and glucose), although absorbed from the ileum, do not leave the lumen nearly as easily as in the duodenum.

Bicarbonate is rapidly absorbed from the upper small intestine, but its equilibrium concentration is lower than in plasma and as expected Cl^- is higher. This indicates an exchange of HCO_3^- out for Cl^- into the lumen. Another possibility, although not as plausible, is that Na^+ may be absorbed in exchange for H^+ with the latter entering the intestinal lumen and generating CO_2 from luminal HCO_3^-. In the ileum, on the other hand, the reaction of the luminal content is more alkaline than plasma. Here there appears to be a net secretion of HCO_3^-. The Cl^- ion seems, in consequence, to be much more rapidly absorbed than HCO_3^-. Chloride is absorbed with Na^+ and in addition Cl^- exchanges for HCO_3^- and since this mucosa is not as leaky as that of the jejunum the net movement is Cl^- out of and HCO_3^- into the lumen. The net movement of HCO_3^- into the lumen depends on the presence of luminal Cl⁻ as one would expect. If Cl⁻ is replaced by SO_4^{2-}, which is not absorbed, then net secretion of HCO_3^- into the ileum ceases and the reaction of the secretion will no longer be alkaline (fig. 10/8).

In most treatises on this subject a great deal is made of the potential difference between the serosal and mucosal surfaces of the cell. The serosa is almost always positive with respect to the mucosal surface. In the usual cell the whole surface is equally positive with respect to the interior because the Na^+ pump mechanism moves the Na^+ and K^+ equally over the whole surface. In the enterocyte by contrast the Na^+ extruding mechanism seems to be confined to the laterobasal or serosal part of the cell, thus although the inside of the cell is still negative to the exterior it is less negative on the mucosal than on the serosal side. Anything which increases the movement of Na^+, e.g. mucosal Na^+, glucose, galactose or l-amino acids, will increase the potential difference across the enterocyte. Na^+ and other positively charged particles have simply to follow the electrical gradient to get into the cell, but they must be pumped up hill against a gradient to get out. The only such known pump is that for Na^+. It seems that actively transported sugars and amino acids are moved in, in combination with Na^+, indirectly by this same pump. The measurement of transmucosal or cellular PD has captured the imagination of physiologists following the work of *Ussing and Zerahn* [62]. It has remained high fashion ever since and would seem to be a secondary phenomenon although the student of the voluminous literature which has followed *Ussing* might be excused for considering it of primary importance. In summary, Na^+ can be moved actively alone by virtue of the laterobasal Na^+ pump or with glucose, galactose or amino acids also by virtue of the pump. These two mechanisms are electrogenic. Cl⁻ always accompanies the Na^+ and water accompanies both the nonelectrolytes and the NaCl. Na^+ can exchange across the enterocyte for H^+; this produces neither electropotential nor net water movement. Chloride can leave the lumen with Na^+, it can leave by exchanging with bicarbonate and it can leak back or be secreted back into the intestine with Na^+.

The absorption of K^+ has not received much consideration. The consensus now is that it is passive following electrochemical gradients into the plasma.

Intestinal Surface Absorption and Digestion

Since pancreatic amylase does not hydrolyse either external 1–4 glycosidic linkages, or points of branching (fig. 7/11) these are the endpoints of starch (amylopectin) or glycogen digestion by amylase. They are produced by pancreatic amylase in almost equal proportions.

Proteins are present as free amino acids to some extent, but mostly as di-, tri-, and tetrapeptides. Of these digestion products only amino acids and monosaccharides are found in portal blood draining the small bowel. Further digestion evidently takes place before absorption.

Intestinal secretion does contain enzyme activity capable of hydrolysing di- and higher saccharides and a variety of peptides. This activity is not free in solution, but associated with the cellular debris. Centrifugation removes all peptidase and saccharidase activity. Based on information of this sort *Ugolev* [61] in Leningrad developed the idea of surface digestion. He was invited to present his work at the International Physiological Congress in Tokyo in 1965, but did not appear. Since then he has virtually disappeared from sight and mind. The latter is unfortunately not uncommon. *Ugolev* presented evidence for the idea that certain enzymes are bound to cell surfaces and hydrolyse only those substrate molecules which come in contact with them.

Carbohydrates [12, 20]

The enzymes concerned in sugar hydrolysis are in what appears, by electron microscopy, to be a filamentous extension from the outer surface of the microvillus called the glycocalyx (fig. 10/3B). This is firmly held to living cells but easily removed from dead ones. These *saccharidases* are all large (MW 200,000) glycoproteins. The large hydrophilic part protrudes from the cell membrane (fig. 10/5). It bears all the enzymic sites and is anchored to the cell and membrane by a small lipophilic base. The enzyme activities present are: *lactase,* which hydrolyses lactose to glucose and galactose; *sucrase,* which hydrolyses sucrose to glucose and fructose; and *maltase,* which hydrolyses maltose to two glucose molecules. The tetra- and penta- 1–6 linked glucoses are cleaved at the 1–6 linkage by an enzyme known either as α-*dextrinase* or as *isomaltase;* glucose molecules are removed sequentially from the nonreducing end of these by maltase and also by sucrase, which is active against both maltose and sucrose. Maltase is often called *glucoamylase* in modern texts, because its activity is not confined to the dissacharide maltose and also because sucrase must also be classified as a maltase. The action of all the saccharidases is quick except for lactase, which may be rate limiting. In many normal adults of nonmilk-drinking ancestry, e.g. Africans, Mediteranean and Asiatic people, lactase activity often becomes so low that unhydrolysed, and, therefore, unabsorbed lactose remains in the intestine. Due to its osmotic activity and to the action of bacteria in such people, milk or milk products will produce diarrhaea and colic.

All of the saccharidases are large glycoproteins with an active half-life much less than that of the cells to which they adhere. In all likelihood, during digestion they fall victim to pancreatic proteases and are replaced from within the cell. In conformity with this idea is the evidence that during digestion the half-life of sucrase is reduced to about 1/10 of that in the fasting intestine.

Of the monosaccharides now in the intestine only glucose and the almost identical galactose are absorbed actively. Fructose, another common monosaccharide is passively absorbed by diffusion. Its problem is simpler than that of glucose, since it is not present in the plasma as glucose is. When glucose concentration is higher at the surface of the surface epithelial cells than in the plasma it can diffuse also, but glucose absorption from the small bowel is almost 100% and must, therefore, be effected at times against a very steep gradient. The postulated mechanism is that a membrane protein acts as a carrier for glucose and galactose. These two sugars seem to have an affinity for the same site since they will compete with one another if presented simultaneously to the surface epithelial cells. Either one, in high concentration, can saturate the mechanism which then reaches a maximum rate. The protein binding site takes up two Na^+ ions with each molecule of either glucose or galactose. In the absence of Na^+ glucose and galactose absorption is greatly reduced and vice versa. The luminal carrier is restricted to the intestinal surface.

Exit from the cell seems to depend in part on an Na^+-independent serosal carrier, which is distinct from the entrance carrier, and also on simple diffusion. Phlorizin, a classical tool for the study of sugar transport for half a century, blocks the entrance carrier, but not the exit one. The absorption of sugars, active or passive, is much higher in the proximal small intestine than in the distal because the surface area is greater: there are more villi and, therefore, more cells and more entrance transporting sites for Na^+ and glucose and/or galactose. Whether or not the exit carrier is energy demanding or simply assisted diffusion down a concentration gradient is unknown. It is argued that the energy of the Na^+ pump fuels the transport of glucose, i. e. the pump expels Na^+ from the base of the cell and more Na^+ at the surface is pulled down the resulting electrochemical gradient to take its place dragging glucose or galactose with it.

The commonest derangement in carbohydrate digestion is lactase deficiency mentioned above. Sucrase-isomaltase deficiency also occurs as a familial disorder in which surface epithelial cells are unable to manufacture the requisite enzyme protein. Since several enzymes have maltase activity

10 Absorption

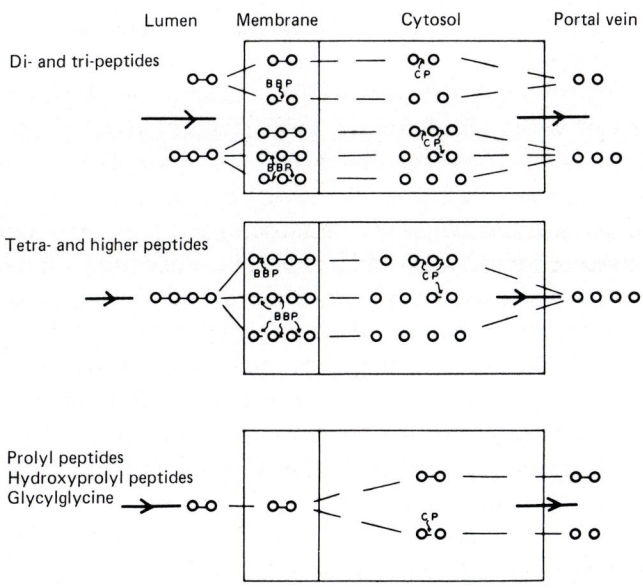

Fig. 10/9. The entry and movement of amino acids and peptides through the enterocyte. BBP = Brush border peptide hydrolase; CP = cytosol peptide hydrolase. Reproduced, with permission, from *Kim* et al. [32].

this deficiency is not seen except in generalized diseases or atrophy affecting the intestinal mucosa in which all saccharidase activity will be diminished.

Peptides [14, 32]

Dietary proteins enter the portal blood as free amino acids, but the products of intraluminal digestion by gastric and pancreatic proteases are di-, tri-, and tetrapeptides and some amino acids. Therefore, on the enterocytes or within them these peptides must be hydrolysed further. There is no other source of peptidases. These cells are rich in peptidase activity, but it is very low in the luminal fluid. Within mature enterocytes peptidase activity is found mainly at the brush border and in the cytosol (fig. 10/9). These two sets of enzymes are quite different proteins. Brush border enzymes have a special affinity for longer peptides (up to 8) but can hydrolyse dipeptides, except those with an N-terminal proline. Those of the cytosol can hydrolyse dipeptides only. Thus it is that the long peptides are hydrolysed on the surface of the cell while dipeptides may gain entrance to the cell as such. Until very recently it was felt that only single amino acids could cross the cell wall.

The cytosol dipeptidases are not specific to enterocytes. Almost identical enzymes can be obtained from a great range of tissue cells. The membrane peptidases are, however, unique to this location. These enzymes are not found in the glycocalyx beyond the cell membrane, but are actually in the membrane. This may explain why it is that peptides for the most part are more rapidly absorbed than single amino acids, i.e. the enzyme protein is not only an enzyme but also a transporter. Some di- and tripeptides can get through to the cytoplasm: those with an N-terminal proline must. These are hydrolysed in the cytosol before leaving the cell as free amino acids. Traces of dipeptides have been reported from time to time in rat portal blood.

All of these enzymes whether in the brush border or cytosol are aminopeptidases, that is in peptides with more than two amino acids they hydrolyse only the N terminal amino acids.

There is evidence, although not as much as for glucose, that the transport of peptides into the enterocyte is facilitated by Na^+ ion and vice versa and again the implication is that the Na^+ pump is the energy source.

Single amino acids are absorbed from the intestinal lumen, the l-isomers actively. Again, Na^+ is necessary for optimal rates and the pump is described as the energy source, just as for glucose.

Four distinct 1-amino acid transport systems are described, for: neutral, dibasic, dicarboxylic and the fourth for proline and hydroxyproline (imino acids) and glycine. These channels do not seem to be the same as those for peptides since in two hereditary disorders, *Hartnup's disease,* and *cystinuria* in which neutral and dibasic amino acids, respectively, are not absorbed; dipeptides containing them are. This is the likely explanation for the puzzling observation that severe protein mulnutrition is uncommon in these patients. Another practical consequence of this new data is that in preparing elemental diets (diets in a form which can be absorbed promptly without further luminal digestion) solutions containing peptides should be much more effective in maintaining protein nutrition than single amino acids. Partially hydrolized proteins have the added advantage of being much less expensive than solutions of single 1-amino acids.

Nitrogen Balance. The adequacy of protein nutrition can be gauged by nitrogen balance. A healthy adult will excrete as much nitrogen per day in his urine, on average, as he consumes in digestable from in his diet. A growing animal will excrete less in its urine because it is forming new body protein and is, therefore, in positive nitrogen balance. An animal receiving insufficient dietary protein will utilize body protein to manufacture essential body proteins such as pancreatic and other metabolic enzymes. This animal will excrete more nitrogen than it eats and will, therefore, be in negative nitrogen balance.

In order to remain in nitrogen balance it is not enough for the healthy adult to consume sufficient protein, but the protein must be of adequate quality; that is, the mixture of proteins eaten must together supply sufficient of the amino acids that the organism cannot make for itself *(essential amino acids)* and moreover these proteins must be sufficiently digestable so that the essential amino acids enter the portal blood simultaneously in the correct proportions. The correct proportions for man are those found in human tissue. The proportions found in hen's egg albumin are close and are used as a reference standard. If a disproportionate surge of a single amino acid or group of amino acids enters the portal blood negative nitrogen balance will result. If the deficit is made good a few hours later by a compensatory surge of the deficient amino acids this will simply accentuate the negative balance. Most cereals are deficient in either lysine, tryptophane or both. This deficit cannot be repaired by adding the calculated compensatory amount of crystalline amino acid to the diet since the added amino acid will be absorbed more quickly than the constituent acids of the cereal protein and produce just such a surge as mentioned above. A deficit of this sort can be corrected by adding a protein to the diet which contains the missing acids.

Antigens and Antibodies. The foregoing discussion assumed that all protein is digested in the intestine, becoming victim in turn to pepsin pancreatic proteases and, finally, to brush border peptidases, but despite this and the selective permeability of the surface epithelial cells which excludes large peptides, it is clear that large immunologically active proteins can cross from gut lumen to blood. There is clear immunological evidence that the infant rodent and ruminant absorb globulin from colostrum and milk. This is effected by pinocytosis but only in infant animals. In the rat within 20 days postpartum it ceases.

In the human fetus passive immunity is gained in utero. Human colostrum and milk contain IgA, which functions only on epithelial surfaces. It is quite likely that were γ-globulin present, the human infant intestine could absorb it. It has been suggested that some passive immunity is acquired through intestinal absorption or γ-globulin from swallowed amniotic fluid.

There is excellent evidence for the presence of antibodies to food proteins in the plasma of human infants. Antibodies to bovine serum albumin fed children over a year old cannot be detected in their serum, but can be in those less than 3 months old. Between these two ages the phenomenon known as 'closure' has occurred, but even in the adult there is good immunological evidence for the absorption of intact antigenic proteins. Special cells *(M cells)* overlying Peyer's patches are said to permit the entry of antigenic protein to the antigen manufacturing lymphoid tissue of the patch but not to the bloodstream (free absorption of such proteins could produce extensive allergy). This mechanism, however, appears to be protective in that intestinal secretions as a result contain *IgA* (immunoglobulin A) which will precipitate antigenic proteins and prevent adherence of bacteria and viruses to the gut surface. There is very good evidence that in celiac disease a deficiency of IgA permits the absorption of the protein gluten from wheat, allergy to which is responsible for the disorder. It is incredible that intact proteins can reach the intestinal surface epithelial cells, but they evidently do. Within the cell itself these large molecules are subject to the action of lysosomal proteases, but even so, from time to time enough of them escape into the bloodstream in some people to constitute a hazard.

Fats [6, 55, 58]

Lipases are enzymes that have a preference for the digestion of water insoluble substances. They do not act well on molecules in solution, but at interfaces. Most dietary fats are water insoluble, forming droplets or films in water. Lipases will bind to and act at the surfaces of these droplets. If agents are employed which solubilize droplets the activity of lipase falls. *Pancreatic lipase* is a typical lipase. For many years it was felt that bile salts had a detergent action which resulted in an enhancement of the activity of lipase. This does happen if lipase activity is studied in pancreatic juice. The surface activity of the bile salts brings about a reduction in the droplet size which increases the surface area available for lipase combination, but if pure preparations of pancreatic lipase are used, bile salts actually reduce lipase activity because they displace the enzymes from the interface. Both bile salts and lipase are negatively charged molecules and repel one another. Pancreatic juice contains a protein *colipase* which effects the binding of lipase to its substrate even in the presence of bile salt. A lipase, colipase, bile salt complex actually binds to the interface. Even without bile salts, colipase

Fig. 10/10. The sites of action of pancreatic lipase and phospholipase. **A** Typical triglyceride. **B** Lecithin, a typical phospholipid. Phospholipase hydrolyzes the ester bond in position 2 to form lysolecithin. R_2 is usually an unsaturated fatty acid; in lecithin of human origin it is most often linoleic acid.

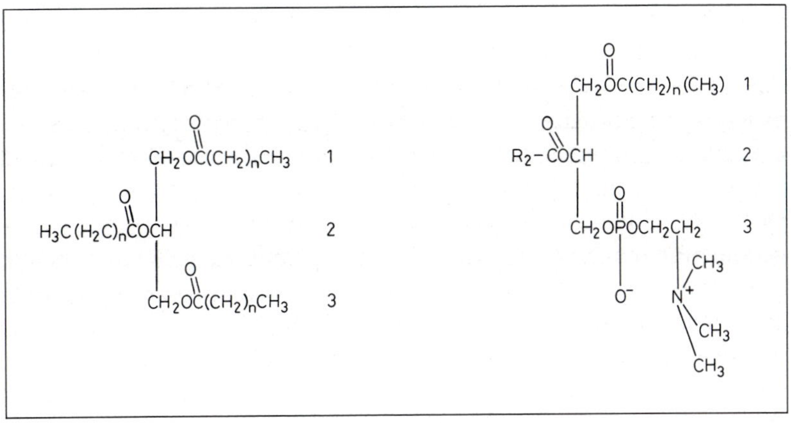

increases lipase activity by promoting substrate binding. It does not increase the intrinsic activity of the enzyme, however.

Colipase is a polypeptide with a molecular weight of about 10,000. It binds to lipase at a pH of 6, which is approximately that of the duodenal lumen. Lipase, by itself, is maximally active at pH9. Since all digestive enzymes are proteins they are subject to proteolytic digestion by pancreatic proteases; however, colipase confers considerable resistance on lipase once they have formed a complex.

A notable characteristic of pancreatic lipase is that it is a *primary ester hydrolase*. This means that the products of its action on going to completion are free fatty acids and 2-monoglycerides, i.e. it does not hydrolyse the middle ester linkage in a triglyceride (fig. 10/10). In vitro, lipase activity, even with optimal concentrations of bile salt and colipase, does not go to completion, but reaches equilibrium since it promotes both hydrolysis and esterification. The velocity of the reactions depends on the surface concentration of lipase and the accessibility of the substrate.

Lipid Classification. Lipids have been classified by *Small* [56] into: (a) *nonpolar,* which do not concern us much here, since they are rare in the diet and unabsorbable because they are completely insoluble in the aqueous chyme – their molecular attraction is so great that water molecules are forced out and they form drops or films in water, and (b) *polar lipids* – triglyceride, diglyceride, cholesterol and long chain fatty acids belong to this class. These are strongly lipophilic and weakly hydrophilic. Molecules of this type will orient themselves such that the hydrophilic parts of the molecules form the boundary between lipid and water. Water is unable to penetrate between the molecules making up such an interface. The bulk lipid phase, either the drop or layer is an oil, the molecules within it existing without any regular orientation. Less polar lipids such as those mentioned above can exist in solution within this oil phase as can the biologically important relatively nonpolar lipid cholesterol. Still within this group are those which are more polar than the above. Amongst these are monoglycerides and lecithins. These are called *swelling insoluble amphophiles,* i.e. they have strong hydrophilic and lipophilic groups in the same molecule. They are swelling because they arrange themselves in bilayers in water, hydrophilic out and lipophilic in, with water between adjacent hydrophilic ends. It is the water between which causes the swelling. Monoglycerides for example have two hydrophilic hydroxyl groups (see fig. 10/10) and an ester linkage plus the lipophilic paraffin fatty acid tail. The latter form the interior of the lamellae while the former the surface (the exterior). Some of the nonswelling polar lipids mentioned above, such as cholesterol, are able to dissolve within the lamellae of these swelling amphophiles. It will be understood that monoglycerides and lecithins are not polar enough to exist in simple monomolecular solution in water except in trace concentrations.

Micelles. Still more polar lipids are those such as bile salts, soaps of long chain fatty acids and lecithins which have had a fatty acid removed by lecithinase leaving a free hydroxyl *(lysolecithins)* (see fig. 10/10). These are capable of existence in substantial concentrations in monomolecular solution. When concentrations of these molecules in solution increase to critical levels then their amphophilic properties become evident in that they form aggregates, again with their hydrophilic heads outwards and their lipophilic tails inwards. The concentration at which this happens is called the *critical micellar concentration.* It varies from compound to compound. These micelles are not static, but exist in equilibrium with the molecules in solution. The bile molecules are flat plates with one side polar and the other not, thus they arrange themselves as plates rather than head and tail. In simple solution the lipophilic surface of bile salts will complex with fats or oils which may be present. They will cover such interfaces lowering surface tension and, in the process, allowing lipids to break up into small droplets. They will also complex with the insoluble swelling amphophiles (e.g. phospholipids, monoglycerides) mentioned above. This greatly increases their detergent properties. Aggregates of this sort are so small that they appear to be water soluble, i.e. they transmit light without interference. These aggregates are themselves amphophiles and as such are able to aggregate further with insoluble lipids of the first group of very slightly hydrophilic lipids (tryglycerides, diglycerides and cholesterol).

The above properties apply only to ionized bile acids. Free bile acids in acid solution (nonionized) are highly insoluble in water. The glycine and taurine conjugates are almost completely ionized in the slightly acid pH of the upper duodenum. Free acids if they occurred physiologically would not be as ionized as their pKa is about 6 as opposed to 2–4 for the conjugates.

When fat enters the duodenum from the stomach, in all likelihood as oil droplets, it encounters bile salt, colipase and pancreatic lipase, as well as the phospholipids and cholesterol which may be in the food and in the bile and intestinal secretions. Thus, it immediately encounters an excess of amphophiles which will reduce surface tension and produce an emulsion. More amphophiles are generated by the action of lipase, which is fixed to the surface of these droplets by its colipase, because its action is the generation of monoglycerides and free long chain fatty acids, which at the pH of the duodenum are likely to be present as soaps. The consequence of this is that detergent action increases and droplets become smaller eventually reaching the proportions of micelles. These micelles will contain all the lipids except those fatty acids short enough to be water soluble. In vitro, this action of lipase eventually reaches equilibrium. This does not happen in the gut as the products of digestion are absorbed.

Micellar Absorption. This is the extent of lipolysis and now we are left with the problem of getting the micelles across the surface epithelial cell. Modern wisdom has interposed a new barrier between the lumen of the gut and the wall of the epithelial cells. This is the unstirred water layer which is said to exist against the walls of any pipe through which liquid is flowing. On the face of it, the formation of micelles would seem to be a hindrance rather than a help since large particles will not diffuse as readily as will small ones through an unstirred layer and only by pinocytosis or phagocytosis can intact particles gain access to cells. Lipid aggregates have never been seen crossing epithelial cell walls by any sort of microscopy so far employed.

Droplets of fat are seen in smooth endoplasmic reticulum vesicles and cisterns. None are seen near the brush border or beneath it, suggesting that the products of fat digestion are absorbed in monomolecular solution. That the micelles must break up prior to absorption is evidenced by the fact that the bile salts are not absorbed with fat. However, absorption of dietary lipids is much more rapid from micellar than from simple monomolecular solution. A reasonable explanation for this is that equilibrium between intestinal fat and monomolecular solution is more easily and more rapidly obtained with micelles than from any other lipid aggregate, i.e. if the monomolecular concentration is reduced by absorption it will rapidly be made up again by disintegration of micelles which maintain the concentration gradient and increase the diffusion flux. Another even more reasonable possibility is that perhaps there is no unstirred layer barrier at all and that, therefore, the micelles break up in contact with the lipid membranes of the enterocyte thus avoiding the problem of transport through water in solution.

Despite lack of clear information concerning the movement of fats into the epithelial cells and the role of micelles in this it is absolutely clear that micelle formation is crucial to the absorption of lipids and that in vivo pancreatic lipase is normally present in great excess. Virtually all meal fat has been absorbed by the mid-jejunum, but even though the upper small intestine is most active, the lower small intestine can absorb fat and therefore constitutes a rather large reserve.

Intracellular Mechanisms. The fat mixture it seems enters the enterocyte in monomolecular solution. By contrast it is clear that fat leaves these cells in particulate form as low density lipoproteins called *chylomicrons*. All the fragments of triglyceride formed by the action of lipase in the gut (fatty acid mono and diglycerides) are reesterified to triglyceride, except for water soluble short chain acids. Short chain acids are not acted on by the necessary intracellular enzymes or perhaps they do not form the necessary fatty acid Co A complex and as a result pass into the portal blood with remarkable ease and rapidity as free acids associated with albumin (fig. 10/11).

The enterocyte can match the free fatty acids with the number of glycerol hydroxyl groups available in partial glycerides both by manufacturing glycerol and by hydrolysing the ester bonds in partial or triglycerides. The same is true of phospholipids. Cholesterol, which is the substrate for a pancreatic esterase, is also reesterified within the enterocyte and so enters the chylomicron.

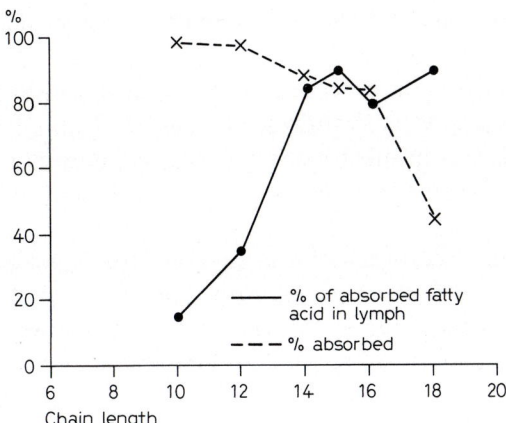

Fig.10/11. The influence of chain length on both the percentage of fat absorbed and its route in 18h following administration. Reproduced, with permission, from *Bloom* et al. [5].

The polar lipids of the chylomicron (phospholipids) coat the surface of the droplet. To this, in the region of the Golgi apparatus, is added a glycoprotein coat. Normal fat transport requires an adequate supply of phospholipid from either the diet or by synthesis in the enterocyte. The latter is diminished in animals deficient in essential fatty acids and in these droplets accumulate within the enterocyte and discharge is impaired. Droplets once discharged into the lateral spaces are taken up by lymphatics after they have penetrated gaps in the basement membrane. All stages of the droplet phase of fat transport within the cell to the lymphatics have been observed microscopically; how they enter lymphatics has not been investigated, but the lymphatic vessels almost everywhere, but especially in the abdominal cavity, pick up large molecules and even microscopic particles, presumably through pores or leaky junctions between cells. Lymph flow increases during fat digestion.

Fecal Fat. Not all the lipid that enters the intestine is dietary; some of it originates in desquamated epithelial cells and from bacteria in the lower intestine. Cellular lipids high up in the intestine are treated just as those of dietary origin, but lower down cellular and bacterial lipids will pass into the colon with those of dietary origin that are indigestible or unabsorbable, e.g. high melting point fats, waxes and plant sterols. On very high fat diets fecal lipids and especially sterols and bile acids do increase even in normal animals.

Normal human fecal fat is of the order of 5 g/day. The fat of endogenous cellular origin has been estimated as 30 g/day in man, of which only 10% is unabsorbed. Fecal fat, whether of dietary origin or not, is radically transformed by the fecal flora. About 50% of it is present as calcium or magnesium soaps.

If the fat content of the feces reaches 50 g/day then *steatorrhea* (fatty diarrhea) is the result. This will affect people with short small intestines, defects in the surface epithelial cells or the villi, defects in the lacteals, reduced concentrations or abnormal bile salts in the intestine or reduced pancreatic lipase.

In addition to lipase, pancreatic juice contains a *cholesterol esterase* and a *phospholipase*. The former may be identical with the nonspecific esterase known to be present in the pancreas. It will hydrolyse water soluble esters as well as cholesterol esters which pancreatic lipase will not do. Bile salt is necessary for its activity, since it will attack cholesterol esters only in micellar form. Bile salts also will protect it from proteolytic digestion. Phospholipase, unlike the other lipolytic enzymes, is stored in and secreted from the pancreas as an inactive precursor. It is activated by trypsin in the duodenum. Once again bile salts are both needed for its action because it is confined to micellar phospholipid and also to protect it from proteolysis. It hydrolyses the 2 ester bond in phospholipid forming acyl lysolecithin and a molecule of free fatty acid (see fig. 10/10). This enzyme is the major phospholipase in the gut.

All of the fat soluble vitamins are highly nonpolar and are absorbed by solubilization in lipid micelles. They leave the enterocytes incorporated in chylomicrons. They are all rather poorly absorbed, no doubt as a consequence of their low solubility.

Bile Salts

Passive absorption of bile salts takes place at all levels of the intestine, but only in the terminal ileum is there an active carrier mechanism. Passive diffusion is diminished by conjugation and accelerated by removal of hydroxyl groups from the sterol nucleus.

Passive absorption of bile salt from the small bowel is substantial in some experimental animals (e.g. 50% in rats). It is the only means of absorption from the colon, but since here bacterial action has deconjugated the salts diffusion is facilitated.

Active ileal absorption is most rapid for conjugated trihydroxy salts (glyco-or taurocholate in man) and less for conjugated dehydroxy salts which in turn exceed all unconjugated acids

Vitamins

Fat-soluble vitamins are absorbed from solution in the fat micelles discussed above. Water soluble vitamins present individual problems; of them, most is known about vitamin B_{12} because of its unique mechanism and the occurrence of pernicious anemia when it fails. Vitamin B_{12} is a large highly polar molecule (1,355 daltons) present in the diet only in microgram quantities. Vitamin B_{12}, after liberation from protein by gastric hydrochloric acid, combines with *intrinsic factor,* a gastric glycoprotein. Intrinsic factor, referred to above under gastric mucosaccharides, is secreted by the parietal cells of the acid and pepsin secreting gastric mucosa. The B_{12} intrinsic factor complex resists intestinal digestion and passes intact down to the terminal ileum, its site of absorption. Another glycoprotein called *R protein* avidly binds B_{12} in the stomach but is digested off in the duodenum, where combination with intrinsic factor can then take place. The R protein complex is not absorbed. This may explain impaired B_{12} absorption in pancreatic insufficiency. The B_{12} intrinsic factor complex combines with surface epithelial cell receptors on a specific carrier protein which evidently recognizes the intrinsic factor [16], as even enormous doses of free B_{12} are virtually unabsorbed. Binding requires calcium ions and a pH above 5.5, both of which conditions are met in the ileum. Even with intrinsic factor absorption is slow and usually incomplete. From a large dose, 10 µg for example, only 1.5 µg is absorbed followed by a mucosal block which persists until the vitamin has traversed the intestinal mucosa and reached the blood (4–6 h). The maximal daily absorption capacity is about 6 µg. Intrinsic factor does not enter the blood stream, but B_{12} circulates there in combination with a specific plasma protein, *transcobalamin II* and a nonspecific one, *transcobalamin I,* which is the R protein of the gut.

Of the other water soluble vitamins, folate presents special difficulties in that it appears in portal blood as pteroylmonoglutamate, but in the diet 95% of it is present as a polyglutamate (usually a chain of seven glutamyl residues). Neither pancreatic proteases nor brush border enzymes are capable of hydrolysing this polyglutamate chain, but an intracellular lysozomal *pteroylpolyglutumate hydrolase* is. The puzzle is in how the large unhydrolyzed molecule gains entry to the cell. The upper jejunum seems to be the main site for folate absorption in man; folate, ascorbic acid, thiamine and riboflavin seems to have saturable transport mechanisms. Ascorbic acid is unique in that its maximum is huge (5g/day). Ascorbic acid is absorbed with sodium by a mechanism which seems to be identical with that of glucose. Pyridoxine seems to be absorbed by simple diffusion and almost nothing is known about nicotinic acid.

Calcium and Iron

The absorption of everything so far considered is determined almost exclusively by the amount eaten; for the two elements Ca^{2+} and Fe^{2+} the homeostatic needs of the body are a decisive factor. Surface epithelial cells can transport calcium against an electrochemical gradient from lumen to blood. The mechanism is saturable, which indicates a carrier dependent on Vitamin D.

Calcium. In vitro intestines from a vitamin D-deficient animal do not absorb calcium well. Addition of vitamin D to the preparation is without effect; however, four hours after administration of the vitamin to deficient animals, absorption by in vitro intestine is increased in a dose-dependent manner. This follows metabolism of vitamin D by the liver and then by the kidney to 1,25-dehydroxycholecalciferol followed by the manufacture of calcium binding protein in the enterocytes. Parathormone injections increase calcium absorption via the vitamin D mechanism. In calcium deficiency, on an adequate intake of vitamin D, the efficiency of calcium absorption increases markedly because the kidney increases its production of dehydroxycholecalciferol. As might be expected calcium absorption is reduced in kidney disease.

Calcium forms many insoluble salts with compounds that can occur in the normal diet. These will impair its absorption. Examples are phosphates and phytates, which are found in wheat bran and high extraction flour. Calcium is much more likely to be in soluble form at an acid pH than at an alkaline one. It is, therefore, much more readily absorbed from the duodenum than from the ileum.

Iron. Iron, like calcium, is actively transported. Twenty years ago there was much more certainty about iron absorption than there is now. Iron in the diet occurs as *heme iron* in hemoglobin and myoglobin and as other organic forms such as in chlorophyl and as inorganic iron. Much of the organic iron, under the action of gastric acid and reductases in food, assumes the form of ferrous chloride, but substantial quantities of hemoglobin iron appear to reach the small intestine. Pancreatic proteases split off the globin. The heme portion enters the enterocyte intact, where the iron is removed from it, eventually to leave and be picked up as ionic iron by *transferrin,* the plasma iron-transporting protein. Ionic iron is taken up by the brush border by, it is assumed, an iron transporting protein. Some iron enters the blood within an hour or two but some enters a slowly turning over compartment where it is incorporated into ferritin. Absorption of iron varies inversely with the size of the iron stores and directly as the rate of hemopoiesis.

If large oral doses are given mucosal block occurs. This is of therapeutic importance since, even in the iron depleted, it is impossible to get large and rapid absorption by the oral route. Mucosal block is not complete and so continual high oral intake results in excessive absorption and the pathological condition called *hemochromatosis*; in this, iron is deposited in excessive quantities in liver, skin (thus giving it a bronze color) and pancreas, which can result in diabetes.

Low luminal pH favors iron absorption. As the pH rises, iron complexes with other ions and compounds and the equilibrium moves towards the ferric form which is much less water soluble than the ferrous and which above pH 7 will precipitate. Reducing agents in the chyme (ascorbic acid is the most important) tilt the equilibrium in favor of the ferrous form. Iron forms insoluble phosphates and phytates just as does calcium and these will impair absorption. These accelerating and depressing factors do not influence the absorption of heme iron which, in consequence, is much better abosrbed than iron of plant origin.

Absorptive Adaptation

If the intestine is shortened, the remaining gut undergoes hyperplasia. The villi become longer and the crypts deeper. The total number of cells per villus increases, the production of enterocyte enzymes increases and the renewal rate of enterocytes is accelerated. In vivo, it is evident that the rate of absorption per unit area of mucosa is increased. The muscle of the intestine undergoes hypertrophy. This adaptation accounts, in part, for the decline in weight loss with time in people who have been subjected to jejeunoileal bypass for malignant obesity. The bypassed segment in this operation resembles the atrophied entire small bowel seen in starvation. The walls become thin and the enterocyte count is reduced, the villi become small and short, but do not disappear and the renewal rate is greatly retarded; surface enzyme production is reduced. Enthusiastic refeeding of victims of prolonged starvation will cause malabsorptive diarrhea and may even cause rupture of the frail intestine.

The mechanism by which the intestinal mucosa adapts is probably both nutritional and in response to movement. All muscle undergoes atrophy when under-used. In the empty intestine the stimulus to activity is reduced. The absorbing mucosa derives much of its energy from absorbed nutrients in transit. This is obvious in the case of glucose.

References

1 Altmann, G.; Enesco, M.: Cell number as a measure of distribution and renewal of epithelial cells in the small intestine of growing and adult rats. Am. J. Anat. *120:* 319 (1967).
2 Bayliss, W.: Principles of general physiology, chap. 7 (Longman Green, London 1915).
3 Bernard, C. (1859): Quoted by Gregory, R. [21]; thesis, St. Petersburg (1904).
4 Boldyreff, W.: Fermentforschung *7:* 1565 (1928): Quoted by Gregory [22].
5 Bloom, B.; Chaikoff, I.; Reinhardt, W.: Intestinal lymph as a pathway for the transport of absorbed fatty acids of different lengths. Am. J. Physiol. *166:* 451–462 (1951).
6 Borgstrom, B.: Digestion and absorption of lipids; in Crane, International review of physiology. Gastrointestinal physiology, vol. 12 (University Park Press, Baltimore 1977).
7 Casley-Smith, J.: The identification of chylomicra and lipoproteins in tissue sections and their passage into jejunal lacteals. J. Cell Biol. *15:* 259–277 (1962).
8 Cheng, H.; Merzel, J.; Leblond, C.: Renewal of Paneth cells in the small intestine of the mouse. Am. J. Anat. *126:* 507–526 (1969).
9 Clementi, F.; Paladi, G.: Intestinal capillaries. I. Permeability to perioxidase and ferritin. J. Cell Biol. *44:* 33–58 (1969).
10 Cooke, A. The glands of Brunner, chap. 61; in Code, Handbook of physiology, sec. 6. Alimentary canal, vol. II (American Physiological Society, Washington 1967).
11 Creamer, B.: Paneth-cell function. Lancet *1:* 314–316 (1967).
12 Dawson, A.: The absorption of disaccharides; in Card, Cramer, Modern trends in gastroenterology; 45th ed., pp. 105–124 (Butterworths, London 1970).
13 Dahlqvist, A.: Rat intestinal dextrinase. Localization and relation to the other carbohydrases of the digestive tract. Biochem. J. *86:* 72–76 (1963).
14 Desnuelle, P.; Figarella, G.: Biochemistry; in Howat, Sarles, The exocrine pancreas (Saunders, London 1979).
15 Dobbins, W., III: The intestinal mucosal lymphatic in man. A light and electron microscopic study. Gastroenterology *51:* 994–1003 (1966).
16 Donaldson, R., Jr.; MacKenzie, I.; Trier, J.: Intrinsic factor-mediated attachment of vitamin B_{12} to brush borders and microvillus membranes of hamster intestine. J. clin. Invest. *46:* 1215–1228 (1967).

17 Erlandsen, S.; Rodning, C.; Montero, C.; Parsons, J.; Lewis, E.; Wilson, I.: Immunocytochemical identification and localization of immunoglobulin A within Paneth cells of the rat small intestine. J. Histochem. Cytochem. *24:* 1085–1092 (1976).
18 Goldsmith, G.: On the mechanisms of absorption from the intestine. Physiol. Rev. *1:* 421–453 (1921).
19 Granger, B.; Baker, R.: Electron microscope investigation of the striated border of intestinal epithelium. Anat. Rec. *107:* 423–441 (1950).
20 Gray, G.: Carbohydrate absorption and malabsorption; in Johnson, Physiology of the gastrointestinal tract, chap. 62 (Raven Press, New York 1981).
21 Gregory, R.: Secretory mechanisms of the gastrointestinal tract, III: The small intestine (Arnold, London 1962).
22 Gregory, R.: Nervous pathways of intestinal reflexes associated with nausea and vomiting. J. Physiol. *106:* 95 (1947).
23 Ham, A.: Histology; 7th ed. (Lippincott, Philadelphia 1974).
24 Hanson, W.; Osborne, J.; Sharp, J.: Compensation by the residual intestine after intestinal resection in the rat. Gastroenterology *72:* 692–705 (1977).
25 Harding, J.; Cairnie, A.: Changes in intestinal cell kinetics in the small intestine of lactating mice. Cell Tiss. Kinet. *8:* 135–144 (1975).
26 Hauri, H.; Green, J.: The identification of rat intestinal membrane enzymes after electrophoresis on polyacrylamide gels containing sodium dodecyl sulphate. Biochem. J. *174:* 61–66 (1978).
27 Hogben, C.; Schanber, L.; Tocco, D.; Brodie, B.: Absorption of drugs from the stomach. II. The human. J. Pharmac. Conf. Ther. *120:* 540–545 (1957).
28 Hogben, C.; Tocco, D.; Brodie, B.: On the mechanism of intestinal absorption of drugs. Pharmacol. exp. Ther. *125:* 275–282 (1959).
29 Holdsworth, C.; Sladen, G.: Absorption from the stomach and small intestine; in Duthie, Wormsley, Scientific basis of gastroenterology, chap. 12 (Churchill-Livingstone, Edinburgh 1979).
30 Jackson, M.: Absorption and secretion of weak electrolytes in the gastrointestinal tract; in Johnson, Physiology of the gastrointestinal tract, chap. 49 (Raven Press, New York 1981).
31 Junqueira, L.; Carneiro, J.: Basic histology; 3rd ed. (Lange Medical Publication, Los Altos 1980).
32 Kim, Y.; Nicholson, I.; Curtis, I.: Intestinal peptide hydrolase. Peptide and amino acid absorption. Med. Clins N. Am. *51:* 1397–1412 (1976).
33 Kokas, F.; Ludany, G.: Über das Villkinin. Pflügers Arch. ges. Physiol. *243:* 589–593 (1934).
34 Kurz, H.: Principles of drug absorption in IEPT; in Forth, Rummel, Pharmacology of intestinal absorption: gastrointestinal absorption of drugs, vol. 1, pp. 245–296 (Pergamon Press, Oxford 1975).
35 Lake, A.; Bloch, K.; Sinclair, K.; Walker, W.: Anaphylactic release of intestinal goblet cell mucus. Immunology *39:* 173–178 (1980).
36 Landon, J.; Bradsher, P.; Dragsted, L.: Experimental studies on the secretions of the isolated duodenum. Archs. Surg. *71:* 727 (1955).
37 Leblond, C.; Stevens, C.: The constant renewal of the intestinal epithelium in the albino rat. Anat. Rec. *100:* 357 (1948).

38   Leblond, C.; Messier, B.: Renewal of chief cells and goblet cells in the small intestine as shown by radioautography after injection of thymidine-H into mice. Anat. Rec. *132:* 247 (1958).
39   Lichtenberger, L.; Trier, J.: Changes in gastrin levels, food intake and duodenal mucosal growth during lactation. Am. J. Physiol. *237:* E98–105 (1979).
40   Linderstrom-Lang, K.: Distribution of enzymes in tissues and cells. Harvey Lect. *34:* 214 (1939).
41   MacKenzie, I.; Donaldson, R., Jr.: Vitamin B_{12} absorption and the intestinal cell surface. Fed. Proc. *28:* 41–45 (1969).
42   McMinn, R.; Mitchell, J.: The formation of villi following artificial lesions of the mucosa in the small intestine of the cat. J. Anat. *88:* 99–104 (1945).
43   Messier, B.; Leblond, C.: Cell proliferation and migration as revealed by radioautography after injection of thymidine-H into male rats and mice. Am. J. Anat. *106:* 247–285 (1960).
44   Moog, F.: The lining of the small intestine. Scient. Am. *245:* :154–179 (1981).
45   Moon, H.; Whipp, S.; Baetz, A.: Comparative effects of enterotoxins from *Escherichia coli* and *Vibrio cholerae* on rabbit and swine small intestine. Lab. Invest. *25:* 133–140 (1971).
46   Nasset, E.; Pierce, H.: On the influence of peptones and certain extracts of small intestine on the secretion of succus entericus. Am. J. Physiol. *113:* 568 (1935).
47   Neutra, M.; Padykula, H.: The gastrointestinal tract; in Weiss, Histology: cell and tissue biology; 5th ed., chap. 19, pp. 658–706 (Elsevier Biomedical, New York 1983).
48   Peeters, T.; Vantrappen, G.: The Paneth cell: a source of intestinal lysozyme. Gut *16:* 553–558 (1975).
49   Rodnig, C.; Wilson, I.; Erlandsen: Immunoglobulins within human small-intestinal Paneth cells. Lancet *i:* 984–986 (1976).
50   Rubin, C.: Electron microscopic studies of triglyceride absorption in man. Gastroenterology *50:* 65–77 (1966).
51   Sandstrom, B.: A contribution to the concept of brush border function. Observations in intestinal epithelium in tissue culture. Cytobiologie *3:* 293–297 (1971).
52   Schanker, L.: Passage of drugs across body membranes. Pharmac. Rev. *14:* 501–530 (1962).
53   Schedl, H.: Water and electrolyte transport: clinical aspects. Med. Clins. N. Am. *58:* 1429–1448 (1974).
54   Schwalbe, G.: Beiträge zur Kenntnis der Drüsen in den Darmwandungen, insbesondere der brunnerschen Drüsen. Arch. Mikroskop. Anat. *8:* 92–140 (1872).
55   Simmonds, W.: Absorption of lipids, chap. 10; in Jacobson, Shanbour, MTP international review of science. Gastrointestinal physiology, ser. 1, vol. 4 (University Park Press, Baltimore 1974).
56   Small, D.: A classification of biologic lipids based upon their interaction in aqueous systems. J. Am. Oil Chem. Soc. *45:* 108 (1968).
57   Sullivan, M.; Hulse, E.; Mole, R.: The mucus-depleting action of bile in the small intestine of the irradiated rat. Br. J. exp. Path. *46:* 235–244 (1965).
58   Thomson, A.; Dietschy, J.: Intestinal lipid absorption. Major extracellular and intracellular events; in Johnson, Physiology of the gastrointestinal tract (Raven Press, New York 1981).

59 Thuneberg, L.; Rastgaard, J.: Motility of microvilli. A film demonstration. J. ultrastruct. Res. *29:* 578 (1969).
60 Trier, J.; Madara, J.: Functional morphology of the mucosa of the small intestine; in Johnson, Physiology of the gastrointestinal tract, chap. 35, pp. 925–961 (Raven Press, New York 1981).
61 Ugolev, A.: Membrane (contact) digestion. Physiol. Rev. *45:* 555–595 (1965).
62 Ussing, H.; Zerahn, K.: Active transport of sodium as the source of electric current in the short-circuited isolated frog skin. Acta physiol. scand. *23:* 110–128 (1951).
63 Van Genderen, H.; Engel, C.: On the distribution of some enzymes in the duodenum and the ileum of the rat. Enzymologia *5:* 71–80 (1938/39).
64 Verzar, F.; McDougall, J.: Absorption from the small intestine (Longmans, Green, London 1936).
65 Williams, P.; Warwick, R.: Gray's anatomy; 36th British ed. (Saunders, Philadelphia 1980).
66 Williamson, R.: Intestinal adaptation. Structural, functional and cytokinetic changes. New Engl. J. Med. *298:* 1393–1402 (1978).

11 The Colon and Defecation

The Large Intestine

General Structure and Function

Beginning at the ileocecal valve, the large intestine consists of an initial blind pouch, the *cecum* (and its small terminal extension, the *vermiform appendix*); the *ascending, transverse, descending* and *sigmoid colon*; and the *rectum*, ending at the external anal orifice (fig. 11/1A). It is usually some 1.5–1.8 m in length in the adult human, its length not exceeding one-fourth that of the small intestine. Its caliber is greatest at its commencement (7.5 cm) and diminishes gradually as it courses distally until it reaches a minimum (2.5 cm) at the rectosigmoid junction. The caliber then increases to form the *rectal ampulla,* a dilatation which narrows abruptly at the tonically contracted anal canal. Throughout its caliber is greater than that of the small intestine, but it is capable of great increase in circumference by distension.

Once again, this portion of the gut largely conforms with the 'General Structure of the GI Canal' (see corresponding paragraph, chapter 1), except for a few modifications related to the functional demands placed upon it, i.e.

Fig. 11/1. The structure of the large intestine. **A** Gross anatomy and major regions of the colon. **B** Macroscopic anatomy of the colonic wall. **C** Microscopic anatomy of the mucosa. Note the solitary lymphatic follicle occupying lamina propria, muscularis mucosa and submucosa on the lower right side. **D** Schematic diagrams of the ultrastructure of cells of the epithelial lining. The cells illustrated in white in **C** which are not represented in **D** are undifferentiated epithelial cells (see text for details). Reproduced, with permission, from *Williams and Warwick* [84].

11 The Colon and Defecation

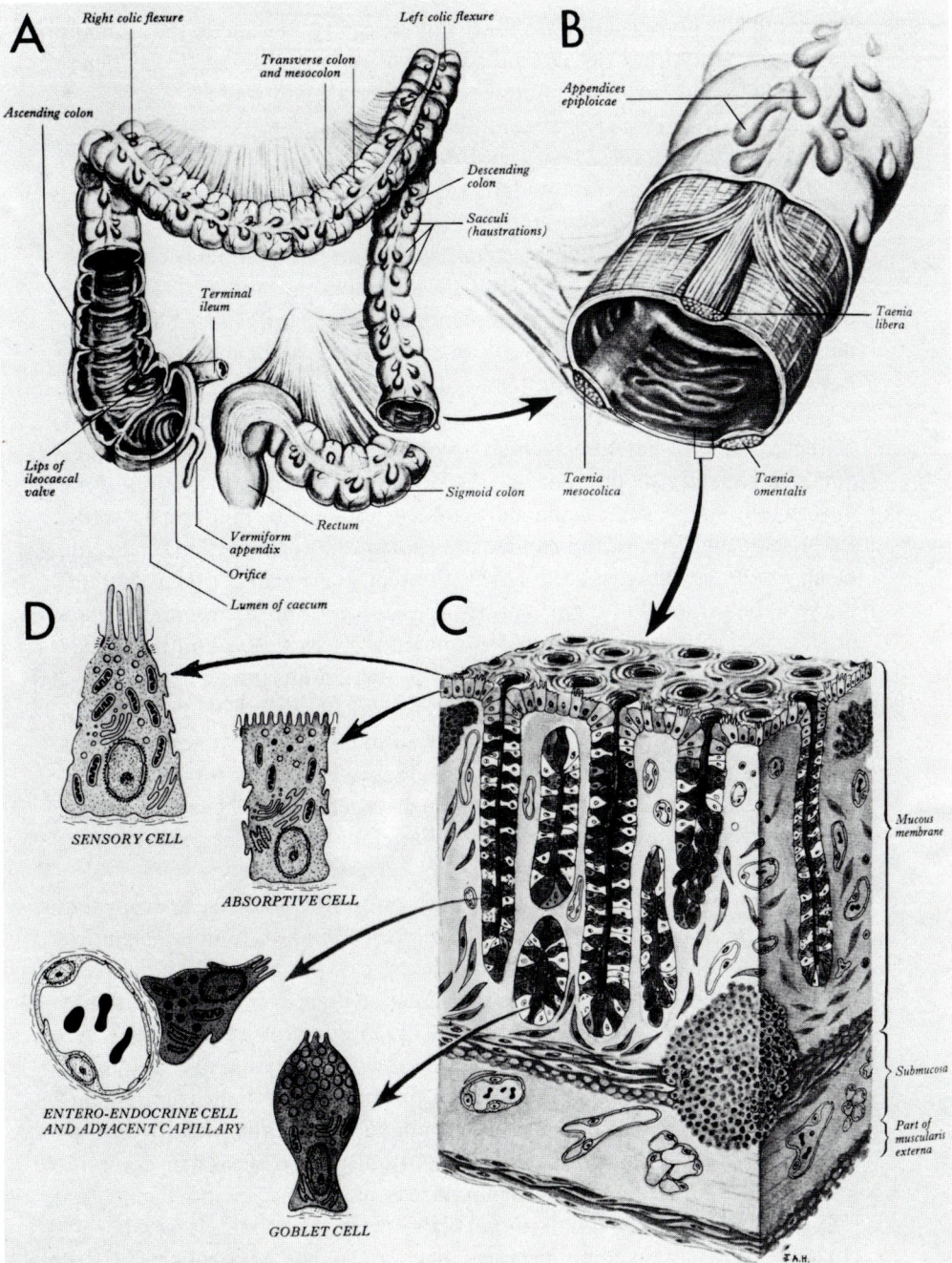

water and Na⁺ conservation (primarily in the right half), and in the left half, formation, temporary storage and passage of the normally relatively-dehydrated and solid fecal mass (undigested materials and waste).

Musculature of the Large Intestine

The well-developed muscularis externa of the large intestine includes an outer longitudinal and inner circular layer of smooth muscle fibers as elsewhere. However, that of the colon and cecum has a uniquely characteristic arrangement in which most of the longitudinal muscle fibers of the outer layer coalesce into three conspicuous, equidistantly placed longitudinal bundles known as *taeniae coli* (taeniae: (Greek) 'bands' (fig. 11/1B). Between them, the remaining longitudinal fibers of the intertaenial areas form a much thinner and sometimes incomplete layer. The extreme thinness of the colon wall between taeniae accounts for the great and sometimes amazing capacity of this part of the bowel to undergo distension when obstructed, the cecum being particularly notable in this respect [40]. Because of their tonus, the taeniae are shorter than the portion of the gut along which they are disposed; hence, they cause the wall of this part of the bowel to be gathered into sacculations or *haustra*. If the taeniae are severed or stripped from the wall, the colon immediately elongates and the sacculated appearance is lost. The taeniae and consequent haustra constitute two of three external anatomic characteristics by which the surgeon may identify a loop of colon, clearly distinguishing it from small intestine. The third is the *appendices epiploicae:* small, fatty appendages covered with peritonium studding the external surface of the colon, especially near the taeniae. They are relatively flat in the proximal colon but elongated and pedunculated in the sigmoid.

Separating adjacent sacculations are crescentic folds, seen from the internal surface as *plicae semilunares* that project into the lumen. The inner circular layer of muscle fibers forms a complete coat that is thicker between sacculations (i.e. within the plicae semilunares) than over the sacculations. The ileocecal valve is said to consist of an opening between two parallel and adjacent plicae semilunares with perhaps some modification of the circular muscle fibers contained therein. Another finding unique to the colon is that bundles of longitudinal fibers from the taeniae frequently blend with the circular coat (fig. 11/1B). Thus, the continuity of the circular layer is interrupted and different intertaenial areas may contract independently [58]. The deviation of longitudinal fibers from the taeniae towards the circular layer may, in some instances, account for the haustration [61].

On approaching the rectum, the three taeniae spread out and fuse to form a muscle coat that completely encircles the rectum, but is thicker on the anterior and posterior aspects than on the sides. These broad anterior and posterior bands of longitudinally disposed muscles are somewhat shorter than the rectum itself, and so again a type of sacculation occurs in which the lateral rectal wall bulges inward to form two to three transverse shelves, the *plicae transversales* of the rectum. These are said to aid in bearing the weight of the fecal mass and thus make the work of the anal sphincters less arduous.

In the anal canal the circular muscle coat is thickened to form the *internal anal sphincter*. The anal sphincters, internal and external, will be discussed under 'Primary Mechanisms of Fecal Continence', this chapter.

The Ileocolic Junction

Approaching sideways from the left, the terminal ileum invaginates the medial wall of the colon, the consequent protrusion into the cecal lumen – designated the ileocecal valve – anatomically demarcating cecum from right (ascending) colon (fig. 11/2). This phenomenon commonly occurs directly posterior to the point where a line connecting anterior superior iliac spine and umbilicus crosses the lateral border of the rectus abdominis muscle (McBurney's point). While the demarcation of cecum and right colon is convenient for descriptive purposes, physiologists studying colonic function (mobility and absorption) have yet to make any distinction between the two.

When observed in cadaveric material, a slit-like ileal aperture is bounded at least 60% of the time by two transversely-directed folds or lips, the lateral commissures of which form mucosal ridges or frenula which resemble the plicae semilunares seen elsewhere within the colon (fig. 11/2). As a consequence of this bilabial appearance, it was assumed that the structure functioned as a 'flap' or 'flutter-valve'. Direct observation of the 'valve' in the living through a cecostomy, however, has shown that the ileal projection assumes a papillary shape in vivo as it surrounds a stellate aperture (fig. 11/2). The appearance of the papilla has been compared to the appearance of the cervix uteri projecting into the vagina. The papillary appearance is more suggestive of sphincteric activity [23].

In most animals, there is a thickening of the musculature of the last few centimeters of ileum. In several, e.g. human and cat, the circular and longitudinal muscle coats of the terminal ileum continue into the valve or papilla and, as they do so, they are surrounded by colonic musculature (fig. 11/2). As a consequence of the overlap of ileal and colonic musculature,

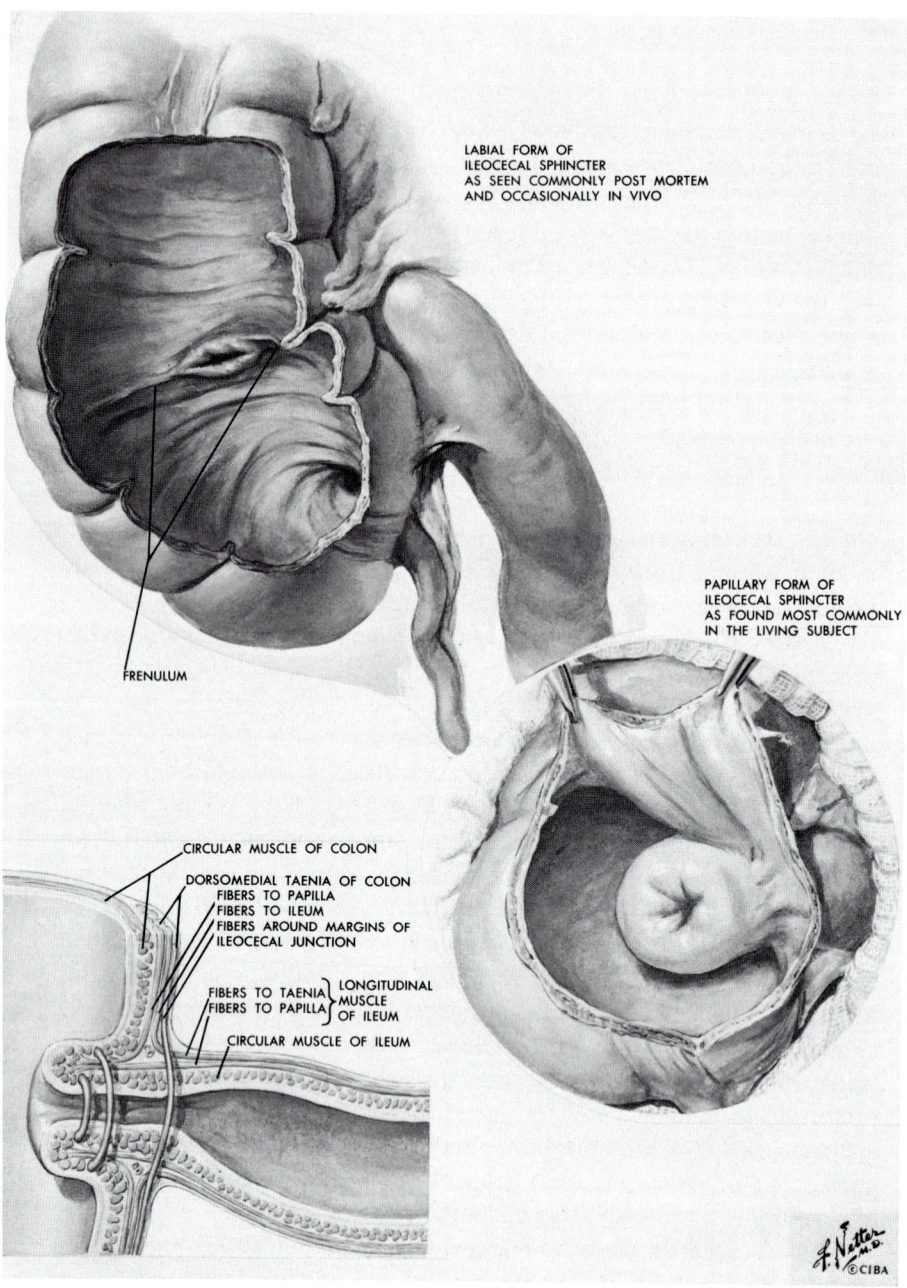

Fig. 11/2. The ileocecal sphincter. [Reprinted, with permission, from The Ciba Collection of Medical Illustrations, illustrated by *Frank H. Netter,* MD. © Copyright 1962, Ciba Pharmaceutial Company, Division of Ciba-Geigy Corporation.]

some of the data on this area is difficult to interpret. The alleged ileocolic sphincter is probably the least studied portion of muscle in the entire gastrointestinal canal. Although hard data are meager, it is held that the myenteric plexus extends into the junction and that extrinsic innervation is supplied by the vagi (parasympathetic) and by fibers from the superior and inferior mesenteric ganglia (sympathetic) [83].

Both function and relative importance of the structure remain uncertain. It has been maintained since the 1920s [3] that the function of the 'sphincter' is twofold: (1) preventing reflux of colonic contents from cecum into ileum, and (2) regulating flow from ileum to cecum, preventing the ileal contents from passing too rapidly; thus contact with the terminal ileal mucosa could be prolonged to maximize intestinal absorption. More recently, however, it has been doubted that it behaves as a sphincter at all because surgical removal and reimplantation of the ileum into the cecum is not followed by any derangement in transit. The junction is reputed to prevent reflux from colon to small bowel, but following X-ray examination by barium enema contrast medium is often seen to regurgitate into the small intestine.

Fifty or 60 years ago the region was studied by direct visual examination in persons and dogs with cecal fistulae. In the fasting state resistance to the passage of a tube or a finger has been described. A very high closing pressure (66 cm H_2O) has been described by some, but not by others. The high closing pressures were reduced or abolished by ileal distension. Recently, isolated strips of circular muscle from this region have been claimed to show a resting tone which does not exist in the muscle on either side. Transmucosal electrical stimulation caused relaxation, but tetrodotoxin did not [18]. The only consistent result of extrinsic nerve stimulation is that the sympathetics cause contraction.

Those who have observed the intact ileocecal opening in vivo have stated that as soon as food is eaten or placed in the stomach through a fistula the sphincter becomes active. They call this the *gastroileal reflex*. Within 1–3 min gushes of intestinal juice and food residues entered the cecum at a rate of 10–20 ml/min. The pattern of activity seems to be like that at the pyloric sphincter, i.e. contraction only when a wave reaches it and quiescence in between. Does this prevent regurgitation? Evidently, it does as in fed dogs, in contrast to those fasted, barium from an enema did not reach the ileum. The patient undergoing diagnostic barium enema is usually fasted and purged. Regurgitation here as in the stomach occurs if the bowel is empty.

A reassessment of the activity of this region is necessitated by the rediscovery of the interdigestive activity. This might clear up the problem of resting tone. The gastroileal reflex is undoubtedly the beginning of the intestinal feeding pattern. An understanding of this region requires information about the pressure and motility of the cecum as it fills and empties. This is lacking.

Motility of the Colon [15]
Right vs. Left Colon
Physiologists studying colonic motility and absorption have yet to make any distinction between cecum and right colon. The cecumward direction of activity gradients in the right colon suggests that the cecum is the major storage and absorptive site in the large bowel.

Several studies have provided us with evidence that the cecum, like the stomach, undergoes receptive relaxation as ileal fluid enters it through the ileocecal sphincter. *Alvarez* [3] concluded from his observations that the motility pattern of the ileum did not carry over to the cecal muscle and certainly modern descriptions of the slow waves show them to be distally directed (i.e. towards the cecum) in the ileum and proximally directed (also toward the cecum) in the right colon. However, at least one modern study using electrodes has described spike potentials which continued from ileum across the sphincter to the cecum. Thus, the question of ileocecal peristaltic continuity is uncertain.

The best studies of the motility of the intact colon were made between 50 and 80 years ago by X-ray [3]. After he had put an X-ray contrast medium into the colon of a cat, *Cannon* [12] noted that the usual movement in the ascending and transverse colon was antiperistaltic, i.e. in the wrong direction – towards the cecum. The peristaltic movements occurred with a frequency of 5.5 times/min and usually travelled all the way to the cecum at 1–2 mm/s. Contrast medium did not enter the ileum, but when ileal contents entered the cecum it contracted driving fluid up the ascending colon which in turn drove it back towards the cecum.

The left side of the colon he found behaved differently. Here contrast medium is divided into globular masses by ring contractions which move slowly towards the rectum and the contents are, of course, no longer fluid. He found the rectum to be quiescent. Movement of material from the right side of the colon to the left may take days; in fact, passage from cecum through transverse colon is the slowest segment of journey from mouth to anus.

11 The Colon and Defecation

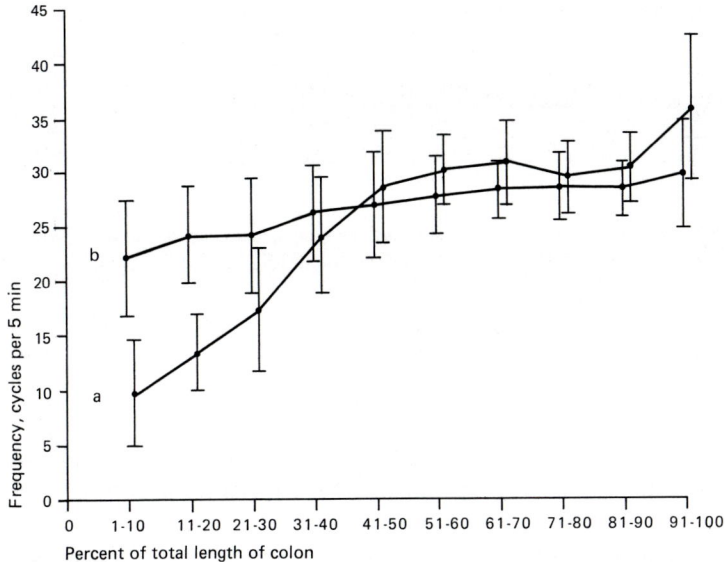

Fig. 11/3. A plot of the average frequency of slow waves from the circular muscle of 14 cats' colons in vitro. b = Intact colon; a = after complete transection of the colon between each electrode. Zero on the abscissa represents the ileocecal junction and 100 represents the anal verge. Reproduced, with permission, from *Christensen* [16].

Embryologically and functionally, the colon may thus be divided into two parts. The right colon, like the small intestine, arises from the embryological midgut, receives its arterial supply from the superior mesenteric artery and its parasympathetic innervation from the cranial outflow (Vagus). Like the small intestine, the right side of the colon has important mixing and absorbing functions. The left colon, derived from hindgut, receives its arterial supply from inferior mesenteric artery, and its parasympathetic innervation from the sacral outflow (pelvic splanchnic nerves). The distal colon is the site of fecal storage, having no essential digestive or absorptive function, and the rectum and anus the site of the defecation mechanism.

Electrical Activity

We have already noted slow waves or a basic electrical rhythm (BER) in the longitudinal muscle of both the stomach and the small bowel. Similar waves have been seen in the proximal part of the large intestine (fig. 11/3),

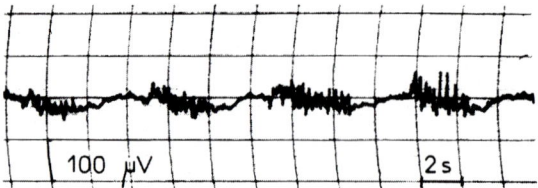

Fig. 11/4. Spike potentials and slow waves recorded from the human colon in situ. Reproduced, with permission, from *Coutourier* et al. [20].

but here they originate in the circular muscle and spread to the longitudinal [13]. Like the stomach, in cats at least, the proximal colon seems to be dominated by a single pacemaker situated at the middle of the ascending colon and from it the waves of depolarization spread towards the cecum, but not into it. If the colon is cut between the pacemaker and the cecum, the rate of the slow waves will fall in the proximal segment, because the dominant pacemaker has been cut off and now a slower proximal one has become dominant (fig. 11/3).

In the cat slow waves are found throughout the colon. In the distal colon, however, slow wave *spread* is not obvious. In other words, there may be many pacemakers with a very limited extent of influence just as in the small intestine. Spiking at slow wave crests is correlated with contraction of the circular muscle so that the slow wave frequency, just as in other intestinal muscle, sets the maximal rate [37].

There has been little consistency in the descriptions of slow waves recorded from the human distal colon undoubtedly because of the difficulty of insuring good mucosal electrical contact (fig. 11/4).

Mass Movements

Radiologists have from time to time described what has been called the *mass movement*. It is usually assumed that this occurs once or twice a day and that it is responsible for filling the rectum which in turn sets off the defecation reflex. No satisfactory experimental model for mass movement has been discovered nor have any myogenic or neurogenic basic mechanisms for this been unearthed. *Alvarez* [3] states that he saw 'mass movements' several times and describes them as follows: 'The haustrations in the middle of the transverse colon disappear and the fecal material runs into a sausage-shaped bolus some 15 cm long. A contraction then appears which in the next 4 or 5 seconds pushes the mass into the descending colon or on into the rectosigmoid.' Does such a movement involve the ascending colon? If it does, unless the gradient of the right colon is reversed from time to time, solid fecal matter will accumulate in the cecum rather than in the transverse colon. The point of transition between the left and the right colon presents a mechanistic mystery.

The Gastrocolic Reflex. Alvarez [3] felt that the mass movement was a manifestation of the 'gastrocolic reflex'. That defecation often occurs following feeding is common knowledge. The baby is changed after its bottle and the dog is put out following its dinner. This has been called the gastrocolic reflex. It has been assumed to be nervous (cholinergic) in mediation. It affects the entire colon ileum and jejunum. Only the effect on the small intestine has been the subject of vigorous investigation. Feeding, sham feeding and gastric distension all increased the motility of the innervated jejunum, but not of the denervated viscus. The effect on the colon is almost uninvestigated. It has been claimed that the effect is cephalic, that it occurs only if food enters the stomach, that it is seen after complete gastrectomy, after bilateral vagotomy and even after complete transection of the spinal cord in the thorax. In short, the effect is obvious, but the afferent receptors, the pathways and the ultimate stimulus to the colon are all unknown; colinergic mediation, a depression of sympathetic inhibitory tone and gastrointestinal hormones have all been suggested as the ultimate mechanism. In the light of today's rediscovery of interdigestive intestinal activity and onset of the feeding pattern this reflex needs to be restudied. It will be remembered from chapter 9 that many of the characteristics of the gastrocolic reflex noted above fit with much of what is known about the feeding pattern; for example: it is not seen in autotransplanted loops of small intestine, it is not affected by vagotomy and it is seen when food enters the intestine. The only serious gap is that no one seems to have studied feeding patterns in the large bowel. Gastrin is certainly capable, and probably CCK is also, of interrupting interdigestive periodic activity and substituting the feeding pattern for it. Gastrin and CCK will both increase intestinal and colonic activity in dogs and man, but gastrin and CCK remain elevated throughout digestion and the gastrocolic reflex is a transitory affair, probably because it can only be manifest when the distal colon contains feces.

Innervation of the Colon
Motor Nerves

In general colonic activity is increased in experimental animals by stimulation of the parasympathetic nerves. It is usually decreased by sympathetic stimulation, but increases have been seen also. The responses to the corresponding neurohumoral agents and their antagonists (atropine and guanethidine) in vivo are equivocal, but in vitro they correspond to the effects of nerve stimulation.

The vagus nerve by the time it reaches the splenic flexure has come to the end of its long journey. From here on its place as an efferent nerve is taken by the sacral parasympathetic outflow via the pelvic splanchnic nerve. These, like the vagi to the esophagus, contain both excitatory and inhibitory fibers to the muscle. The sensory or afferent vagal innervation continues further to the sigmoid colon, where again it is replaced by afferents, including pain afferents, which travel in the pelvic splanchnics. These are the afferent fibers responsible for the defecation reflex. The sympathetic supply is to the smooth muscle of the colonic vasculature; in addition it is largely inhibitory to the colonic musculature by virtue of its action on the cells of Auerbach's plexus as in the small intestine. Sympathetic stimulation, here also, diminishes the release of acetylcholine; it has long been suspected that cholinergic receptors are scarce here. This view is substantiated by the failure of metoclopromide, an acetylcholine releaser, to increase colonic motility and by the finding that the stimulatory action of TRH is blocked only by atropine and a 5-hydroxytryptamine blocker given together (see chapter 8).

Pain

Colonic pain results from spasm, mucosal edema and inflammation. The latter are themselves potent causes of spasm and hyperactivity. Distension of the colon, unlike the small bowel, does not cause pain, but only a sensation of fullness. This may be because distension develops slowly here. In colonoscopic examination rapid distension or stretching is painful. This may also explain why it is that even in the massive distension of Hirschsprungs' megacolon the activity of the small bowel is not inhibited, i.e. the intestino-intestinal inhibitory reflex is inoperative.

The occurrence of *Hirschsprung's disease* or *aganglionosis* of the colon makes it very obvious that the intrinsic nerves of the colon are essential for its normal function just as was seen to be the case in the small intestine. It is likely, although there is little evidence, that spiking depends on stimuli external to the muscle itself and is dependent on the intrinsic innervation. *Bayliss and Starling* [6] described a peristaltic reflex in the cat's colon and *Hukuhara and Miyaka* [47] have described it in the anesthetized dog. The existence of an inhibitory component is supported by the finding that the application of tetrodotoxin to the colon increases its motility. It is likely, therefore, that the scheme of Hirst referred to above (see chapter 8) applies also to the colon. This is supported by Hirschsprungs' disease which is caused by spasm in the aganglionic segments, usually rectum, accompanied by great dilation of the otherwise normal colon proximally.

Lining of the Large Intestine

The submucosa and muscularis muscosae of the large intestine do not vary significantly from that of the small intestine. Many scattered solitary lymphatic follicles are present in the lamina propria. They are often so large that they project through the muscularis mucosae to displace the crypts and lie in the submucosa (fig. 11/1C). They are most abundant in the cecum and vermiform appendix, but are irregulary distributed throughout. They have epithelial M cells, specialized for antigenic transport, associated with them [60]. Many lymphocytes, macrophages, and plasma cells that locally produce IgA are also found within the lamina propria. The profusion of lymphoid cells and follicles found associated with the mucosa here is a reflection of the hundreds of bacterial species (over 500 types of anaerobe alone) present in the lumen of the large intestine of the average healthy human.

The mucous membrane of the large intestine differs from that of the small intestine in several other respects (see fig. 11/1C). In postnatal life, the entire large intestine is devoid of villi. The intestinal glands (crypts of Lieberkuhn) are larger, more numerous and more densely packed than those in the small intestine. They are also longer (deeper), increasing in length distally to reach a maximum depth (0.7 mm) in the rectum. Consequently, the mucosa itself is thicker, especially distally and in the rectum. The glands of the large intestine are characterized by a great abundance of goblet cells (up to one goblet cell for every four columnar cells; their broad shape conveys the illusion that they constitute a majority) and a small number of entero-endocrine cells of the two major types. Except in the young, they contain no Paneth (zymogenic) cells. The minute orifices of the glands give the intestinal surface a cribriform appearance. The mucous membrane of the cecum and colon is pale and is raised into numerous plicae semilunares between sacculi as mentioned above; that of the rectum is more vascular, of a darker color and more loosely connected with the muscular coat.

The lining of the luminal surface of the colon continues to be a simple, columnar epithelium wherein mature absorptive cells with striated borders predominate, with goblet cells interspersed among them (fig. 11/1C, D). The epithelium also lines the walls of the intestinal glands (crypts) which, being large and numerous, provide a large surface area for mucous secretion and water and sodium absorption. Thus, lubrication is provided for the passing feces and many of the substances secreted into the GI tract in the buccal cavity, stomach and small intestine are reabsorbed here.

The epithelium lining the crypts and luminal surface here also is a continuous sheet that is constantly being renewed via proliferation and sloughing, as described under 'The Mucous Membrane of the Small Intestine', chapter 10. The immature-looking presumptive stem cells found in the bases of the crypts of the ascending colon closely resemble their counterparts in the small intestine, i.e. they appear as young (small and undeveloped) absorptive cells. However, those of the descending colon and rectum include secretory vacuoles in their apices and, consequently, have frequently been referred to as *vacuolated cells*. As these cells migrate towards the glandular orifice, they first become distended with secretory vascuoles but, before reaching the luminal surface, they discharge their vacuolar contents and become typical absorptive cells with microvilli packed into a striated border [14]. The secretion provided by such cells appears to be mucoid in nature, but is also rich in antibodies of the IgA group which provide a measure of protection against invasion by microorganisms [72]. Other columnar cells, described as demonstrating an apical tuft of especially long microvilli, have been suggested as providing a type of sensory ending [74] (fig. 11/1D).

Intestinal crypts cease to exist at the anorectal junction. Here, *circumanal glands* occur. These glands have a stratified columnar epithelium and are of the branched tubular type, but do not seem to be actively functioning. It is probable that they are a vestige of the actively functioning circumanal glands of certain mammals.

The *anorectal junction* (pectinate line) is the demarcation between that which is 'visceral' and that which is 'parietal'. At this point, the simple columnar epithelium of the gut is replaced by a nonstratified squamous epithelium which then lines the anal canal (a distance of about 2 cm) before it becomes continuous with the keratinized epidermis of the skin.

Absorption [21]

All that is left of the chyme, which formed the contents of the upper small intestine, by the time it reaches the ileum is the unabsorbable constituents of the diet such as dietary fiber, lignin, waxes, high melting point fats (which are not very common), insoluble or unabsorbable salts, and the bodies of the saprophytic bacteria, which inhabit the lower small bowel, plus about 1.5–2 liters of water per day containing Na^+ in concentrations almost isotonic with plasma. As noted above the small bowel, although it can easily absorb 20 liters of water per day, absorbs it as an almost isotonic solution. The large intestine on the other hand usually absorbs all the fluid and Na^+ that enters it, except for the 100 or so ml of fluid lost per day in the feces and about 4 mmol of Na^+. If the large bowel is removed the patient will exist on the edge of dehydration and dangerous Na^+ depletion. On the other hand, if in excess of 3 liters of water or 450 mmol of Na^+ per day enters the cecum, the upper limit of the colon's absorptive capacity will be exceeded. This means that the colonic reserve capacity is very small. As in the small bowel the absorptive capacity of the large bowel diminishes from above down, i.e. cecum to the anus, where it is zero. The decrease in absorptive capacity occurs as absorptive cells diminish in number towards the rectum, being replaced by goblet cells. In the distal colon, the epithelium consists primarily of goblet cells. For this reason, surgical removal of the cecum and ascending colon will generally cause diarrhea.

Sodium. The most important substance moved from the colonic lumen to the portal blood, sodium is actively transported against both electrical and concentration gradients [27, 28]. The mucosa is not as effective an absorber as is the duodenal mucosa of glucose, but it can reduce the $[Na^+]$ of colonic fluid to 15 mmol/l. In the small bowel, it will remembered that Na^+ was absorbed in association with both glucose and actively absorbed amino acids. In the colon neither amino acids nor glucose are actively absorbed nor, therefore, do they influence the rate or efficiency of Na^+ absorption. Of the three methods by which Na^+ is moved in the ileum only the electrogenic mechanism seems to be operative in the colon [45]. Na^+ movement seems to be responsible for the transmucosal PD across the colonic mucosa, the luminal side of which is negative as expected.

Water. Water moves passively with Na^+ and Cl^-, but neither the movement of these ions nor of water is unidirectional or isotonic. Na^+ and its accompanying anion and water diffuse fairly easily back into the lumen, but the net flow is usually in the direction of active absorption into the portal blood. This is normally 4 times as rapid as back diffusion into the lumen.

Chloride. The absorption of Cl^- is passive. It follows Na^+ down the electrical gradient from lumen to blood. There is less Cl^- in the fecal water than one would expect if Cl^- moved in tandem with Na^+ only. The explanation for this is the Hamburger chloride shift (fig. 10/7). The colonic mucosa manufactures and secretes HCO_3^- into the lumen where it exchanges with Cl^-. In a study of rat intestinal and gastric mucosa the highest carbonic anhydrase activity was found in the colonic mucosa. Bicarbonate secretion into the colon is diminished if luminal Cl^- is replaced by SO_3^- and by conditions which increase luminal HCO_3^-. Acidification accelerates Cl^- absorption.

Potassium. The large bowel secretes K^+ into its lumen where it can reach 15 mmol/l. Above this K^+ back diffusion into the blood stream can balance secretion. Movement of K^+ in both directions is said to be passive, in the luminal direction the transmucosal Na^+ potential is advanced as the explanation. It is, however, difficult to explain the very high concentrations often found in colonic contents on this basis alone. Concentrations four or five times those of plasma are commonplace and several groups of workers have reported fecal concentrations 10–12 times higher than in plasma. The most likely explanations are either active transport of K^+ or progressively declining back permeability as the anus is approached (fig. 10/7).

As elsewhere the adrenocortical hormone aldosterone accelerates the active absorption of Na^+ and increases the Na^+ transmucosal potential. This may be the explanation for the accompanying increase in luminal K^+.

Diarrhea. In diarrhea three or more liters of fluid per day may be lost from the intact gut. In diarrheas of small intestinal origin the composition will come to resemble more that of the jejunal contents, that is the electrolyte concentrations will approach those of plasma and because of the rapid flow there will be insufficient time for colonic modification. In diarrheas which result from colonic irritation and infection not only is reabsorption impaired, but secretion, which is characteristically high in K^+, will be increased. A rare congenital form of watery diarrhea is attributed to active secretion of Cl^- by the colonic mucosa or to uptake of HCO_3^- in exchange for Cl^-. In diarrhea of colonic origin, although normal colonic secretion is scanty, liters of fluid can be lost daily.

The large bowel is not essential for life, but in its absence an ileostomy opening will lose the amount of fluid which normally enters the cecum, i.e. about 1.5 liters/day and 200 mEq of Na^+. In long-standing ileostomies these figures may decline to about half, but even so the patient's fluid and electrolyte balance may be in a precarious state and is dependent solely on renal regulation. Extra fluid loss, such as sweat, may precipitate acute dehydration.

Colonic Digestion

From examination of modern literature one might easily get the impression that the sole colonic function is the movement of Na^+, Cl^- and water. This again is the modern preoccupation with cyclic AMP, the Na^+ pump and transmucosal electrical potential differences. We have dealt already with bile salts and acids, but we must consider many colonic constituents of great biological importance. Some of these are unabsorbed remnants of the diet, some originate from desquamated mucosal cells and from secretions, while others are plasma constituents which have diffused into the colon. All are subject to bacterial transformation.

In the latter category are creatinine, uric acid and urea [87]. There is no known mechanism by which creatinine is destroyed in the body yet it does disappear. ^{14}C from orally administered creatinine appears in the expired air. Ileal effluent contains creatinine and urate levels equal to those in plasma, yet the stool contains none because fecal bacteria convert creatinine to sarcosine, to methylamine and to 1-methyl guanidine, which has been promoted as a toxic agent in uremia. Some at least of these degradation products are absorbed by simple diffusion to be degraded further by the healthy liver, or to accumulate in the plasma if the liver is defective. Uric acid is degraded to allantoin, urea and ammonia.

In animals and man treated with antibiotics the urea in colonic contents seems to be in equilibrium with that of plasma. The colonic mucosa is not very permeable to urea in either direction, so most of the urea must originate from the small bowel, its breakdown to ammonia must, however, take place in the colon, as the bacterial population is several orders of magnitude greater there than in the ileum. The amount of urea degraded and the resulting ammonia production are very large. It is reckoned that 20% of the daily urea production is degraded in the large bowel. This produces 200 mmol of ammonia, which is a prodigious amount [22, 86].

These are not the only sources of ammonia since there are other nitrogenous substances of dietary origin and the bodies of the bacteria themselves from which ammonia can be produced.

Only about 1/100th of the ammonia produced in the colon reaches the feces, the difference being absorbed. Absorption seems to be by passive diffusion of the non-ionic form, NH_3 which is predominant at high pH, therefore bicarbonate assists in ammonia absorption. As noted above the colon produces a great deal of bicarbonate.

Analysis of feces has shown that lower fatty acids, acetate, propionate, butyrate and, to a lesser extent, Krebs cycle acids (all present as salts) are abundant [86]. These are similar to the acids found in rumena of sheep and cattle and are of bacterial origin. They are all readily absorbable and metabolizable by colonic mucosa. It is frequently stated and assumed that sugars are not absorbed from the colon, however, except after antibiotic therapy it is likely that the situation never arises. *Bond* et al. [10] have shown that almost all of sucrose or glucose on entering the colon is converted by bacterial action to the volatile lower fatty acids mentioned above, all of which are highly diffusible. The presence of unabsorbed sugars could produce dangerous diarrhea, but bacterial action reduces this. Diarrhea from carbohydrate malabsorption does occur, but largely because the ileal effluent is increased in volume.

Fecal Fat. In normal people there is less than 5 g fat/day in the stool. Anything above this is considered to be abnormal. Virtually all of this is endogenous [52]; that is, it is not unabsorbed dietary fat but is the fat of desquamated cells, bacteria and those dietary lipids and waxes which are undigestible. Addition of extra meat fat (lard) to the diet, within physiological limits (93–183 g/day) does not increase the daily fecal fat. In experimental animals diversion of bile and or pancreatic juice from the intestine (fig. 11/5) increases the amount of dietary fat in the stool, but does not alter the level of fecal fat in animals on fat-free intakes. If bile is diverted from the intestine between 42 and 54% of lard fat is absorbed. Of the extra fat in the stool in this preparation virtually all is present as soaps of fatty acids. From this one would conclude that the major action of bile salt is to facilitate the absorption of fatty acids and that hydrolysis by lipase and the absorption of partial hydrolysates is relatively unimpaired. By contrast, when pancreatic secretion is impaired neutral fat is found in the stool, but so also are fatty acids. In the total absence of pancreatic juice about 60% of beef lard is absorbed. If the bicarbonate of the lost pancreatic juice is replaced by

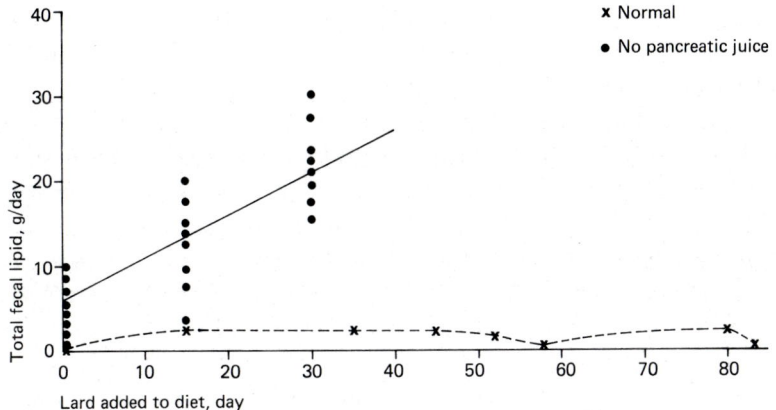

Fig. 11/5. The effect of excluding pancreatic juice from the duodenum on the fecal fat in dogs fed diets of varying fat content. Drawn from *Annegers* [4].

intraduodenal infusion much more fat is absorbed because in the more alkaline medium which results, most fatty acids will form amphophilic soaps (see 'Absorption', chapter 10). The presence of free fatty acids or their soaps are not a good diagnostic measure of the causes of steatorrhea as fecal bacteria are capable of hydrolysing neutral fat. The presence or absence of neutral fat is a better measure.

Fecal Bulk. As the amount of an ordinary mixed meal is increased so the amount of fecal fat increases. In the past this was related to the dietary fat and a straight line relationship was obtained. As noted above additions of fat to a standard diet do not increase fecal fat. The discrepancy arises in that the relationship is not really to the fat content of the diet but to its fiber. As the intake of a mixed diet increases the daily intake of all the contained foodstuffs increases and so of course does the fiber. Fiber, obviously, increases fecal bulk and weight and fecal fat increases with both [68]. This is all endogenous fat and represents about 0.05 g for each gram of dietary fiber.

In adults on the usual European or North American diet 133 g stool/day is considered to be the upper limit of normal. As fecal bulk increases the frequency of bowel movements will also increase. The average frequency for people on a normal free choice diet is one movement each 1.1 days. As fiber is removed the frequency decreases. On elemental fiber-free diets one movement every 5 days has been recorded. As a general rule, however, the weight of the stool and its bulk, whether it occurs everyday or every five days are remarkably constant. In view of the importance of distension in triggering the defecation reflex this is hardly surprising. As stool weight increases, transit time, mouth to anus, decreases. The measurable difference is almost entirely colonic since it is here that there is the greatest delay. Strangely, pressure measurements taken from the human colon which are an indication of activity, are reduced by additions of bran fiber to the diet and this is now recommended for diverticular disease. In this disorder pulsion diverticuli are pushed through the enveloping muscle at weak points, i.e. where blood vessels enter. High luminal pressures would favour the formation of such blind saccules in which stagnant fecal material can accumulate. Such high intracolonic pressures are found in people who habitually consume low residue diets. However, any relationship between consumption of fiber and diverticular disease is speculative at the moment.

Vitamins. It is an orthodox view that the organisms of the lower ileum and colon are capable of synthesizing a variety of vitamins both fat and water soluble [53]. It is generally assumed, in the absence of any evidence pro or con, that these can be absorbed. If they are it must be by simple diffusion. In the case of water soluble vitamins this must be slight as these are all large molecules and they must cross lipid cell membranes. In the case of the fat soluble, none of the special small bowel transport mechanisms exists. Vitamin K is the one most often thought to be absorbed from the colon. It certainly is present in substantial quantities in the colonic contents and feces of most animals. From time to time newborn children suffer from vitamin K deficiency. It has been assumed that the sterility of the newborn bowel and consequent lack of vitamin K from this source is an important contributing factor in this disorder.

Flatus [50, 51]

The normal human GI tract contains approximately 100 ml of gas distributed through stomach, small and large bowels. Between 0.2 and 2 liters of this per day is discharged per anus as flatus depending on the individual and his/her diet. Most of this gas is swallowed air. During normal eating and drinking ingestion of air is unavoidable and some people swallow air as a habit *(aerophagia)*. Most of this air is almost immediately eructated. In the supine position it may be difficult to do this as the opening of the esophagus may be below the gastric fluid level. This may account for the periodic large expulsion of gas often with gastric content in the suckling infant. Of the air that remains in the gut the predominant constituent is N_2; O_2 is rapidly absorbed; both it and CO_2 come into equilibrium with the plasma. The body fluid N_2 is in equilibrium with the atmosphere, therefore none is absorbed from the gut. In the duodenum a great deal of CO_2 is generated from the neutralization of gastric juice by the pancreas, but this is all rapidly absorbed.

Whenever there is bacterial growth in the gut H_2 and CH_4 will be produced. This is normal in the large bowel but can occur in the esophagus, stomach or intestine if there is stasis. In the colon H_2 is produced from unabsorbed or unabsorbable carbohydrates. The amount produced provides a fairly exact measure of the amount of substrate present. Since this gas is absorbed and can be measured in the breath it can be used as a measure of carbohydrate malabsorption. Methane (CH_4) also is produced from colonic carbohydrate, but this cannot be used diagnostically even though, like H_2, it can be detected in the breath, because production varies greatly from person to person; indeed, it seems to be familial and must mean that the concerned organisms unlike H_2 producers show a familial preference. Some foods such as beans are notorious flatus producers because they are rich in undigestible saccharides. Foods high in fiber may also result in gas production since cellulase containing organisms are present in the human gut. Gas will be formed from the resulting sugars. Hydrogen is usually about 10–15% of flatus and N_2 about 80%. This means that flatus is normally inflammable.

All the gases mentioned above are odorless, but information on those responsible for the odor of flatus/feces is scarce. Trace amounts of H_2S are common especially after eating sulfur-containing foods such as eggs or cabbage. Ammonia is produced, as previously discussed. The others are usually listed simply as indoles, skatols and mercaptans. All of these do enter the blood in small quantities, but are removed by the liver. They contribute to the *foetor hepaticus,* i.e. foul breath of patients with advanced liver disease.

Colonic Secretion (see fig. 10/8) [30]

Information on the regulation and composition of colonic secretion is scanty, except for the electrolytes, especially K^+ which have already been discussed. *Wright and Florey* [85] almost 50 years ago published the last systematic study of secretion from the colon of anesthetized cats.

If one makes a Thiry loop of the colon in a dog nothing will run out. The cannula and the lumen of the loop will, however, become filled with white, dried mucus. After mechanical stimulations or irritation more profuse secretion can be obtained but even then the amount is small. The record so far is 10 ml/h. This 'irritation juice' is a watery fluid, alkaline in reaction (pH 8.5), containing lumps of mucus. It contains almost equal concentrations of bicarbonate and chloride varying between 85 and 95 mEq/l. It is isotonic with plasma. Sodium is the main cation but potassium is higher than elsewhere in the intestine. It is likely as already noted that potassium is secreted into the lumen. Chloride and bicarbonate are the main anions. A great array of enzymes has been claimed to exist in colon secretion: amylase, proteases of various sorts, lipase and sucrase. Enterokinase has not been found. It is far from certain that any of these enzymes exist; their occurrence has frequently been denied. None of the work in their favor is modern and, in a region where desquamation and bacterial contamination is so likely and important, it is probable that in pure colonic secretion none exists.

The most certain stimulant of colonic secretion is local irritation, whether it be the result of chemical or mechanical stimulation or from a disease process. Evidently, the goblet cells are stimulated directly and as a result discharge their mucus. Neither cocaine nor atropine will interfere with this response. A surprising feature of colonic secretion is that no matter how secretion is stimulated the cells which show most evident loss of their mucus are those at the bottoms of their crypts. Those on the free surface appear relatively unchanged. Rapid secretion occurs through the sequential exocytosis of large numbers of intracellular mucous granules, called *compound exocytosis* and results in deep cavitation of the apical cell surface. During secretion, the plasma membrane remains intact, and the cell can rapidly replenish its store of mucin [58]. The Oxford workers have shown that nerves can play a part in secretion since Faradic stimulation of the pelvic nerves will increase the secretion of the distal half of the cat colon about fivefold. A clear mucoid secretion was the result. Reflex secretion has been described, stimulation of the central end of the cut pelvic nerves on one side causing an augmentation of secretion. Separated segments with intact innervation show increased secretion at the time of defecation. The response to nerve stimulation can be reproduced by pilocarpine, while both drug and nerve stimulation effects are abolished by atropine. Both pilocarpine and pelvic splanchnic nerve stimulation increased colonic motility but the augmentation of secretion observed cannot be ascribed entirely to motility change.

Sympathectomy produces no change in colon secretion but stimulation of the sympathetic supply while the pelvic splanchnic nerves are being stimulated will reduce the secretory volume in response to pelvic splanchnic nerve stimulation. The only enzyme in the juice collected as a result of nerve stimulation which could not be accounted for by cellular debris was a trace of amylase which may have been pancreatic in origin since this enzyme is very resistant to proteolytic destruction.

The ultimate fate of the colonic secretion is not known in its entirety. Certainly the water and electrolytes are, for the most part, absorbed but the fate of the 0.7% organic material is unknown. Is there a mucolytic enzyme? This has not been claimed, but a thorough investigation has not been conducted. Perhaps the inspissated mucin is lost with the feces and helps to bind the fecal masses together.

The Anorectal Region

General Structure and Function

Anatomically, the anal canal is defined as the terminal portion of the large intestine extending from the level of the superior surface of the pelvic diaphragm to the anus. Clinically, however, the anal canal is commonly considered to be limited to that portion inferior to the pectinate line (fig. 11/6). The pectinate (dentate) line is the demarcation between that which is 'visceral' and that which is 'parietal'. At this point, visceral efferent (parasympathetic) and afferent innervation (via the splanchnic nerves) ceases and parietal (spinal nerve) innervation begins (via the pudendal nerve) (fig. 11/6). Here, also, the blood ceases to be supplied via the inferior mesenteric artery (its superior rectal branch) and to drain into the portal system as supply via the internal iliac artery (via the internal pudendal artery) and drainage into the caval venous system begins. The lymphatic drainage also changes here, as the deep route of drainage to pelvic nodes is replaced by superficial drainage to the superficial inguinal nodes. A change occurs in the epithelium as well, but only with regard to the thickness of the squamous epithelium (it becomes stratified); the transition from simple columnar mucous membrane to non-stratified squamous mucocutaneous epithelium occurs approximately at the beginning of the anatomically defined canal [59].

Although only 3 cm or so in length, this short segment of the GI tract is of great importance clinically both because of its function in fecal continence and defecation and because it is prone to certain disorders [40]. thus, it is perhaps appropriate that it should receive consideration disproportionate to its length.

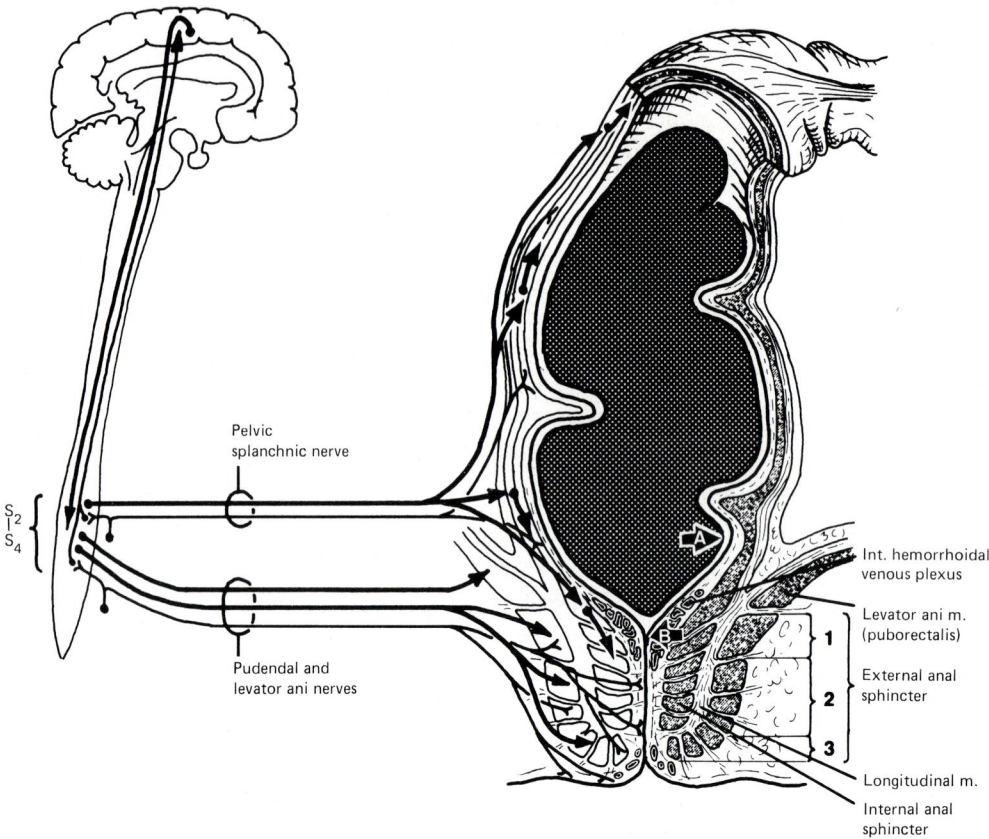

Fig. 11/6. Schematic illustration of the anorectal mechanism and its innervation. A fecal mass occupies the rectal ampulla. Anatomically, the anal canal commences at the superior surface of the pelvic diaphragm (arrow A); clinically, it is commonly considered to begin at the pectinate line (arrow B). Traditionally, the external anal sphincter has been subdivided into three parts: deep (1), superficial (2) and subcutaneous (3). *Shafik* [77] describes the external sphincter as consisting of a series of three loops [deep (1), intermediate (2) and base (3)] but includes the puborectalis as part of the deep loop rather than as part of the levator ani as shown. Others dispute the subdivisions, believing the external sphincter to be a single, continuous muscle sheet or mass. The pelvic splanchnic nerve conveys afferent fibers stimulated by stretching of the wall of the ampulla and presynaptic parasympathetic (visceral efferent) fibers. The pudendal nerve conveys proprioceptive and tactile afferent fibers and somatic efferent fibers. Not illustrated is the presumed intramural link between the stretch receptors of the rectum and the postsynaptic (enteric) motor fiber to the internal sphincter, believed to be responsible for the relaxation reflex. Redrawn from *Netter* [57].

11 The Colon and Defecation

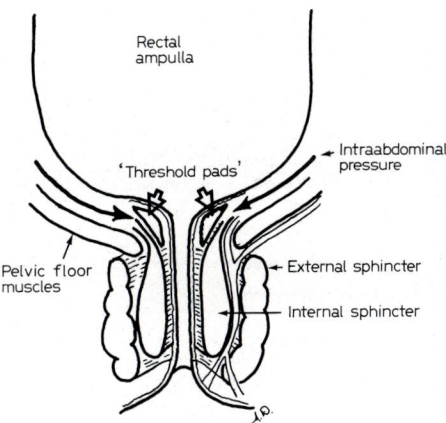

Fig. 11/7. Schematic diagram of the 'flutter-valve' mechanism proposed by *Phillips and Edwards* [67]. The 'threshold pads' described by *Thomson* [82] consist of the engorged veins of the internal hemorrhoidal venous plexus (see fig. 11/6). They augment anal closure and lend support to the 'flutter-valve' concept. Redrawn and modified from *Duthie and Wormsley* [26].

Primary Mechanisms of Fecal Continence

As the alimentary tube passes through the striated musculature of the pelvic diaphragm it not only changes its name (from rectum to anal canal) but also its shape and direction. The circular and relatively capacious lumen of the rectal ampulla abruptly collapses into a slit (fig. 11/7, 11/8) directed anteroposteriorly due to firm attachment to the perineal body anteriorly and the coccyx posteriorly, with the pliable ischiorectal fat bodies located laterally. The slit-like configuration of the anal canal passing through the pelvic floor has been compared to a 'flutter-valve' which has transmitted intra-abdominal pressure applied on its lateral aspects just proximal to its perforating the pelvic diaphragm (fig. 11/7) [67]. The lumen is held closed (and therefore empty) not only by the tonic contraction of the surrounding and overlapping internal and external anal sphincters (and any medially-directed force resulting from pressures applied to the fat bodies lying lateral to them) but also by the adherence of the opposed moist mucocutaneous surfaces (fig. 11/8). *Duthie* [24] claims that, even in vitro, with no sphincteric activity or fatty packing, force is required to separate the apposed canal walls, and that the force required is proportional to the surface tension of the mucus.

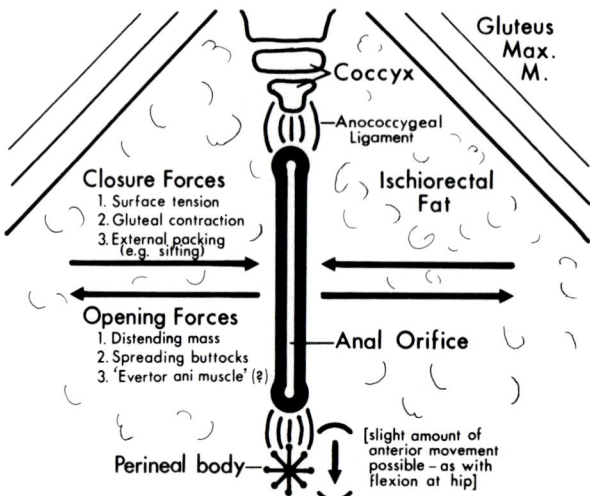

Fig. 11/8. Schematic diagram of the anal region. With the pliable ischiorectal fat pads located laterally and firm attachments anteriorly and posteriorly, the anal canal (and its orifice) is drawn into an anteroposterior slit when empty. The (nonsphincteric) forces which tend to open and close the canal (orifice) are indicated.

Perineal Flexure

Concurrent with the reduction in shape and volume, the alimentary tube negotiates a sharp bend, there being an angle of about 80° between the anteroinferiorly directed axis of the rectal ampulla and the posteroinferiorly directed axis of the anal canal (fig. 11/9). This angle, the so-called *perineal flexure,* is maintained by the tonic contractions of a U-shaped muscular sling formed by the puborectalis, which passes posterior to the canal, drawing the anorectal junction anteriorly toward the pubis to which the ends of the sling are attached (fig. 11/9).

The perineal flexure is generally accepted as being a major factor in controlling the passage of solid rectal content [24, 40], the canal thus being 'kinked' as when one folds a garden hose on itself to shut off or reduce the flow of water. During peristalsis of the filled descending colon, sigmoid colon and rectum (when sensations of urgency may be perceived), the puborectalis muscle is further contracted by reflex and/or voluntary contraction to maintain continence, or if conditions are appropriate, reflex contraction is voluntarily suppressed to allow passage of the fecal mass (defecation) which causes a temporary reduction of the angle of flexure [40,

11 The Colon and Defecation

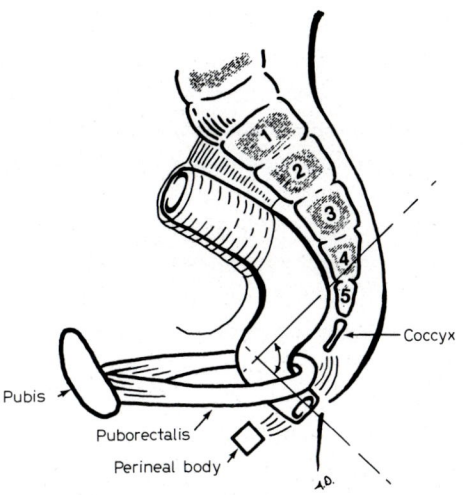

Fig. 11/9. Diagram of the perineal flexure, the 80° angle between the axes of the rectal ampulla and the anal canal. The flexure is maintained by the tonic contraction of the muscular 'sling' formed by the puborectalis. Voluntary and/or reflex contraction of the muscle increases the angle or maintains it during increased intrarectal pressure. During defecation, a lack of voluntary contraction and inhibition of the tonic contraction of the puborectalis allows temporary reduction of the angle to facilitate passage. [Modified and redrawn from Basmajian, J. V.: Primary anatomy; 7th ed. (Williams & Wilkins, Co. Baltimore 1976).]

63, 67]. The anal canal is often 'straight' in incontinent persons and restoration of the flexure leads to anal continence in a majority of these cases [64].

Positional Effects: Squatting. It is common experience that the squatting position facilitates the process of defecation. The radiological studies of *Tagart* [79] have shown that flexing the hips 90° or more reduces the angle of perineal flexure by some 12–20°. Since the rectum is anchored inside an inflexible bony framework while the joints at which hip and lumbar flexions occur lie outside, any consequent straightening of the flexure would have to involve movement of the relatively mobile anal canal; the movement presumably mediated by the skin and subcutaneous fascia between the thighs/buttocks and the anal pore which becomes increasingly taut as squatting proceeds (fig. 11/8). The posterior wall of the anal canal is, however, firmly attached to the tip of the coccyx via the anococcygeal ligament, which would become taut with any anterior movement of the anal canal. Thus, as was shown by Tagart, the anterior wall of the anal canal/anus undergoes more anterior excursion on squatting than does the posterior wall, the result being an increase in the anteroposterior diameter, i.e. an opening of the canal which, it is assumed, would also aid the defecation process. Since *Tagart's* studies were performed with a flexible but sizeable radiopaque tube extending from the rectum through the flexure and out via the anus – which could not help but produce some degree of rectal and anal stimulation – it is difficult to estimate what the effects of voluntary and/or reflex-mediated contraction of the levator ani and the puborectalis, in particular, were on altering the angle of flexure. The study did show, however, that the angle was further reduced by some 5°–14° by 'straining, as in defecation' (it is assumed this meant voluntarily increased intraabdominal pressure while attempting to relax the perineal musculature).

It seems to us that perhaps an equally important benefit of the squatting position is that when the hips are flexed 90° or more the inferior portions of the gluteus maximus muscles no longer overlap the ischial tuberiosities (which now become palpable) nor the medially located ischiorectal fat pads, as they do in the erect posture. Thus, any medially directed force of the gluteal muscles on the anal canal mediated by the fat bodies is removed, and the resilience of the fat pads can now facilitate the lateral expansion of the canal as the fecal mass passes. The same effect is achieved by spreading the buttocks as one sits on a toilet seat (fig. 11/8).

It is interesting to note that dogs (which can also be taught socially acceptable patterns of anal continence and defecation) possess neither perineal flexure nor buttocks to compress ischiorectal fat, yet also assume a squatting position for defecation. In so doing, although there is no straightening of a flexure, the tail and hind-limbs are moved in opposite directions so that there is a tendency to draw the dorsal and ventral walls of the anal canal/anus apart and so that musculature attached to the coccyx (rectococcygeus) can act as a fixator for the anal canal [54]. Here also, the position permits maximum pliability of the ischiorectal fat pads. Perhaps the greatest advantage gained from the squatting position is that legs and feet are kept out of the way of the free-falling feces!

The Anal Sphincters

The Internal Anal Sphincter. The sphincteric region proper is a rather complicated combination of both smooth and striated muscle. The internal anal sphincter is composed entirely of smooth muscle and is continuous superiorly with the inner circular layer of the gut (rectum); it is, in fact, generally considered to be merely an approximately fourfold thickening of this layer. Inferiorly it ends with a well-defined (palpable) rounded edge 6–8 mm above the anal orifice [40]. It is traversed by fibers from the longitudinal layer which terminate in the submucosal layer, thus anchoring the overlying mucosa to the sphincter itself (fig. 11/6).

The internal anal sphincter is normally tonically contracted when the rectum and anal canal are 'at rest' – i.e., when the rectum is 'empty' (less than 20 ml of feces, fluid or flatus) or, if not, when it has not received any additional, distending material for 30s or more and is not distended to the point of conscious awareness (no feeling of fullness are perceived nor is there any desire to defecate) [24]. The mechnism by which its tone is maintained is unknown, but as a general rule smooth muscle has both motor and inhibitory innervation. In fact, it has been concluded that the internal sphincter has two excitatory innervations (cholinergic and α-adrenergic) and two inhibitory ones (β-adrenergic and an unknown). The relative importance of these four kinds of nerves in the operation of the internal anal sphincter remains to be determined, however [15].

The tonic contraction usually maintained by the internal sphincter is interrupted, however, in response to distension of the lower sigmoid and rectum ('rectal filling') which promptly produces relaxation [26, 35, 43, 73] (see fig. 11/10). This is the initial event in the chain of events resulting in defecation, and which may be collectively referred to as the *defecation reflex*. Rectal filling can be experimentally imitated through distension of an inserted balloon – a moving balloon being much more effective than a stationary one. The distal rectum and proximal anal canal ('rectal ampulla') are more sensitive to distension than more proximal sites. The consequent decrease in internal sphincter tone *(relaxation reflex)* [71] occurs even at small distending volumes and pressures (20–50 ml, 6–11 mm Hg) [48] and occurs whether or not any sensation is consciously perceived [67, and many others], but the degree of relaxation is determined by the degree of distension [49]. Some have suspected that the receptors for the afferent limb of the reflex lie within the mucosa, and others that they are in muscle layers. In cats local anesthetics have been claimed to abolish the reflex; in dogs and in man reflex is said to be present after removal of the rectal mucosa. In man it is only necessary to have 6 cm of the distal rectum in order to preserve the defecation reflex.

It is often stated that the rectum is empty except during the initiation of the defecation reflex. This is not true of the human or dog rectum.

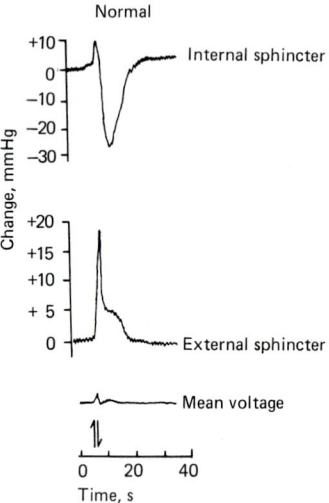

Fig. 11/10. Normal (continent) response to rectal distension. The upper graph demonstrates the 'relaxation reflex' of the involuntary musculature of the internal anal sphincter (*and* the surrounding *tonically* contracted voluntary muscle fibers) in response to inflation of a rectal balloon (arrows at bottom of chart). The lower graph demonstrates the nearly simultaneous 'inflation reflex' of the *phasically* contracted voluntary muscle fibers (of puborectalis as well as of external anal sphincter) in the continent individual. The 'inflation reflex' is actually a conditioned response learned during toilet training. Incontinent individuals often fail to recognize the onset of the relaxation reflex and thus to invoke the inflation reflex. Reproduced, with permission, from *Alva* et al. [2].

Rectal examination usually reveals feces. The rectum, like the stomach and cecum, relaxes receptively. Intrarectal pressure rises following distension, and then falls quickly to preinflation levels. An urge to defecate may be experienced, but it can be suppressed voluntarily and passes off rapidly as the ampulla adapts to the size and pressure of the contents (this process requires a minute or more and has been termed the *accommodation response* [24]) (see fig. 11/11). When the pressure has decreased, the tone of the internal sphincter is restored. As additional gas, liquid or solid enters the distal sigmoid and rectum, this process is repeated until the contents reach a critical amount or until the intrarectal pressure becomes excessive during peristalsis (in excess of 50 cm H_2O), at which time the internal sphincter enters a state of constant relaxation, its tone not being restored until the pressure has fallen (as it does following evacuation of the ampulla) [48].

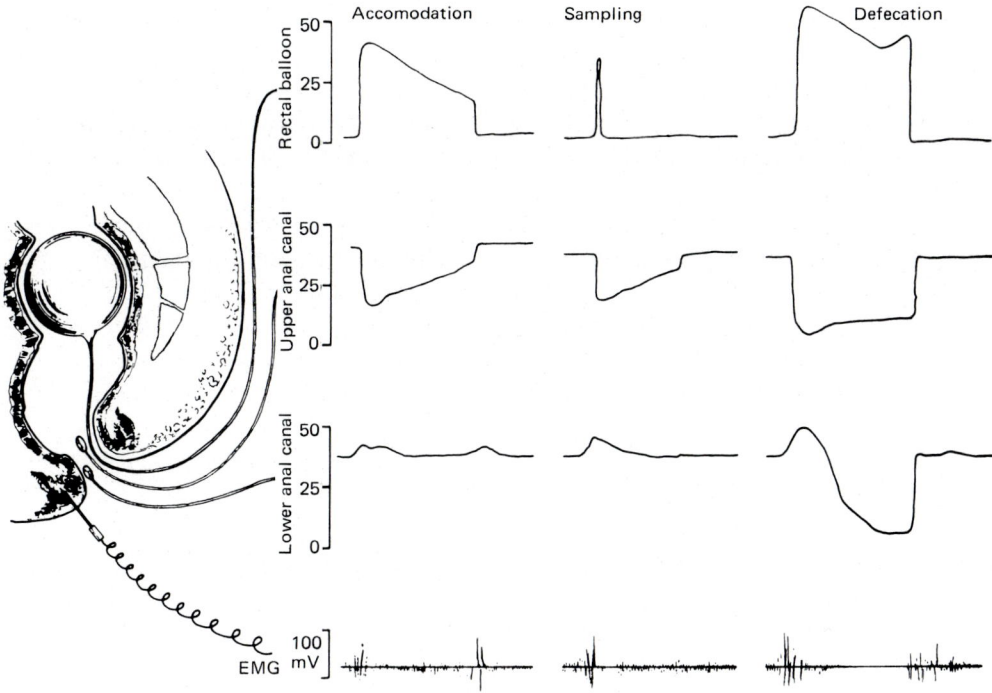

Fig. 11/11. Activity of upper and lower anal canal musculature (as demonstrated by changes in intraluminal pressure) and electromyography (EMG) in response to rectal filling. During accommodation (first panel) intra-ampullary pressure gradually diminishes as the initially resistive rectal wall relaxes, adjusting to the suddenly increased volume. The responses of the upper and lower anal canal are identical to those depicted more precisely in figure 11/10, except that the EMG also demonstrates a burst of activity when the rectal balloon is suddenly deflated toward the end of the accommodation response. In the sampling response (center panel) the rectal balloon has been rapidly inflated and deflated to initiate the response. In practice, the response is initiated either by sudden rectal filling or by suddenly increasing intra-abdominal pressure to stimulate baro (stretch) receptors in the terminal rectum. The relaxation reflex of the upper anal canal musculature thus invoked allows a portion of the distending substance to enter the anorectal junction where specialized nerve endings are able to distinguish solid, liquid and gas. During defecation (right panel) inflation of the rectal balloon above the threshold of accommodation (alternately achieved in practice by increasing intra-abdominal pressure, i.e. 'straining') results in sustained intra-ampullary pressure and thus sustained inhibition of the tonically contracted musculature. Phasic contraction of the musculature of the lower anal canal initially resists passage (in practice, to facilitate the desired increase in intra-abdominal pressure). Subsequent cessation of the phasic contraction then allows expulsion. Increased EMG activity occurs as levators and sphincters return the prolapsed and distended anal canal to normal position. Reproduced, with permission, from *Goligher* [40].

The viscoelastic properties of the ampullary wall are important to both defecation, in terms of triggering the defecation reflex, and to continence, in terms of providing a capacious and functional reservoir. In chronic ischemia of the rectosigmoid, the wall becomes fibrous and thus unyielding. The reservoir capacity is lost; a small amount of fecal matter triggers a disproportionate sense of urgency which must be responded to promptly. Conversely, a congenitally inelastic colon will fill to incredible volumes without triggering a defecation reflex. Ability to maintain continence following surgical ileo-anal anastomoses is directly correlated with the compliance and reservoir capacity of the neorectum (as well as frequency and strength of peristaltic contractions) [7]. In Hirschsprung's disease, the accommodation response does not occur. Neither will it occur in the normal individual when distension is unusually abrupt; in which case voluntary suppression of the defecation response may be impossible. This is called *urgent defecation*.

The mechanism of relaxation of the internal anal sphincter has been extensively studied (see *Duthie* [24] for details). It has been shown that the relaxation reflex persists when the spinal cord has been transected, when the pudendal nerves are severed or blocked, and during low and high spinal anaesthesia. Relaxation occurs if the presacral sympathetic (hypogastric) nerves are stimulated [34]. The relaxation reflex does not occur in Hirschsprung's disease, the tonic state being maintained at all levels of filling as well as during defecation [1]. It is currently held that the relaxation response is neither sympathetic nor parasympathetic [34], but is conditional by an intramural (locally mediated muscular) reflex and is initiated by the propulsive contraction (peristalsis) generated upon distension of the rectum [71]. No stimulus has been found which can induce reflex *contraction* of the internal sphincter. The only response normally available to this muscle is relaxation. This suggests that the internal anal sphincter is contracted at or very near maximal levels during the 'resting state' of the rectum/anal canal [2].

From the foregoing, it seems reasonable to conclude that the internal anal sphincter does *not* function to prevent passage of: (a) a distending mass or volume of gas approaching from within (as during initial rectal filling); (b) a mass or volume of gas already in the ampulla during the propulsive contraction generated by further filling, nor (c) a mass of or volume of gas that has reached a critical level (approximately 130 ml) through continued filling. During these moments, which might be considered as the critical moments in terms of maintaining anal continence (yet which, with the exception of (c), can hardly be categorized as moments of emergency), the internal sphincter is in a state of relaxation. Hence, it appears that the internal sphincter functions primarily during the 'resting state' when it acts (a) to prevent leakage of fluid or flatus from a distending mass or volume of gas to which the ampulla has already adapted or which is so small (1 ml or less) that its approach and presence is undetected, and (b) to guard against invasion from without.

External Anal Sphincter. Although studies of the organization and fine details of the structure of the remainder of the sphincteric complex date as far back as 1715, it has defied consistent and consensual anatomic description. The existence and disposition of subdivisions of the external anal sphincter and the extent and arrangement of longitudinal muscular and fascial planes in the sphincteric complex are still subjects for study [26].

As initially described by *Santorini* [70] in 1715 and further amplified by *Milligan and Morgan* [56], it has become customary to consider the external anal sphincter in three parts (deep, superficial and subcutaneous). A recent elaboration of this concept by *Shafik* [75] describes the external sphincter as consisting of a series of three muscular loops – top, intermediate and base (fig. 11/6, 11/12) – the disposition of which, he contends, facilitates the sphincter's role in maintenance of fecal continence. In agreement with earlier workers, he found the deepest portion of the sphincter ('top loop') to be ill defined and intimately associated with the puborectalis portion of the pelvic floor muscles. He goes further, however, in proposing that the puborectalis is actually a part of the external anal sphincter rather than of the levator ani [76]. Other recent investigators disagree with both his subdivision and loop arrangement [5, 19, 40, 42], regarding the external anal sphincter as one continuous, circumferential sheet or muscle mass. *Shafik's* claim regarding the puborectal sling has also been disputed [5, 59].

Regardless of how they may appear to be subdivided or fused or otherwise disposed, the striated muscle of the sphincteric complex (external sphincter and puborectalis) act as a unit as shown by electromyography [48, 69]. They are innervated by a branch or branches of the pudendal nerve (either from its perineal division or via the inferior rectal nerve) and by the perineal branch of the fourth sacral spinal nerve. Electromyography also shows that, in contrast with most other striated muscles, these muscles exhibit continual detectable activity (*Floyd and Walls* [32], any many

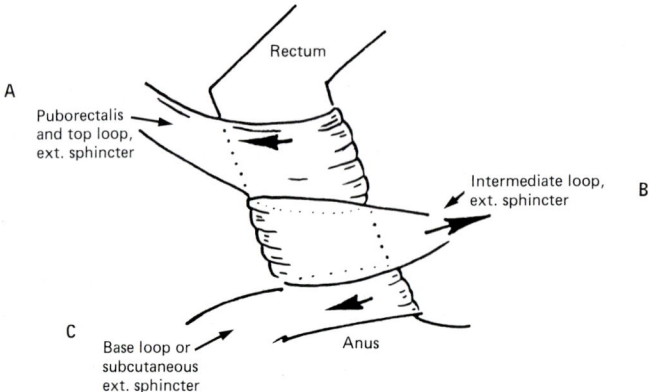

Fig. 11/12. Illustration of the 'triple-loop' arrangement of the external anal sphincter as proposed by *Shafik* [75]. **A** Top loop fused with puborectalis, attached to and pulling anteriorly toward the pubis. **B** Intermediate loop, attached to coccyx and pulling posteriorly. **C** Base loop attached to perineal body and skin, also pulling anteriorly.

others). Actually, they demonstrate two types of electrical activity: a basic (tonic) rhythm and another (phasic) rhythm which occurs upon reflex or voluntary contraction of the muscles [48].

It has long been known that, on the basis of histological appearance and speed of contraction, there are at least two types of striated muscle fibers: one smaller and slower, the other larger and faster; the higher myoglobin content of the former endows them with a reddish appearance in contrast to the 'white' appearance of the latter [11, 49]. It has been determined that in the external anal sphincter both types exist; the red fibers serving to maintain the tonic contraction of both the external sphincter and the puborectalis [49]. This activity is always present (except under extraordinary circumstances to be explained shortly), so that even at rest the external sphincter and the puborectalis have some tone. The level of tonic contraction varies somewhat with bodily position, being highest in the upright position. Thus these muscles are in harmony with the postural (antigravity) muscles which demonstrate similar tonus. The tone in the striated muscles of the anal region is amplified or inhibited via spinal reflexes (see below) and is not affected by transverse section of the spinal cord [64]; it disappears during sacral anaesthesia. The tonic activity is inhibited upon pronounced distension of the rectal ampulla [69].

As mentioned previously, this tonic activity helps to maintain the 80° angle of the perineal flexure. The contribution of this constant activity to the intraluminal pressure of the anal canal inferior to the flexure during the 'resting state' can be assessed by measuring the decrease which results when the striated muscles are paralyzed by muscle relaxants or pudendal block anaesthesia. *Frenckner and von Euler* [33] observed a 15%-decrease in the intraluminal pressure after pudendal block, a finding which generally agrees with what was described as 'only a small decrease' in earlier studies using muscle relaxants [25]. These findings lend further support to the conclusion implied earlier that, apart from the role played by the tonus of the striated muscle in maintaining the perineal flexure (and the contributions, perhaps, of some passive, 'mechanical' factors), it is the tonic contraction of the internal anal sphincter that is primarily responsible for maintaining anal continence when the rectum is 'at rest'.

In contrast to the internal sphincter, however, the activity detected in the striated muscles during the resting state does not represent maximal contraction, but only the base-level tonus maintained by the red muscle fibers. When we are being toilet trained, we learn to recognize the onset of the relaxation reflex, and to simulataneously contract the voluntary musculature to prevent unwanted defecation. This markedly increased ('phasic') activity emanates from the white muscle fibers. We also learn to contract these fibers defensively in response to perianal stimulation *(anal reflex)*, and during increases in intra-abdominal pressure (as during deep inspiration, sneezing and the Valsalva maneuver) if we do not wish to defecate or pass gas [17, 24]. The intensity of this activity is regulated voluntarily. Thus, one may conclude that the external anal sphincter, while normally relatively insignificant in maintaining closure of the anal canal inferior to the perineal flexure during the resting state, actively resists what is detected as imminent passage from within as well as from without.

Contraction of the external sphincter in response to rectal distension *(inflation reflex)* [48], produced either by the initial fillings or by increases in intrarectal pressure due to subsequent filling or peristalsis, is of particular importance with regard to maintaining continence since it occurs in unison with the relaxation reflex of the internal sphincter. The pudendal block studies of *Freckner and von Euler* [33] found the external sphincter to be decisive in maintaining continence immediately following distension, the (primarily phasic) activity of the muscle accounting for 60% or more of intraluminal pressure depending on the abruptness and level of distension. However, recall that these white muscle fibers become rapidly fatigued. In fact, maximal contraction of the striated muscle in this region can be sustained only for a few seconds and, despite continued effort, is followed by a steady and progressive relaxation (evidenced by both declining electromyographic activity and intraluminal pressure) until the base level of tonus is reached. Phasic activity cannot be maintained in this region for more than about two minutes [69] and the length of time during which increased intraluminal pressure can be voluntarily sustained in the anal canal is only 40–60 s [38, 63, 69, 80] (see fig. 11/10).

Synchrony of the Sphincters. Contraction and relaxation (inhibition) of the two sphincters must therefore be in synchrony with each other and be correlated with rectocolic peristalsis as well if continence is to be maintained. Correlation of sphincteric contraction was demonstrated by *Gaston* [38]. By means of saline enema, the colon was distended sufficiently to initiate peristalsis. The changes in intrarectal pressure due to peristalsis were recorded simulataneously with variations in anal intraluminal pressure. The recordings demonstrated that during each peristaltic wave the increased pressure exerted against the outlet is countered by an appropriate degree of contraction of the external anal sphincter. (Recall: Increases in intraluminal pressure must be attributed to the external sphincter since the internal sphincter, maximally contracted during the resting state before peristalsis, is inhibited by increases in intrarectal pressure.) As the peristaltic wave recedes and pressure against the sphincter diminishes, relaxations of the sphincter occurs so that at any given moment contraction is maintained just sufficient to effectively counter the peristaltic thrust. Normally, peristaltic waves last less than a minute, with the peak of each wave enduring for only a few seconds. Thus maximal contraction of the external sphincter is ordinarily required for only a few seconds at a time and its endurance is usually sufficient for the demands normally placed upon it.

The synchrony of the two anal sphincters is evident both in the way in which the relaxation and inflation reflexes occur in unison in response to rectal distension and in the way in which the previously inhibited tone of the internal sphincter returns as the phasic fibers of the external sphincter either fatigue or relax as the peristaltic wave recedes or, if filling has occurred, as ampullary adaptation occurs.

During the 'resting states' between the peristalsis stimulated by the initial filling and the time at which maximal filling or pressure is reached, the tonic activity of the external sphincter is augmented so that the intraluminal pressure in the anal canal is increased over that maintained during the 'resting state' preceding the initial filling (85% of which was contributed by the internal sphincter). Thus, between peristalses when the rectal ampulla is substantially – but not maximally – filled (i.e. to the level of conscious awareness and the need to defecate is building), *Frenckner and von Eular* [33] found the two sphincters more nearly sharing the responsibility for continence. They found that during such times the tonic contractions of the internal and external sphincters contributed to the intraluminal pressure in the ratio of 65:35.

Another subtlety of the synchrony of the sphincters occurs between initial and maximal filling. This is called the *sampling response* [24] (see fig. 11/11). Following each increase in the distension of the ampulla, as receptive relaxation is occurring but prior to restoration of the tone of the internal sphincter (i.e. during those moments when continence is primarily determined by the level of contraction of the voluntary musculature), one may voluntarily and abruptly increase intra-abdominal pressure (Valsalva maneuver) sufficiently to force a small amount of the distending substance into the partially relaxed upper anal canal while increasing contraction of the lower anal canal. Thus the distending matter may stimulate sensory nerve endings located in the anal epithelium approximately 1.5 cm above the anal valves. Nervous impulses from these endings enable higher centers to distinguish gas, liquid and solid. If the recent distension is determined to be due to gas, it may be allowed, through voluntary relaxation of the lower anal canal, to pass; by contrast, if liquid or solid is detected (or if passage of gas is deemed inappropriate), one may elect to increase contraction of the voluntary musculature sphincter sufficiently to push the mass back into the ampulla to await completion of the accommodation response.

If the need to defecate is not satisfied, however, and content and/or pressures are allowed to reach maximal levels – or if filling occurs very abruptly – the internal sphincter is pushed to the level of constant relaxation and the tonic activity of the striated musculature of the region is inhibited as well. Thus, the burden of maintaining both the perineal flexure and the intraluminal pressure (i.e. of maintaining continence) falls on the phasic muscle fibers, and an emergency situation exists where their endurance may be sorely tested. During such emergencies, extraordinary measures may be taken, such as (a) standing to maximize the angle of the perineal flexure; (b) contraction of the gluteus maximus muscles and/or sitting to 'pack' ischiorectal fat against the lateral walls of the anal canal (fig. 11/8), or (c) lying supine in an attempt to do both (a) and (b) simultaneously. Such emergencies may even produce a sympathetic response, which is said to aid sphincter contraction [84] via direct innervation (recall, however, that direct stimulation of the presacral sympathetic nerves was found by *Freckner and Ihre* [34] to elicit the relaxation reflex of the internal sphincter) and indirectly via the circulated adrenalin.

The findings of the laboratory investigations cited above have provided considerable insight into the roles typically assumed by the two sphincters and their relative importance in maintaining continence in the *normal* individual – i.e. that they *both* play significant roles at various moments and under certain circumstances, and that in actual fact it is the reciprocity of the combined tonic contraction of both sphincters with the phasic contraction of the voluntary musculature that is the cardinal factor enabling complete continence in the normal individual. The findings of clinical studies go somewhat further, however, in determining which sphincter is the sine qua non for maintaining either absolute or relative continence.

Postoperative studies of patients who have undergone internal sphincterotomies [8, 9, 37, 41, 67] have found that, in spite of the consequent reduction in intraluminal pressure during the resting state, full continence is common in the absence of internal sphincter activity; no cases of gross incontinence were reported. 'Minor defects' in continence were attributed to scarring or areas of anaesthesia at the surgical site. In studies of patients with gross incontinence suffered for a variety of reasons, *Alva* et al. [2] found the tonic contraction and the relaxation reflex of the internal sphincter present but the tonic and phasic (inflation reflex) contraction of the external sphincter absent. In all the grossly incontinent patients studied by *Parks* et al. [65] the pathology was detected in the striated musculature, particularly the external sphincter.

The preponderance of clinical findings support the conclusion [stated by *Bennett and Goligher,* 9] that continence normally depends in greatest measure on the voluntary musculature of the anal region – the external sphincter and the puborectalis. The role of the internal sphincter seems to be concerned with the refinements of anal continence (i.e. the completeness of the gas- and water-tight seal during the resting state), and in many people its role can be assumed by the voluntary muscles.

Because they do not differ greatly in size in humans (see fig. 11/6), the gross appearance of the two sphincters offers no clues as to their relative importance. However, it is of interest (if little else) that in the dog referred to earlier as an animal that, despite the lack of a perineal flexure, is capable of learning 'socially-acceptable' patterns of fecal continence, the size of the internal anal sphincter – particularly its width – is much less than that of the external anal sphincter [54].

The Defecation Reflex

The voluntary and controlled act of defecation in the normal mature individual is a learned behavior, much influenced by social and environmental factors. The infant and the domestic animal must first be taught to recognize the connection between the afferent signal (urge) and the subsequent automatic elimination (the 'defecation reflex'), and then may learn the ability voluntarily to suppress the reflex (i.e. to maintain fecal continence) within limits. As stated previously, the initial event triggering the defecation reflex is the peristaltic propulsion of a fecal bolus into the rectum – often after arising *(orthocolic reflex)* or following ingestion of food or beverage *(gastroileal/gastrocolic reflexes)*. The rectal distension produced causes the relaxation reflex to occur, and the urge to defecate is experienced. In the trained subject, conscious decisions are made depending on the intensity of that urge. If mild to moderate, one may choose to suppress the reflex, taking no further action at that time. Moderate intensity will usually motivate the individual to seek a toilet in addition to the temporary suppression. In either case, the inflation reflex will counter the relaxed internal sphincter until the accommodation reflex ensures. Urgent defecation – which may allow neither decision making nor suppression – has already been discussed.

When conditions are appropriate, the sitting/squatting posture will be assumed. If the intraluminal pressure has not reached the urgent-defecation stage (approximately 50 cm Hg – in which case mere cessation of whatever voluntary contraction one has managed to maintain will result in immediate automatic elimination), then efforts are made to increase intraabdominal pressures via the *Valsalva maneuver* (bearing down). The diaphragm is contracted and fixed in position and the glottis is closed (i.e. a breath is drawn and held), followed by contraction of the abdominal wall and pelvic floor muscles. This is often accompanied by a variety of facial muscle contractions as well. The increased pressure may coincide with or may stimulate peristaltic contraction of the gut, the concomitant forces raising the intraluminal pressure sufficiently to relax the internal anal sphincter (relaxation reflex) and inhibit the tonic contractions of the external sphincter (see fig 10/11). As propulsive (mass) contractions of left colon, sigmoid and rectum develop, all inhibitory influences exerted by higher (cortical) centers cease and phasic contraction of the voluntary musculature stops as well. Relaxation of the puborectal sling allows straightening of the perineal flexure of the anal canal as the pelvic floor descends, and the anal canal prolapses and dilates to allow passage of the fecal mass. Following evacuation, intra-abdominal and intraluminal pressures drop precipitously to normal levels, the pelvic floor ascends and the anal canal is retracted via reflex contraction of the levator ani muscle; the tonic contractions of the puborectal sling and anal sphincters resume, restoring the normal perineal flexure and closure of the anal canal.

The defecation reflex will be abolished if the peripheral receptors are destroyed, if the afferent or efferent nerves are severed (the fibers of both are conveyed by pelvic splanchnic nerves), or if the sacral cord is destroyed. Following spinal transection above the sacral region the reflex becomes completely automatic (once spinal shock has worn off); consciousness of receptor stimulation is lost, and, consequently, the ability to suppress defecation, just as in the untrained infant.

Pharmacological studies which should have told us a great deal about this reflex have yielded comparatively little, in part because very few have been carried out and also because the results in the light of current knowledge have been hard to interpret. Atropine, for example, seems to be without influence on the reflex in vivo and ganglionic blockade has an equivocal effect on internal sphincter relaxation produced by pelvic splanchnic nerve stimulation.

Other Factors Effecting Continence and Defecation

Longitudinal Anal Musculature. Disagreements similar to those concerning the anatomical details of the external anal sphincter and puborectalis exist among investigators [42, 56, 63, 76, 77] concerning the organization and disposition of the longitudinal muscle fibers in the anal region and its functional role. It is basically accepted, however, that the main layer of longitudinal muscle (the 'conjoined longitudinal muscle') [57]: (a) is a downward continuation of the outer longitudinal layer of smooth muscle of the gut (rectal) wall joined by striated muscle of the levator ani and mixed with elastic tissue, and (b) extends into a plane between the internal and external anal sphincters from which a number of fine septae (fibrous and/or muscular (?)) diverge both internally and externally to traverse both sphincters and insert into the submucosa and dermis respectively (the latter forming the *corrugator cutis ani* 'muscle') (see fig. 11/6). Some disagreement remains as to the organization and composition of that plane and the septae (ligaments) extending from it [31, 40, 63, 77] and as to whether the corrugator cutis ani is a muscle at all [29, 55, 77]. The longitudinal muscle (or at least its striated component) may play a minor role in anal competence in that it may help to resist the prolapsing force of the fecal mass (fig. 11/13). It is common experience that the anus is retracted when the external sphincter is actively constricted during moments of urgency. (The apparent retraction results in part, however, from the simultaneous voluntary contraction of the gluteus maximus muscles referred to earlier.) *Shafik* [77] describes an additional role of the longitudinal muscle in stating that the corrugations of the anal and perianal skin (caused by the fibrous septa extending through the tonically contracted base-loop of the external sphincter) help the subcutaneous portions (base-loop) of the external sphincter to induce an airtight anal occlusion.

It is generally agreed, however, that the major role of the longitudinal muscle occurs during defecation when it (a) facilitates passage of the fecal mass by shortening the anal canal, and (b) fixes the anal canal (and – if the muscularis mucosae is competent – its mucocutaneous lining), resisting the tendency of the increased intra-abdominal pressure and the passage of the (possible abrasive) fecal mass to prolapse the canal. Its continued contraction following passage of the mass normally retracts the canal, restoring its position from whatever degree of temporary prolapse that may have occurred. *Shafik* [77] also assigns an additional role here, claiming that – through the portion of the longitudinal muscle derived from the levator ani – a laterally-directed pulling force is applied to the canal walls, widening the canal and tending to evert (open) the anal orifice. He believes this is, in fact, the main function of the longitudinal musculature and has suggested the name 'evertor ani' muscle for it.

Threshold Pads. One more mechanism proposed as playing a role in continence and defecation seems worth mentioning. Substantial venous plexuses lie between the mucosa and skin and the internal and external anal sphincters immediately above and below the pectinate line (see fig. 11/6). It has long been held that varicosities of these anal canal veins result in hemorrhoids or 'piles'. As a result of dissecting 95 cadaveric anorectal specimens, however, *Thomson* [81] confirmed what was known to 19th century French and German anatomists, that is, that these veins are *normally* dilated and tortuous ('varicose') in appearance. The submucosa is normally extraordinarily thickened due to the presence of the saccular veins and may resemble cavernous tissue on section. *Claude Bernard* is credited with having suggested that the function of the submucosal venous plexus was

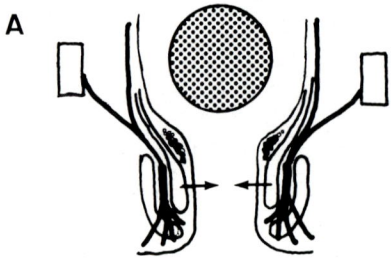

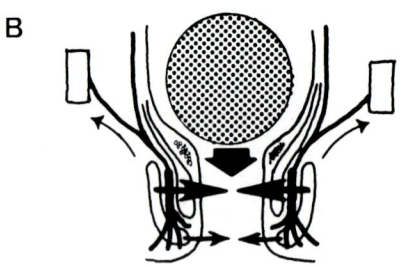

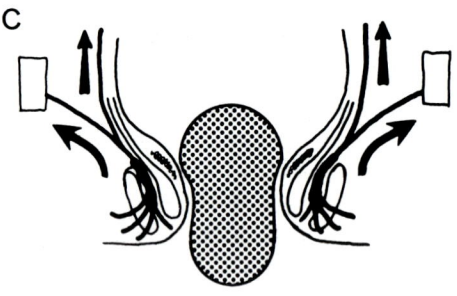

Fig. 11/13. Schematic illustration of a proposed mechanism of action of the conjoint longitudinal musculature of the anorectal region. **A** At rest. **B** During increased intraluminal pressure (increased fecal mass or peristalsis). **C** During defecation. In the latter, the more vertically oriented smooth muscle component (derived from the longitudinal outer coat of the gut) and the more horizontally oriented striated muscle component (derived from the levator ani), act together to shorten the anal canal and to resist the tendency toward proplase (i.e. they serve as fixators for the anal canal). The same occurs during **B** when attempting to maintain continence during peristalsis, but is combined with voluntary contraction of the external sphincter. However, during defecation, the relaxed sphincters allow the more horizontally disposed muscle fibers to widen the canal (and, according to *Shafik* [76], to evert the anal orifice). Adapted from *Shafik* [77].

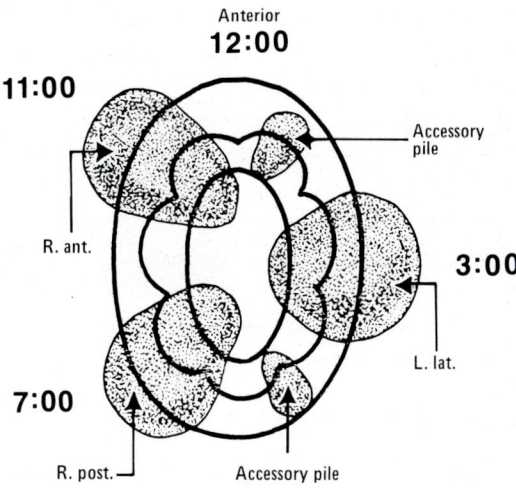

Fig. 11/14. Diagram of the anal region demonstrating the positions of the 'corpora cavernosa recti' [78] or 'threshold pads' [82]. Pathologically, these venous masses become hemorrhoids or piles. Their positions are frequently described in terms of a clock-face, with 12:00 situated anteriorly. Modified from *Goligher* [40].

to assist in anal closure. *Thomson* [82] has been the most recent advocate of this theory. Believing that it would be difficult for the anal canal to both dilate sufficiently for defecation and yet constrict tightly enough to maintain continence (for both gas and liquid as well as solid) were the sphincters lined with only a thin mucosal layer, he proposes that pliable 'threshold pads' on which the sphincters can squeeze are necessities. He refers to, and has confirmed, the work of *Stelzner* et al. [78] wherein direct arteriovenous anastomoses were demonstrated between the submucosal arteries of the anal region and the sacculated veins, providing a mechanism whereby the anal submucosa could become actively engorged and hence be variably pliable and turgid. The sacculated veins and arteriovenous anastomoses are found even in neonates. The presence of variably pliable and turgid threshold pads just proximal to the slit-like lumen of the anal canal supports the 'flutter-valve' concept of *Phillips and Edwards* [67] (see fig. 11/7).

It is well known that the submucosal hemorrhoidal veins occur in distinct clusters (referred to by *Stelzner* as pads or 'corpora cavernosa recti'); in fact, the clusters of the apparently dilated veins are often described as resembling 'bunches of grapes' in appearance. The location of the clusters (there are usually three, although 'accessory clusters or piles' do occur) is determined by the way the three terminal branches of the superior rectal artery are distributed: one to the left of the rectum and two to the right in anterior and posterior positions (see fig. 11/14). Clinicians described the position of the *anorectal cavernous bodies* ('piles' when enlarged and problematic) by considering the anus as a clock face with 12 o'clock directed anteriorly and with the venous masses occurring at 3, 7 and 11 o'clock. However, because patients often lie on one side or assume a knee-elbow position for anorectal examination, the clock face becomes rotated or inverted causing confusion for some students [44]. *Thomson* presumes that the 'threshold pad' is broken up into separate 'cushions' to better accommodate changes in anal circumference and believes the mucosal adaptability is further enhanced by the superimposition of vertical corrugations (the rectal columns of Morgagni).

Thomson, claiming that the histological appearances of hemorrhoidectomy specimens do not differ substantially from that of normal anorectal submucosa, supports the views of others [36, 46, 62] concerning the etiology of hemorrhoids, i.e. that rather than being due to abnormal veins and venous congestion per se, they are due simply to a prolapse of the lining of the anal canal – including the submucosal cushions – as a result of stretching or fragmentation of the muscularis mucosae [82]. This view has been disputed 'on clinical and operative grounds' [40].

References

1 Aaronson, K.; Nixon, H.: A clinical evaluation of anorectal pressure studies in the diagnosis of Hirschsprung's disease. Gut *13:* 138 (1972).
2 Alva, J., Mendelhoff, A.; Schuster, M.: Reflex and electromyographic abnormalities associated with fecal incontinence. Gastroenterology *53:* 101–106 (1967).
3 Alvarez, W. C.: Movement of the colon, chap. XXV; in An introduction to gastroenterology (Hoeber, New York 1940).
4 Annegers, J.: Functions of pancreatic juice and of bile in assimilation of dietary triglycerides. Archs intern. Med. *93:* 9 (1956).
5 Ayoub, S.: Anatomy of the external anal sphincter in man. Acta anat. *105:* 25–36 (1979).
6 Bayliss, W.; Starling, E.: The movements and the innervation of the large intestine. J. Physiol., Lond. *26:* 107–118 (1900).
7 Beart, R., Jr.: Rectal continence: anatomic and physiologic implications of ileo-anal anastomoses. 1st. Ann. Meet. Am. Ass. Clinical Anatomists, Mayo Clinic, Rochester 1984.

8   Bennett, R.; Duthie, H.: The functional importance of the internal anal sphincter. Br. J. Surg. *51:* 355–357 (1964).
9   Bennett, R.; Goligher, J.: Results of internal sphincterotomy for anal fissure. Br. med. J. *ii:* 1500–1503 (1962).
10 Bond, J.; Currier, B.; Buchwald, H.; Levitt, M.: Colonic conservation of malabsorbed carbohydrates. Gastroenterology *78:* 444–447 (1980).
11 Buller, A.: The physiology of the motor unit; in Watson, Disorders of voluntary muscle, pp. 17–28 (Churchill, London 1969).
12 Cannon, W.: The movements of the intestine studied by the roentgen rays. Am. J. Physiol. *6:* 251–277 (1907).
13 Caprilli, R., Onari, L.: Origin, transmission and ionic dependence of colonic electrical slow waves. Scand. J. Gastroent. *7:* 65–74 (1972).
14 Chang, W.; Leblond, C.: Renewal of the epithelium in the descending colon of the mouse. Parts I–III. Am. J. Anat. *131:* 73, 101, 111 (1971).
15 Christensen, J.: Motility of the colon; in Johnson, Physiology of the gastrointestinal tract, chap. 14, pp. 445–471 (Raven Press, New York 1981).
16 Christensen, J., Anuras, S., Hauser, R.: Migrating spike bursts and electrical slow waves in the cat colon: effect of sectioning. Gastroenterology *66:* 240–247 (1974).
17 Collins, C.; Duthie, H.; Shelley, T.; Whittaker, G.: Force in the anal canal and anal continence. Gut *8:* 354–360 (1967).
18 Conklin, L.; Christianesen, J.. Local specialization of the ileo-cecal junction in the cat and the opossum. Am. J. Physiol. *228:* 1075–1081 (1975).
19 Courtney, H.: The posterior subsphincteric space: its relation to posterior horseshoe fistula. Surgery Gynec. Obstet. *89:* 222–226 (1949).
20 Coutourier, D.; Roze, C.; Coutourier-Turpin, N.; Debray, L.: Electromyography of the colon in situ; an experimental study in man and in the rabbit. Gastroenterology *56:* 317–322 (1969).
21 Cummings, J.: Absorption and secretion by the colon in symposium on colonic function. Gut *6:* 323–327 (1975).
22 Dawson, A.: Regulation of blood ammonia. Gut *19:* 504–509 (1978).
23 Didio, L.: Dados anatomicos sobre o 'piloro' ileo-ceco-colico. (Com observacao direta in vivo de 'papila' ileo-ceco-colica.) (English summary); thesis, Sao Paulo (1952).
24 Duthie, H.: Dynamics of the rectum and anus. Clin. Gastroent. *3:* 467–477 (1975).
25 Duthie, H., Watts, J.. Contribution of the external anal sphincter to the pressure zone in the anal canal. Gut *6:* 64–88 (1965).
26 Duthie, H., Wormsley, K.: Anorectal region; in Scientific basis of gastroenterolgy (Churchill-Livingstone, Edinburgh 1979).
27 Edmonds, C.: Transport of sodium and secretion of potassium and bicarbonate by the colon of normal and sodium depleted rat. J. Physiol. *193:* 589–602 (1967).
28 Edmonds, C.: Symposium on colonic function. Electrical potential difference of colonic mucosa. Gut *16:* 315–318 (1975).
29 Ellis, G.: Demostrations of anatomy; 8th ed., p. 420 (Smith, Elder, London 1878).
30 Field, M.: Intestinal secretion. Gastroenterology *66:* 1063–1084 (1974).
31 Fine, J., Lawes, C.: On the muscle-fibres of the anal submucosa, with special reference to the pectin band. Br. J. Surg. *27:* 723–727 (1940).

32 Floyd, W.; Walls, E.: Electromyography of the sphincter ani externus in man. J. Physiol. *122:* 599–609 (1953).
33 Frenckner, B.; Euler, C., von: Influence of pudendal block on the function of the anal sphincters. Gut *17:* 482–489 (1975).
34 Frenckner, B.; Ihre, T.: Influence of autonomic nerves of the internal anal sphincter in man. Gut *17:* 306–312 (1976).
35 Garry, R.: The response to stimulation of the caudal end of the large bowel in the cat. J. Physiol. *78:* 208–224 (1933).
36 Gass, O., Adams, J.: Hemorrhoids: etiology and pathology. Am. J. Surg. *47:* 40 (1951).
37 Gaston, E.: Fecal continence following resections of various portions of the rectum with preservation of the anal sphincters. Surgery Gynec. Obstet. *87:* 669–678 (1948).
38 Gaston, E.: Physiological basis for preservation of fecal continence after resection of the rectum. J. Am. med. Ass. *146:* 1486–1489 (1952).
39 Gillespie, J.: The electrical and mechanical responses of intestinal smooth muscle to stimulation of their extrinsic parasympathetic nerves. J. Physiol. *162:* 76–92 (1962).
40 Goligher, J.: Surgery of the anus, rectum and colon; 4th ed. (Bailliere Tindall, London 1980).
41 Goligher, J.; Hughes, E.: Sensibility of the rectum and colon: its role in the mechanism of anal continence. Lancet *i:* 543 (1952).
42 Goligher, J.; Leacock, A.; Brossy, J.: The surgical anatomy of the anal canal. Br. J. Surg. *43:* 51–61 (1955).
43 Gowers, W.: The automatic action of the sphincter ani. Proc. R. Soc. *26:* 77 (1877).
44 Green, J., Silver, P.: An introduction to human anatomy (Oxford University Press, New York 1981).
45 Hawkens, P.; Mashiter, K.; Turnberg, L.: Mechanisms of transport of NaCl and K in the human colon. Gastroenterology *74:* 1241–1247 (1978).
46 Hughes, E.: Surgery of the anus, anal canal and rectum (Livingstone, Edinburgh 1957).
47 Hukuhara, T.; Miyaka, T.: The intrinsic reflexes in the colon. Jap. J. Physiol. *9:* 49–55 (1959).
48 Ihre, T.: Studies on anal function in continent and incontinent patients. Scand. J. Gastroent. *9:* suppl. 25 (1974).
49 Kerremans, R.: Morphological and physiological aspects of anal continence and defecation (Arscia, Belgium 1969).
50 Levitt, M.; Bond, J.; Levitt, D.: Gastrointestinal gas; in Johnson, Physiology of the gastro-intestinal tract, chap. 52, vol. 2 (Raven Press, New York 1981).
51 Levitt, M.: Intestinal gas production: recent advances in flatology. New Engl. J. Med. *302:* 1474–1745 (1980).
52 Lewis, G.; Partin, H.: Fecal fat on an essentially fat free diet. J. Lab. clin. Med. *44:* 91 (1954).
53 Mickelson, O.: Intestinal synthesis of vitamins in the non-ruminant. Vitams Horm. *14:* 1–82 (1956).
54 Miller, M.: Anatomy of the dog (Saunders, Philadelphia 1964).

55 Milligan, E.: The surgical anatomy and disorders of the perianal space. Proc. R. Soc. Med. *36:* 365 (1943).
56 Milligan, E.; Morgan, C.: Surgical anatomy of the anal canal, with special reference to anorectal fistula. Lancet *ii:* 1150–1156, 1213–1217 (1934).
57 Netter, F.: The CIBA collection of medical illustrations; Digestive system; Part II; Lower digestive tract, Ciba Pharmaceutical Co. (1962).
58 Neutra, M.; Padykula, H.: The gastrointestinal tract; in Weiss, Histology: Cell and tissue biology; 5th ed. (Elsevier Biomedical, New York 1983).
59 O'Rahilly, R.: Basic human anatomy: A regional study of human structure (W. B. Saunders, Philadelphia 1983).
60 Owen, R.; Nemanic, P.: Antigen processing structures of the mammalian intestinal tract: an SEM study of lymphoepithelial organs; Scanning Electron Microscopy, Vol. II (SEM Inc., O'Hare, Ill., 1978).
61 Pace, J.: Stereoscopic micro-anatomy of human colonic mucosa and its blood vessels. J. Anat. *103:* 602P (1968).
62 Parks, A.: The surgical treatment of hemorrhoids. Br. J. Surg. *43:* 337 (1955).
63 Parks, A.; Porter, N.; Melzak, J.: Experimental study of the reflex mechanism controlling the muscles of the pelvic floor. Dis. Colon Rect. *5:* 407–414 (1962).
64 Parks, A.: Post-anal perineorrhaphi for rectal prolapse. Proc. R. Soc. Med. *59:* 477 (1967).
65 Parks, A.; Swash, M.; Urich, H.: Sphincter denervation in anorectal incontinence and rectal prolapse. Gut *18:* 656 (1977).
66 Patey, D.: Aetiology of varicosity (Letter to the editor). Br. med. J. *ii:* 712 (1972).
67 Phillips, S., Edwards, A.: Some aspects of anal continence and defecation. Gut *6:* 396–406 (1965).
68 Pinter, K.; Mclean, A.: Correlation between weight, nitrogen and the fat content of stool. Am. J. clin. Nutr. *21:* 1310–1312 (1968).
69 Porter, N.: A physiological study of the pelvic floor in rectal prolapse. Ann. R. Coll. Surg. *31:* 379–404 (1962).
70 Santorini (1715): Quoted in Gabriel, W.: The principles and practice of rectal surgery; 5th ed. (Lewis, London 1963).
71 Scharli, A.; Kiesewetter, W.: Defecation and continence: some new concepts. Dis. Colon Rect. *13:* 81–107 (1970).
72 Schofield, G.; Atkins, A.: Secretory immunoglobulin in columnar epithelial cells of the large intestine. J. Anat. *107:* 491–504 (1970).
73 Schuster, M.: The riddle of the sphincters. Gastroenterology *69:* 249–262 (1975).
74 Silva, D.: The fine structure of multivesicular cells and large microvilli in the epithelium of the mouse colon. J. Ultrastruct. Res. *16:* 693–705 (1966).
75 Shafik, A.: A new concept of the anatomy of the anal sphincter mechanism and the physiology of defecation. I. The external anal sphincter: a triple-loop system. Investve Urol. *13:* 412–419 (1975).
76 Shafik, A.: A new concept of the anatomy of the anal sphincter mechanism and the physiology of defecation. II. Anatomy of the levator ani muscle with special reference to puborectalis. Investve Urol. *13:* 175–182 (1975).
77 Shafik, A.: A new concept of the anatomy of the anal sphincter mechanism and the physiology of defecation. III. The longitudinal anal muscle: anatomy and role in anal sphincter mechanism. Investve Urol. *13:* 271–277 (1976).

78 Stelzner, F.; Staubesand, J., Machledit, H.: Das corpus cavernosum rectidie. Grundlage der inneren Hämmorrhoiden. Arch. klin. Chir. *299:* 302–312 (1962).
79 Tagart, R.. The anal canal and rectum: their varying relationship and its effect on anal continence. Dis. Colon Rect. *9:* 449–452 (1966).
80 Taverner, D.; Smiddy, F.: An electromyographic study of the normal function of the external anal sphincter and pelvic diaphragm. Dis. Colon Rect. *2:* 153 (1959).
81 Thomson, W.: The nature of haemorrhoids. Br. J. Surg. *65:* 542–552 (1975).
82 Thomson, W.: Hemorrhoids: their true nature and best management. Res. Staff Phys. *104:* 20s–27s (1976).
83 Weisbrodt, N.: Motility of the small intestine; in Johnson, Physiology of the gastrointestinal tract, vol. 1, pp. 411–444 (Raven Press, New York 1981).
84 Williams, P.; Warwick, R.: Gray's anatomy; 36th Br. ed. (Saunders, Philadelphia 1980).
85 Wright, R.; Florey, H.: The secretion of the colon of the cat. Q. Jl exp. Physiol. *28:* 207–212 (1938).
86 Wrong, O.; Metcalfe-Gibson, A.; Morrison, R.; Ng S.; Harvard, A.: In vivo dialysis of feces as a method of stool analysis. Technique and results in normal subjects. Clin. Sci. *28:* 357–375 (1965).
87 Wrong, O.: Nitrogen metabolism in the gut. Am. J. clin. Nutri. *31:* 1587–1593 (1978).

12 The Gut and the Diffuse Endocrine System

Introduction and Historical Perspective

A great deal has been said up to this point about the gut hormones but relatively little about their cells of origin, the widespread distribution of these cells, and the system(s) which they compose or of which they are a part (if indeed such exist). Enteroendocrine cells have been described in relation to 'The Gastric Mucosa' (see corresponding paragraph, chapter 5) and their presence was noted in the small intestine. Their crucial role in the coordination and control of GI function and current interest in them as subjects of investigation warrant a separate and more detailed treatment here.

Gut Hormones

Secretin has the distinction of being the first hormone or blood-borne chemical messenger postulated to exist [3; for an eyewitness account of its discovery, a logical consequence to experimentation on the afternoon of January 16, 1902 see *Martin*, 16]. Although the existence of gastrin seemed likely in 1911 following the experiments of *Edkins* (the difficulties and pitfalls of confirmation were referred to earlier, see chapter 5), it was decades later before definite evidence of other gut hormones (gastrin and CCK) became available. Although the science of endocrinology experienced marked development during the interim, the gut was left behind in the advance of hormone physiology and biochemistry because its hormones are not secreted by cells amassed into discrete glands but frequently by singly occuring, diffusely scattered cells. Apart from the pyloric antrum, where the greatest density of enteroendocrine cells occurs (fig. 12/1), it was difficult to locate the presumed source of any of the suspected hormones. For the same reason, tissue extracts were difficult to prepare. The only certain way to designate the cells of origin is by immunofluorescence; this was possible, of

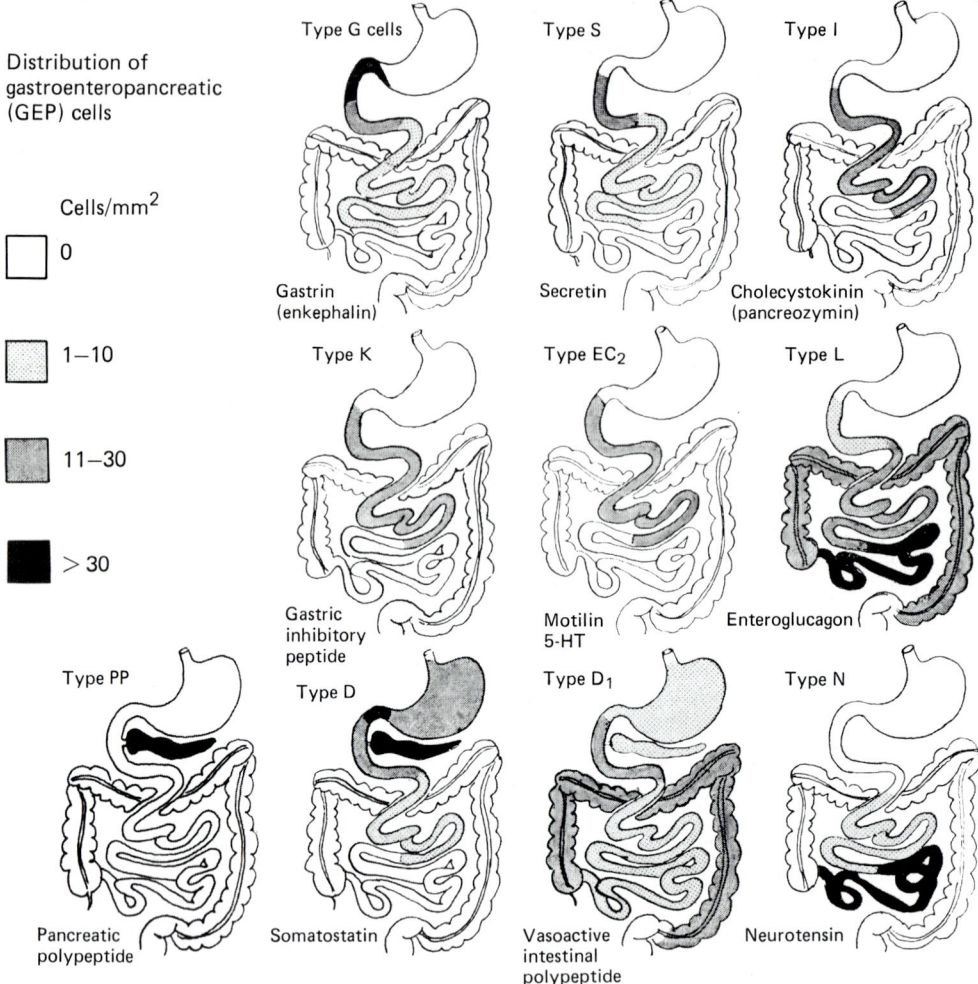

Fig. 12/1. Approximate quantitative distribution of ten types of gastroenteropancreatic (GEP) cells. Modified from Krause, W.J.; Cutts, J.H.: Concise text of histology, p. 280 (Williams & Wilkins, Baltimore 1981); based on *Bloom and Polak* [4].

course, only if the hormone had been isolated previously, which required tens of thousands of pig stomachs and intestines. The development of chromatographic separation techniques and of the technology for amino acid sequencing have made the task much easier. Indeed, today the problem is one of the finding roles for the many gut peptides and amines which have been isolated – a complete reversal of the state of affairs existing only 20 years ago. As a result of immunofluorescence identification, the morphological cell types responsible for most of the more than 20 gut peptides and amines isolated – whether of proved physiological function or not – are now known. Interestingly enough, they have many features in common.

Chromaffin Cells

As early as 1870, *Heidenhain* observed that certain cells with apparently clear cytoplasm (by light microscopy) were selectively stained by chromium salts. Because of this affinity, they were called 'chromaffin' or *EC (enterochromaffin) cells*. In 1914, *Masson* discovered such cells also have the capacity to reduce silver and recognized their endocrine-like nature. On the basis of their staining reaction with silver, they were categorized as (a) 'argentaffin cells' if they reduced silver salts without special pretreatment, and (b) 'argyrophilic cells' if they require exposure to an external reducing agent before they react. *Erspamer* and co-workers [6, 33] identified the heavy metal salt-reducing substance stored in their secretory granules as 5-hydroxytryptamine (5-HT or serotonin).

APUD Cells and the DES

The presence of other endocrine-like cells in the mucosa of the gut which did not contain serotonin was postulated during this time, but proof awaited the development of proficient electron microscopy [32] and immunohistochemical techniques [18]. Soon several types of non-EC cells were identified and designated as the cells of origin of one or another of the rapidly growing family of gut endocrine peptides. In view of a number of histological, histochemical and ultrastructural characteristics common to both the EC and the non-EC cells able to produce peptides and/or amines active as hormones or neurotransmitters (as well as an apparently erroneously postulated common derivation to be explained shortly), they have been collectively referred to as *'APUD' cells* [22] and are considered to comprise the *'diffuse endocrine* (neuroendocrine by some) *system'* (DES) [8, 9]. 'APUD' *(amine precursor uptake and decarboxylation)* refers to their ability to process biogenic amines [23]. These cells are found throughout the gut, but peptide- and/or amine-producing cells occurring in such distantly placed and diversified tissues as the pineal gland and pituitary, the thyroid and parathyroid, the carotid body, the lungs, the pancreatic islets, the urogenital tract and the placenta are also APUD cells of the DES system [24].

GEP Cells

That population of endocrine (APUD or DES) cells associated with the GI tract in particular has been specified as the *gastroenteropancreatic (GEP) cells* [31]. GEP or *enteroendocrine cells* are widely dispersed in the epithelial lining of the stomach, small and large intestine (glands, crypts and villi); the epithelia of the appendix, distal esophageal glands, and the ducts of the pancreas and liver; and among the cells of the acini and islets of the pancreas. Of the specific types of GEP cells recognized thus far in man, only one type (B) is restricted to the pancreatic islets, and at least three are distributed in both the gastrointestinal mucosa and the pancreas [34, 35] (see table 12/I and fig. 12/1). Were they grouped together, the GEP cells would constitute the largest endocrine gland in the body [21]. Still, the concentration of endocrine cells within the GI mucosa is very low. The concentration is maximal in the pyloric antrum and generally decreases progressively in an aboral direction (fig. 12/1). The greatest number of GEP cells is found associated with the small intestine.

Origin

Because APUD cells are found in such locations as the adrenal medulla ('chromaffin cells') and the paravertebral sympathetic ganglia (SIF cells), and since most of them contain catecholamines or serotonin, it has long been assumed they are neural in origin (as from the neural crests) [2]. *Pearse* [25] proposed that all cells of the APUD series are derived from neuroendocrine-programmed cells of the ectoblast in the early embryo. However, recent graft experiments in embryos showing that transplanted embryonic endoderm produces both gut and pancreatic endocrine cells [5] dispute this. The currently-favoured theory is that at least some APUD cells – specifically the GEP cells – are endodermal in origin, and that in the adult they arise from the pool of cells that produces all of the intestinal epithelial cells [1, 14, 15]. Similar cells (Kulchitsky cells) occurring in the epithelia of the respiratory passages – also developed as an outgrowth of the primitive embryonic gut – are likewise thought to be of endodermal origin (see table 12/I).

Ultrastructure

The ultrastructure of human GEP cells, described in detail by several investigators, has been summarized by *Solcia* et al. [29]. The various types of GEP cells resemble each other ultrastructurally for the main part, and the gastrin or G cells situated at the necks of the pyloric glands can be considered typical (see fig. 5/16). They were among the first to be identified immunocytochemically [19], and because of their importance in the stimulation of gastric acid secretion (see chapt. 5), they have been extensively investigated. (For a thorough treatment of the G cell in particular see *Ito* [11].)

Human GEP cells are small, clear pyramidal cells characterized by a broad basal cytoplasm resting on – and making extensive contact with – the basement membrane and filled with dense, membrane-bound cytoplasmic granules (fig. 12/2). This polarity is a reversal of the situation in the exocrine cells that may surround them (fig. 1/4, 1/5) and is strongly indicative of their endocrine nature. The Golgi apparatus, however, remains supranuclear in position but its cisternae, as well as those of the rough endoplasmic reticulum, are diminished in comparison with most exocrine cells such as the pancreatic acinar cells.

Open and Closed Cells. In the pyloric and intestinal mucosa GEP cells possess a tapering apical cytoplasmic pole, bearing a tuft of microvilli of variable number, shape and size, which usually reaches the lumen (fig. 12/2). This protrusion is presumed to be a receptor surface capable of sampling the changing composition of the local environment and enabling the cell to respond accordingly. In the case of the cells producing gastrin, secretin and cholecystokinin, the physiological conditions which determine their release makes this almost certain. This type of GEP cell is the 'open type'. GEP cells in the fundic (oxyntic) mucosa lack luminal contact and show less polarity. These cells are referred to as the 'closed type'. Of the six types of GEP cell known to occur in the stomach, only two – the G and EC cells – are of the open type.

Basal Secretion. Observation of the polarity imposed by the infranuclear concentration of secretory granules – as well as their proximity to blood capillaries and the fact that many GEP cells lack luminal exposure – strongly suggested that the substance bound within the cytoplasmic granules is released into the bloodstream (or at least into the basal extracellular space, perhaps for continued passage into the circulation) rather than into the lumen of the alimentary canal (see fig. 12/2). Indeed, it has been demonstrated by electron microscopy that the secretory granules are discharged at the basal (or basolateral) surfaces of the cell. The lateral intercellular spaces are sealed off at the apical or juxtaluminal end by junction complexes (gap

Table 12/I. The APUD cells of the DES [from *Williams and Warwick*, 35]

Location	Type	Main secretion	
		peptide	amine
I. APUD cells or neural crest origin			
Thyroid	parafollicular (C)	calcitonin	5-HT, Da
Ultimobranchial body	C	calcitonin	5-HT, Da
Carotid body	type I glomus	–	Da, NA
Sympathetic ganglia	SIF	–	NA
Adrenal medulla	chromaffin	–	Ad
Adrenal medulla	chromaffin	–	NA
Skin	melanoblast	–	promelanin
Urogenital tract	EC	–	5-HT
Urogenital tract	E	–	–
II. APUD cells of placodal or specialized ectodermal origin			
Hypothalamus	N pv	oxytocin, CRF	–
	N so	vasopressin	–
	N sch	–	–
	N dm/vm	TRF	–
	N arc	LHRF	Da
	N ant/post	SRF, CRF	–
	N periv	somatostatin	–
Pineal gland	P	LHRF	5-HT, MT
Parathyroid	chief	PTH	–
Pituitary	somatotroph	somaotropin	Da
	mammotroph	prolactin	Da
	gonadotroph	follitropin	Da
	gonadotroph	lutropin	Da
	corticotroph	corticotropin	–
	M	melanotropin	T
	thyrotroph	thyrotropin	Da
Placenta	endocrine	gonadotropin	–
	endocrine	somato-mammotropin	–
	endocrine	corticotropin	–

Table 12/I (continued)

Location	Type	Main secretion	
		peptide	amine

III. APUD cells of disputed origin (possible endodermal)

Location	Type	peptide	amine
Pancreas	A	glucagon	5-HT
	B	insulin	5-HT
	D	somatostatin	Da
	D$_1$	VIP-like	Da
	P	bombesin-like	–
	PP	pancreatic polypeptide	Da
Stomach	A	glucagon	–
	D	somatostatin	–
	ECL	–	H?
	EC$_1$	substance P	5-HT
	G	gastrin, enkephalin	–
	X	–	–
Intestine	D	somatostatin	–
	D$_1$(H)	VIP	–
	EC$_1$	substance P	5-HT
	EC$_2$	motilin	5-HT
	EC$_n$	–	5-HT
	I	cholecystokinin	–
	K	GIP	–
	L	enteroglucagon	–
	N	neurotensin	–
	S	secretin	–
Lung	Kulchitsky (P$_a$)	–	–

Ad = Adrenalin; CRF = corticotropin-releasing factor; Da = dopamine; GIP = gastric inhibitory peptide; H = histamine; 5-HT = 5-hydroxytryptamine; LHRF = luteotropin-releasing factor (luteinizing hormone-releasing factor); MT = melatonin; NA = noradrenalin; N ant/post = anterior and posterior nuclear 'zones' of hypothalamus; N arc = nucleus arcuatus (nucleus infundibularis); N dm/vm = nucleus dorsomedialis/ventromedialis; N periv = nuclei periventriculares; N pv = nucleus paraventricularis; N sch = nucleus suprachiasmaticus; N so = nucleus supraopticus; PTH = parathyroid hormone; SIF = small intensely fluorescent; SRF = somatotropin releasing factor; T = tryptamine; TRF = thyrotropin-releasing factor; VIP = vasoactive intestinal peptide; – = unidentified.

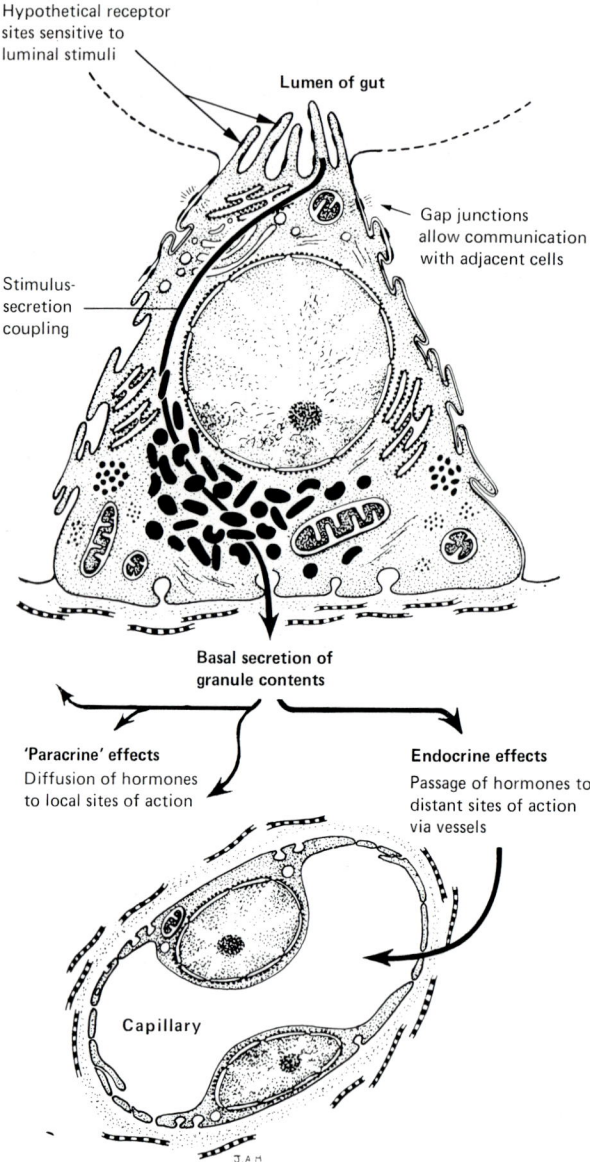

Fig. 12/2. Schematic diagram of the ultrastructure of a typical gastroenteropancreatic (GEP) cell of the 'open type', indicating postulated modes of action. From *Williams and Warwick* [35].

junctions). Apical granular discharge into the lumen has never been observed [28].

Classification. Based on highly-detailed ultrastructural findings (specifically with regard to the staining density, size and shape of the secretory granules) and immunohistochemical studies, at least 15 specific GEP cell types have now been recognized based on the Lausanne classification advocated by specialists in the field [29, 30] (see table 12/I).

Paraneurons? An unexpected and intriguing discovery was the recent finding in neurons of the CNS many of apparently the same neurohormones and neurotransmitter-like peptides known to be produced by GEP cells. In view of this finding and the observation of additional neuron-like features of GEP cells, *Fujita* [9a] has designated them as 'paraneurons'. Among the peptides common to the brain and the mucosa of the GI tract are gastrin, CCK (found in the hypothalamus where some feel it is responsible for satiety), substance P, bombesin, neurotensin, the opiate-like enkephalin, VIP and somatostatin. VIP has also been found in peripheral nerves, and at least one paper [7] has advanced the view that VIP is a vagal mediator in the stimulation of pancreatic secretion. Somatostasin, the hypothalamic hormone which inhibits the release of growth hormone, is found in APUD cells throughout the gut where it also inhibits the release of hormones, but not through the same mechanism. Thus, a new mechanism has been suggested to explain the mode of action of it and other gut hormones and candidate hormones.

GEP Cell Secretions: The Gut Peptides and Amines

Mode of Action
Endocrine Secretion. Of the 20 or so gut peptides and amines isolated, only a few are known to function physiologically in the classical *modus operandi* for circulating hormones, i.e. being secreted into the bloodstream for transport to distant target organs where they exert diffuse effects after which they are rapidly broken down and eliminated, mainly by the kidneys. These are gastrin, secretin and CCK [4]. Enteroglucagon, motilin, GIP and VIP may function in the same way, but currently they fall into the category

Table 12/II. Primary actions of the secretions of GEP cells [Modified from Krause, W.J.; Cutts H.J.: concise text of histology, p. 279 (Williams and Wilkins, Baltimore 1981).]

Secretion	Cell of origin	Primary location of cell type	stomach secretion acid	stomach secretion pepsin	stomach motility	pancreas insulin release	pancreas enzyme secretion	pancreas H_2O/HCO_3 secretion	gallbladder motility (contraction)	gallbladder bile flow	intestine motility	intestine secretion	Summary of primary action
Gastrin	type G	pyloric antrum	+++	++	++	+	+	+	0	0	++	0	stimulates secretion of gastric acid
Secretin	type S	duodenal and proximal jejunal mucosae	--	++	--	+	0	+++	0	0	-	++	controls pancreatic/biliary water/ion secretion
Cholecystokinin (CCK)	type I	mucosae of duodenum and jejunum	+	?	+	+	+++	0	+++	+++	++	0	→ pancreatic enzyme secretion; gallbladder emptying
Enteroglucagon	type L	mucosa of ileum	-	-	-	++	-	-	-	+	-	++	hepatic glycogenolysis; inhibits muscularis externa
Gastric inhibitory peptide (GIP)	type K	mucosae of duodenum and jejunum	--	-	-	+++	0	0	0	0	0	+	inhibits gastric acid secretion
Motilin	type EC_2	mucosa of jejunum	0	+	+	?	+	+	+	+	+	?	stimulates motor activity of gut
Vasoactive intestinal polypeptide (VIP)	type D_1, nerve cells and fibers	gastric, small intestinal and colonic mucosae	-	++	-	?	-	+	--	0	-	++	vasodilation

Table 12/II (continued)

Secretion	Cell of origin	Primary location of cell type	stomach secretion acid	pepsin	motility	pancreas insulin release	enzyme secretion	H_2O/HCO_3 secretion	gallbladder motility (contraction)	bile flow	intestine motility	secretion	Summary of primary action
Somatostatin	type D nerve cells and fiber	mucosae of duodenum and jejunum	−	−	−	−	−	−	−	0	−	−	local inhibition of other GEP cells and GEP secretions
Pancreatic polypeptide	type PP	pancreas (prox. small intestine?)	+	0	+	0	−	−	−	0	+	0	?
Substance P	type EC$_1$ nerve cells and fibers	gastric mucosa	0	0	±	−	±	±	?	−	±	±	?
Bombesin in-like peptide	type P nerve fibers	mucosae of stomach and jejunum	±	±	−	−	+	+	+	?	−	?	?
Neurotensin	type N	mucosa of ileum	−	?	−	−	?	?	?	?	±	?	stimulates contraction of the muscularis externa
Serotonin (5-HT)	type EC	mucosae of small intestine and colon	−	−	±	?	±	±	±	±	±	0	vasoactive agent
Histamine	type ECL	fundic mucosa of stomach	+++	+	0	?	±	±	±	±	0	±	stimulates secretion; vasodilation

0 = No effect; + = stimulation; − = inhibition; ± = stimulation or inhibition depending on animal species; ? = unknown or uncertain.

Table 12/III. Amino acid and sequences of gastro-intestinal hormones. From *Murphy* [20].

Hormone	Amino Acid Sequences
	1 ... 5 ... 10 ... 15 ... 20 ... 25 ... 30
Secretin	His Ser Asp Gly Thr Phe Thr Ser Glu Leu Ser Arg Leu Arg Asp Ser Ala Arg Leu Leu Gln Gly Leu ValNH
Glucagon	His Ser Gln Gly Thr Phe Thr Ser Asp Tyr Ser Lys Tyr Leu Asp Ser Arg Arg Ala Gln Asp Phe Val Gln Trp Leu Met Asn Thr
VIP	His Ser Asp Ala Val Phe Thr Asp Asn Tyr Thr Arg Leu Arg Lys Gln Met Ala Val Lys Lys Tyr Leu Asn Ser Ile Leu AsnNH$_2$
GIP	Tyr Ala Glu Gly Thr Phe Ile Ser Asp Tyr Ser Ile Ala Met Asp Lys Ile Arg Gln Gln Asp Phe Val Asn Trp Leu Leu Ala Gln Gln- 35 40
Cholecystokinin	Lys Gly Lys Lys Ser Asp Trp Lys His Asn Ile Thr Gln 5 10 15 20 25 30 Lys Ala Pro Ser Gly Arg Val Ser Met Ile Lys Asn Leu Gln Ser Leu Asp Pro Ser His Arg Ile Ser Asp Arg Asp Tyr Met Gly Trp- Met Asp PheNH
Gastrin II	Glu Gly Pro Trp Leu Gln Glu Glu Glu Ala Tyr Gly Trp Met Asp PheNH
Motilin	Phe Val Pro Ile Phe Thr Tyr Gly Glu Leu Gln Arg Met Gln Glu Lys Glu Arg Asn Lys Gly Gln
Somatostatin	Ala Gly Cys Lys Asn Phe Phe Trp Lys Thr Phe Thr Cys
Pancreatic polypeptide	Ala Pro Leu Glu Pro Gln Tyr Pro Gly Asp Asp Ala Thr Pro Glu Gln Met Ala Gln Tyr Ala Ala Gln Leu Arg Arg Tyr Ile Asn Met- Leu Thr Arg Pro Arg TyrNH
Neurotensin	Glu Leu Tyr Glu Asn Lys Pro Arg Arg Pro Tyr Ile Leu
Enkephalins	Tyr Gly Gly Phe Leu Tyr Gly Gly Phe Met

of 'candidate hormones' since their pharmacological actions are known but their exact physiological roles are not fully determined.

Paracrine Secretion. GEP cells producing peptides such as somatostatin appear to release their product primarily into the local subendothelial connective tissue, or directly onto adjacent or nearby cells via long basal cytoplasmic processes [21]. This has been termed *paracrine* secretion [37]. Thus, it is speculated that via paracrine secretion GEP cells may be able to exert direct influence on neighboring smooth muscle, endocrine cells or other enterocytes, and on the neurons of the gut. Such phenomena are feasible in light of their isolated dispersal among what may well be their target cells, and in light of the fact that most GEP cells have a luminal border which would allow sampling of the local environment and enable direct response to luminal stimuli. Further, paracrine secretion may possibly explain why, in such phenomena as the modulation of the neural control of gut motility and gastric secretion, there is no direct correlation between plasma hormone levels and functional response.

The DES and the Nervous and Endocrine Systems. Returning to the concept of GEP cells as members of the APUD series composing the diffuse endocrine system, the opinion of *Pearse,* as stated in *Williams and Warwick* [35], is intriguing. He proposes that the cells of the DES be viewed collectively 'as a third division of the nervous system, acting as third-line effectors to support, modulate or amplify the actions of neurons in the autonomic and somatic division and of each other'. The actions of DES cells are of slower onset and longer duration than are those of autonomic neurons; the actions of autonomic neurons, in turn, are of slow onset and longer duration than those of somatic neurons. From another point of view (also stated in *Williams and Warwick* [35], since the products of APUD cells may act upon either (1) contiguous cells; (2) groups of nearby cells, or (3) distant cells (after transport in the blood), they may, perhaps, be considered intermediate between neuron-produced neurotransmitter substances (local and focal in action) and the hormones secreted by discrete endocrine glands (distant and diffuse in action). In light of the apparent interaction of gut hormones with neural mechanisms and with the 'hypothalamic-pituitary axis' in regulating and coordinating the secretion and motility required for digestion, the DES system seems to 'compliment and link the nervous and endocrine systems, all three systems interacting to provide a sensitive mechanism allowing for homeostatic control'.

Cells of Origin/Actions

The accompanying tables and figures list the GEP secretions, identifying their cells of origin and indicating their distribution in the gut. The known physiological or pharmacological actions of specific secretions are listed in table 12/II and are described elsewhere in other chapters when appropriate.

Biochemistry

An odd thing about gut hormones is that many are produced in various sizes. Gastrin, for example, has been isolated in 34, 17 and 13 amino acid sizes (table 12/III). To arrive at the smaller forms, it is evident that biologically inactive N-terminal residual chains of amino acids must have been split off within the cells. CCK likewise has been seen with 39, 33 and 8 amino acids, and once again the inactive N-termini have been found in blood and tissue. Immunoactive forms of glucagon have been isolated from the gut, some with biological activity and some without. Glicentin, an inactive form, seems to have an extension added to both ends of the glucagon chain so that the active molecule is completely buried.

Neoplasia

Another odd thing is that when cells of this system become neoplastic they often produce hormones belonging to other cells in the system. The D cells of the pancreas often produce gastrin when cancerous, giving rise to the Zollinger-Ellison syndrome.

A hypothesis to explain this odd behaviour is that the cell produces a large peptide estimated at 65,000 Daltons. (There is biochemical evidence for this in a variety of 'apudomas', as such tumors are called [26].) In the normal differentiated cell a variety of microsomal proteases prune this polypeptide to the size and shape appropriate to the cell type. Dedifferentiation is a characteristic of all malignant neoplasms, which apudomas usually are, and may explain excessive production of inappropriate biologically active polypeptides.

The common characteristics exhibited by these cells have been invoked to explain the occurrence of multiple endocrine tumors. These seem to represent a genetic defect. They can affect the thyroid, adrenal, and parathyroid or pituitary, pancreas and parathyroids. Individual tumors producing VIP, glucagon, somatostatin, pancreatic polypeptide, and, of course, gastrin are known to occur. Deviations in the relative levels of the secretions of the various types of DES cells have been suggested as possible causes for many of the disorders currently considered psychosomatic [26] or frankly psychotic [36]. If so, increased knowledge of the DES may significantly alter their treatment in the future.

References

1 Andrew, A.: Further evidence that enterochromaffin cells are not derived from the neural crest. J. Embryol. exp. Morphol. *31:* 589–598 (1974).
2 Andrew, A.: The APUD concept: where has it led us? Br. med. Bull. *38:* 221–226 (1982).
3 Bayliss, W.; Starling, E.: The mechanism of pancreatic secretion. J. Physiol. *28:* 325–353 (1902).
4 Bloom, S.; Polak, J.: Gut hormone overview; in Bloom, Gut hormones, pp. 3–18 (Churchill Livingstone, Edinburgh 1978).
5 Edkins, J.: The chemical mechanism of gastric secretion. J. Physiol *34:* 183 (1906).
6 Erspamer, V.; Asero, B.: Identification of enteramine, the specific hormone of the enterochromaffin cell system, as 5-hydroxytryptamine. Nature, Lond. *169:* 800–801 (1952).
7 Fahrenkrug, J.: Vasoactive intestinal polypeptide (VIP) measurement distribution and potential neurotransmitter function. Digestion *19:* 149–169 (1979).
8 Feyrter, F.: Über diffuse endokrine epitheliale Organe. Zentbl. inn. Med. *545:* 31–41 (1938).
9 Feyrter, F.: Über die peripheren endokrinen (parakrinen) Drüsen des Menschen, pp. 1–231 (Maudrich, Wien 1953).
9a Fujita, T.: The gastro-enteric endocrine cell and its paraneuronic nature; in Coupland, Fujita, Chromaffin, enterochromaffin and related cells, pp. 191–208 (Elsevier, Amsterdam 1976).
10 Heidenhain, R.: Untersuchungen über den Bau der Labdrüsen. Arch. mikrosk. Anat. EntwMech. *6:* 368 (1870).
11 Ito, S.: Functional gastric morphology; in Johnson, Physiology of the gastrointestinal tract, pp. 517–550 (Raven Press, New York 1981).
12 Kobayashi, S.; Sasagawa, T.: Morphological aspects of the secretion of gastro-enteric hormones; in Fujita, Endocrine gut and pancreas, pp. 255–271 (Elsevier, Amsterdam 1976).
13 Lauber, M.; Nicolus, P.; Bausetta, H.; Fahz, C.; Beguin, P.; Carrier, M.; Vaudry, H.; Cohen, P.: The M_r 80,000 common forms of neurophysin and vasopressin from bovine neurophysins have corticotropins and endorphin-like sequences and liberate by proteolysis biologically active corticotropin. Proc. natn. acad. Sci. USA *78:* 6086 (1981).
14 Leblond, C.; Cheng, H.: Identification of stem cells in the small intestine of the mouse; in Cairnie, Lala, Osmond, Stem cells of renewing cell populations, pp. 7–31 (Academic Press, New York 1976).

15 Le Douarin, N.: The embryological origin of the endocrine cells associated with the digestive tract. Experimental analysis based on the use of a stable cell marking technique; in Bloom, Gut hormones, pp. 49–56 (Churchill-Livingstone, Edinburgh 1978).
16 Martin, C.: Obituary notice: E. H. Starling. Br. med. J. *1:* 900–904 (1927).
17 Masson, P.: La glande endocrine de l'intestin chez l'homme. C. r. hebd. Séanc. Acad. Sci., Paris *158:* 52–61 (1914).
18 McGuigan, J.: Gastric mucosal intracellular localization of gastrin by immunofluorescence. Gastroenterology *55:* 315–327 (1968).
19 McGuigan, J.; Greider, M.: Staining characteristics of the gastrin cell. Gastroenterology *62:* 959–969 (1972).
20 Murphy, R.: The chemical characterization of gastrointestinal hormones. Clin. Endocrinol. Metab. *8:* 281–295 (1979).
21 Neutra, M.; Padykula, H.: The gastrointestinal tract, chap. 19; in Weiss, Histology: cell and tissue biology; 5th ed., pp. 658–706 (Elsevier Biomedical, New York 1983).
22 Pearse, A.: 5–Hydroxytrytophane uptake by dog thyroid C cells and its possible significance in polypeptide hormone production. Nature, Lond. *211:* 598–600 (1966).
23 Pearse, A.: Histochemistry; 3rd ed. (Churchill, London 1968).
24 Pearse, A.: The cytochemistry and ultrastructure of polypeptide hormone-producing cells (the APUD series) and the embryologic, physiologic and pathologic implications of the concept. J. Histochem. Cytochem. *17:* 303–313 (1969).
25 Pearse, A.: The apudomas; with particular reference to those of gastroenteropancreatic origin; in Yardley, Morson, Abell, The gastrointestinal tract, pp. 206–218 (Williams & Wilkins, Baltimore 1977).
26 Pearse, A.; Polak, J.: The diffuse neuroendocrine system and the APUD concept; in Bloom, Gut hormones (Churchill-Livingstone, Edinburgh 1977).
27 Polak, J.; Bloom, S.: Hormones of the gastro-intestinal tract; Duthie, Wormsley, Scientific basis of gastroenterology (Churchill-Livingstone, Edinburgh 1979).
28 Solcia, E.; Capella, C.; Buffa, R.; Usellini, L.; Fiocca, R.; Sessa, F.: Endocrine cells of the digestive system, chap. 2; in Johnson, Physiology of the gastrointestinal tract (Raven Press, New York 1981).
29 Solcia, E.; Capella, C.; Buffa, R.; Usellini, L.; Fontana, P.; Frigerio, B.: Endocrine cells of the gastrointestinal tract. General aspects, ultrastructure and tumor pathology; in Grossman, Speranza, Basso, Lezoche, Gastrointestinal hormones and pathology of the digestive system, pp. 11–22 (Plenum Press, New York 1978).
30 Solcia, E.; Polak, J.; Larsson, L.; Hakanson, R.; Lechago, J.; Fujita, T.; Rubin, W.; Grube, D.; Falkmer, S.; Grieder, M.; Creutzfeldt, W.; Grossman, M. I.: Human GEP endocrine-paracrine cells: Lausanne 1977 classification revisited; in Grossman et al., Cellular basis of chemical messengers in the digestive system (Academic Press, New York 1981).
31 Solcia, E.; Polak, J.; Pearse, A.; Forssman, W.; Larsson, L.; Sundler, F.; Lechago, J.; Grimelius, L.; Fujita, T.: Gastro-entero-pancreatic endocrine system; in Fujita, Endocrine gut and pancreas (Williams & Wilkins, Baltimore 1974).
32 Solcia, E.; Vassallo, G.; Sampietro, R.: Endocrine cells in the antro-pyloric mucosa of the stomach. Z. Zellforsch. *81:* 474–486 (1967).
33 Vialli, M.; Erpsamer, V.: Sulle reazioni chimiche colarate dell' enteramina. I.

Ricerche su estratti acetonici di mucosa gastro-intestinale. Archs Sci. biol., Bologna *28:* 101–121 (1942).
34 Walsh, J.: Endocrine cells of the digestive system; in Johnson, Physiology of the gastrointestinal tract (Raven Press, New York 1979).
35 Williams, P.; Warwick, R.: Schematic diagram of the ultrastructure of a typical gastroenteropancreatic (GEP) cell of the 'open type', indicating postulated modes of action. Gray's anatomy; 36th British ed. (Saunders, Philadelphia 1980).
36 Webster, K.: The brainstem reticular formation; in Hennings, Hemmings, The biological basis of schizophrenia (MTP Press, Lancaster 1978).
37 Wingate, D.: The eupeptide system: a general theory of gastrointestinal hormones. Lancet *i:* 529–532 (1976).

13 Hunger and Appetite

Hunger vs. Appetite

Hunger and appetite [1] are often used interchangeably, but to the physiologist they are different. Hunger is, in the physiological sense, the urge to eat caused by a caloric deficiency, i.e. an energy deficit. Hunger is experienced in man and animals after a long fast, and it is satisfied after the caloric deficiency has been repaired. Hunger is an absolutely vital sensation since it is the first signal of a caloric deficit.

Appetite is often present without hunger; eating dessert is largely a matter of appetite since hunger has usually been satisfied before dessert is reached. No such confusion exists with regard to drink. No one suggests that whiskey is drunk to alleviate thirst but water usually is. The demarcation line between eating to satisfy hunger and eating for enjoyment has almost disappeared and the word glutton is now rarely encountered despite the fact that sequelae of overeating pose a major medical problem in the western world.

Socioeconomic Factors

The satisfaction of hunger results in the intake of energy (calories) in addition to all the other essential elements, foods and food factors necessary for well-being. There are communities in which all the available food is either of insufficient quantity or of a quality too poor to achieve these ends. There are others in which the consumption of food tends to be excessive, and obesity, coronary disease of the heart, etc., become problems. If the intake of calories customarily exceeds the expenditure, which includes physical work and the energy necessary for our vital processes and warmth, obesity will result. If the converse occurs loss of weight and eventually emaciation will occur.

13 Hunger and Appetite

Calories or energy provided by the food we eat are unique in that there is no automatic mechanisms by which an excess can be disposed of. It is as if we are designed for life in an environment in which calories would always be marginal. For animals in the wild, this is almost always the case, population density often being determined by the availability of energy. For man, of course, this has not been the case for a long time in most northern countries and in consequence appetites rather than hunger have become progressively more important stimuli for the consumption of food. Lunch and dinner are often eaten because the hands of the clock point to the conventional hour for food, and the social occasion rather than hunger is the dictate. Social occasions are rare at breakfast time and because few, except the very thin, are hungry first thing in the morning – this meal is often missed.

In societies in which hard manual work is the rule and food scarce, there is no regulation of food intake but the availability of calories to provide the energy for manual work and the vital processes. If calories become scarce manual labor will diminish and/or the individual will become thinner. If hard manual work becomes an unimportant expenditure of calories, the vital processes are the major expenditure of energy and unless food is consumed in response to the hunger alone, rather than to appetite or social signals, fat will be stored.

The hunger signal will be a long time in returning if the previous meal eaten was a large one and conversely if the meal was a small one. In experimental animals under conditions of controlled activity the quantity of food consumed to maintain constant weight is such that the ratio of the calories consumed to the time elapsed until the next meal will be constant [12]. Since modern sedentary man, unlike caged rats, does not eat to assuage recurring hunger, this generalization does not hold and things are much more complicated.

Hunger and Satiety

It seems fairly obvious that neither man nor experimental animals eat enough simply to repair a caloric deficit. There is always an overshoot. In other words an energy deficit is repaired, but in addition the animal anticipates the next gap between feeds. Feeding is, therefore, terminated when a signal reaching the hypothalamus indicates that enough energy has been swallowed to alleviate hunger and keep it alleviated until the arrival of the next meal.

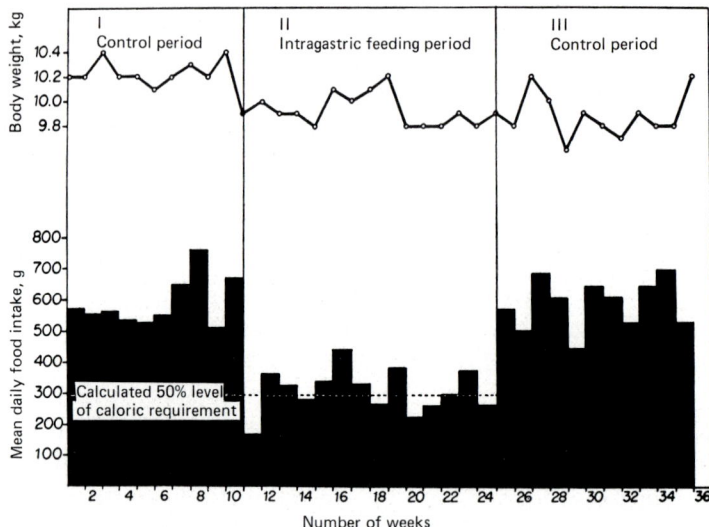

Fig. 13/1. The effect on oral food intake and body weight of introducing 50% of a dog's daily caloric requirement directly into the stomach through a gastric fistula at mealtime. Reproduced, with permission, from *Janowitz and Hollander* [8].

The Effect of Distension/Bulk

How this is monitored is problematical, since the assessment is made before the meal has entered into the body's energy pool. Much of the meal is still in the stomach when the signal for enough reaches the brain. One might argue that simple gastric or abdominal distension provides the signal and clearly this must be considered. If for example food is diluted with non-nutritive bulk in dogs, fed once per day, several days elapse before the animal increases its intake enough to restore its calorie balance. If a diet of this sort is fed ad libitum the interval between feeds will diminish and then gradually lengthen over a few days as the intake at each feed increases. In other words, a new calibration of the abdominal event as a measure of calories has taken place.

Where does this signal come from and what is its nature? Very many have, in the past, assumed it to be simple gastric distension signaled to the satiety center via the vagi. Food intake in hungry dogs can easily be decreased if 20% or more of the volume of the normal meal is introduced through a gastric fistula (fig. 13/1). Introduction 4 h before the test did not diminish the size of the subsequent meal. It was inconsequential whether the

13 Hunger and Appetite

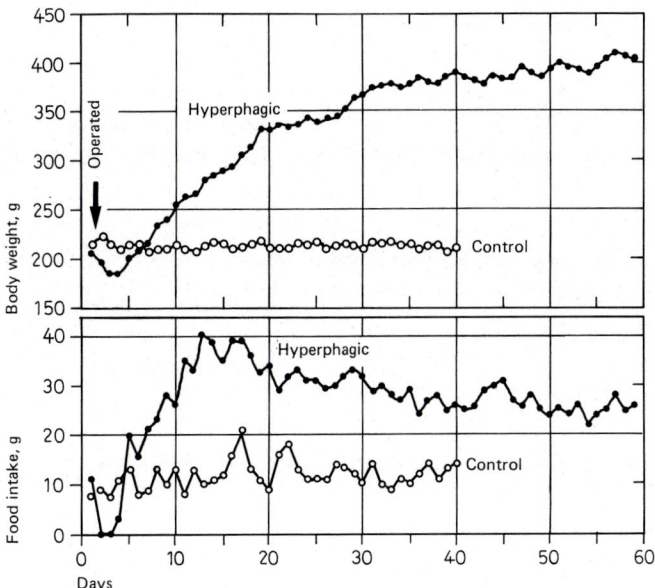

Fig. 13/2. Hypothalamic hyperphagia and its effect on body weight. Reproduced, with permission, from *Teitelbaum* et al. [20].

material was calorigenic or increased bulk, which of course would have passed on into the small bowel within 4 h. A balloon, however, will reduce intake even if inflated 4 h earlier and kept inflated, since it cannot pass on from the stomach. What this means, of course, is that gastric metering by simple bulk does occur, but that by itself it is far from a precise adjustment to caloric needs. Precise caloric adjustment to either the initiation or termination of intragastric prefeeding usually takes 3–5 weeks. Some investigators [16] have found an increase in the electrical activity of the hypothalamic satiety center on distending the stomach.

Hypothalamic Satiety and Hunger Centers

There are two discrete and well-authenticated centers in the hypothalamus concerned with food intake. The lateral one is the hunger center which if destroyed produces anorexia, weight loss, and ultimately death. The medial one is the satiety center which if destroyed partially or completely produces hyperphagia and obesity; the degree of obesity and hyperphagia is directly proportional to the extent of the destruction (fig. 13/2). Stimulation of these centers produces the converse of ablation. The satiety center is dominant: thus, electrical activity in this center means suppression of hunger and the hunger center and the animal will not eat.

Stretch receptors from the stomach and gut which send impulses up vagal afferents have been described by *Paintal* [14], but thorough denervation of the upper GI tract does not alter the precision of gastric monitoring. One must conclude that the distension mechanism is dispensible or else that pathways other than the vagi are involved. It is not recorded whether or not electrical activity on gastric distension occurs in the satiety center following denervation of the stomach.

It is possible that abdominal distension rather than gastric distension or that weight rather than bulk are the signals. These seem not to have been investigated, but it is common knowledge that heavy Christmas pudding is very filling, satisfying, or anorexic depending on one's opinion of plum pudding.

Oropharyngeal Factor

An oropharyngeal component to the satiety signal has been postulated because dogs with esophagastomies through which all the swallowed food falls to the floor do stop eating but after 15 or so min instead of the usual 2–3 min in the intact dog [19]. Bouts of feeding recur at short intervals. This may be oropharyngeal signalled satiety, but it might also be simple fatigue; but as a curb to food intake an oral prefeed in an animal with a gastric fistula is much more effective than an isovolumic intragastric feeding.

Theories Regarding Satiety

Do animals become hungry again as soon as the stomach is empty? Data is slight, but one suspects they do not. Can an animal's hunger be completely or even partially assuaged by intravenous feeding? In dogs intravenous glucose either 30 min before or 4 h before does not diminish the size of the following meal; however, over the course of a week if this is continued food intake will decline but it does not cease altogether.

It is evident that sensations from the digestive tract are not the whole answer to satiety and the cessation of eating. There are other ideas such as the availability of glucose to the neurons in the hunger center, the concentration of free fatty acids or amino acids [6] in plasma, the movement of water into the gut, the heat generated by food and, most recently, elevations in circulating gut hormones; CCK is the favored one at the moment.

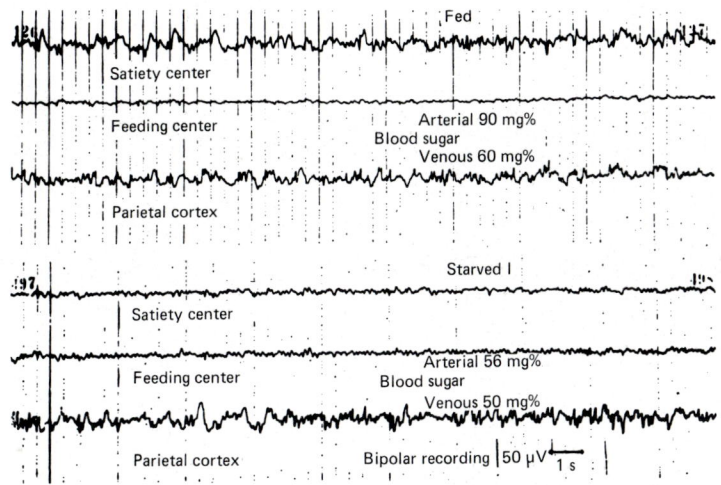

Fig. 13/3. Electrical activity in the satiety center and feeding center during starvation and after feeding. [Reproduced, with permission, from Anand, B.; Subberwal, U.; Manchanda, S.; Singh, B.: Glucoreceptor mechanisms in the hypothalamic feeding center. Indian J. Med. Res. 49: 717–729 (1961).]

The Glucostatic Theory. The glucostatic theory [13] (glucose availability) probably has most support at present. When blood glucose falls to low levels people and animals become hungry. Glucose injected into the cerebral ventricles of rats reduces food intake, but only if the satiety center is intact. *Anand* et al [2, 3] have found that the electrical activity of the satiety center was unique to the brain in that it increased with hyperglycemia and decreased with hypoglycemia (fig. 13/3).

Microinjections of glucose into the center increased satiety center activity and decreased that of the hunger center. Another unique property of these two centers is that their cells are sensitive to the action of insulin. In the rest of the brain, glucose uptake depends only on concentration. Insulin plus glucose greatly increases satiety center activity, but systemic insulin alone produces a transient increase and then a decrease in activity. The initial effect is increased utilization followed by a decrease when the blood glucose concentration falls. It seems clear now that the satiety center increases its electrical activity when glucose utilization is high and conversely when it is low (fig. 13/4). The activity of the hunger center is the converse of the satiety center. This explains why diabetics who have given themselves too much

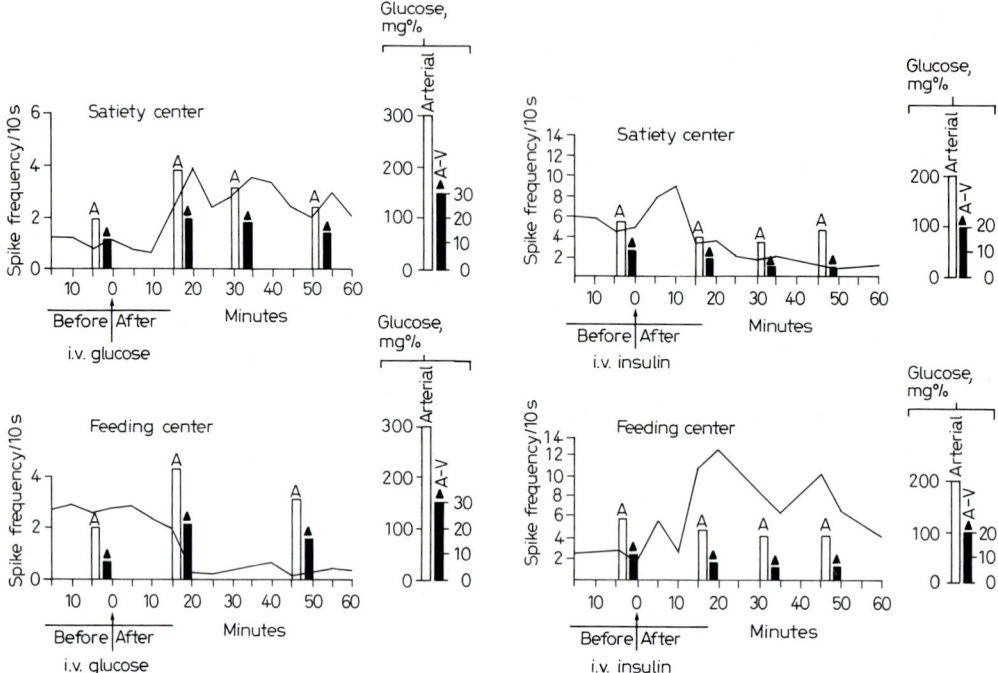

Fig. 13/4. Lines showing spike frequency from the feeding and satiety centers before and after intravenous glucose (left) and insulin (right). The black bars are the arterio-venous glucose difference, the white bars the arterial glucose levels. Reproduced, with permission, from *Anand* et al. [3].

insulin are hungry and those with too little insulin and hyperglycemia are also hungry. In the former the insulin is there, but glucose levels are too low for good utilization; and in the latter glucose levels are high enough but there is insufficient insulin so that hypothalmic utilization again is low.

The Fatty Acid Hypothesis. The fatty acid idea is the converse of the glucostat idea. During fasting when glucose utilization is low, fatty acid utilization will increase. There will be lipolysis in the depots and circulating free fatty acids (FFA) will be elevated and conversely, after feeding under the influence of insulin, glucose utilization will go up and lipogenesis will take place with a concomitant fall in circulating FFA. These levels do correlate well with feelings of hunger and satiety. There is little direct evidence that either fat or amino acids play a role in the short-term regulation of food intake, i.e. in signalling that one should leave the table, but there is considerable evidence for a role for fat in long-term regulation in the maintenance of a stable body weight. Changes in blood amino acids do not increase electrical activity in the satiety center and, moreover, amino acids do not pass the blood-brain barrier easily.

Thermoregulation. Brobeck [5] and *Strominger and Brobeck* [18] have proposed the idea that animals eat to keep warm; during fasting, heat production falls slightly, hence hunger. When food is ingested and its metabolism begins heat production (specific dynamics action) starts and food intake stops to avoid hyperthemia. A difficulty with this idea, if it is to be considered as the sole signal, is that specific dynamic action depends on the nature of the food consumed; but cessation of food intake seems to depend on the volume or perhaps weight of food eaten before significant metabolism has taken place. When the preoptic area in goats is cooled eating has been observed to take place. At a rectal temperature of 41 °C goats will not eat, but if the anterior hypothalamus is directly cooled they can be induced to do so even with higher rectal temperatures. When this area is warmed, it is claimed that eating stops. Others [4], using rats, have stated that warming or cooling either the hunger or satiety centers is without influence on their electrical behavior.

Water Mobilization. The idea that water mobilization may be a factor again is based on digestion rather than on bulk or volume of food eaten. When digestion begins water in the form of gastrointestinal secretions pours into the lumen of the intestine. It is suggested that this withdrawal of water from sensitive areas like the hypothalamus is the signal that hunger is relieved. However, the fact that hunger induces excessive drinking seems to be contrary to this idea, but thirst *is* common following meals. Lesions in the lateral hypothalamus produce both aphagia and adipsia. The water deprivation theory would associate hyperphagia with adipsia.

CCK. A more modern idea is that CCK is the satiety signal [17]. It was found that rats with gastric fistulae open to the exterior feed for much longer than do normals; however, when food was introduced directly into the duodenum the animals stopped eating promptly. Psychological studies to measure aversion indicated that this procedure did not nauseate the animals. *L*-Phenylaline, which is a good releaser of duodenal CCK, was effective in suppressing eating while *d*-phenylaline, which does not release CCK, was not. Finally, CCK itself was found to mimic the complete sequence of events seen in the normal rat. Gastrin had a similar effect, but secretin had not. It is interesting that both CCK and gastrin have been extracted from the brain. This work has been repeated in monkeys in which intravenous doses of between 5 and 20 U/kg were found to be effective. These are very large doses.

Diet and Body Weight

It is virtually the rule that normal healthy adults maintain an almost constant body weight, winter or summer through periods of sloth or activity, whether they be obese or thin. This results from long-term regulation of food intake and must of course be determined by the feeling of hunger and the signals which indicate satiety. The obvious implication is that, somehow, we eat to constant body weight or, since the only factor which can account for the enormous variations encountered is fat, to constant body fat. The *fat set point* must be high in the obese and low in the lean. This idea is known as the *lipostatic hypothesis* [10, 11, 13]. To be valid it must in some fashion or another be linked to a mechanism which signals satiety at individual meals. Fats have been implicated in short-term regulation as FFA (see above) and there is, of course, a connection between glucose metabolism and fat via the hormone insulin [12].

Experimental Data

Rats force fed to excessive body weight will not eat until they are back to normal. Animals in which a part of the satiety center is destroyed eat voraciously until they reach a new elevated body weight plateau and subsequently eat enough to maintain this (see fig. 13/5). If they are fasted for a few days they will, when allowed, eat themselves back to the obese plateau. Further destruction of the center will further raise the plateau. If two rats are sewn together parabiotically so that their circulatory systems mingle and the satiety center in one is then destroyed it will eat up to a new obese plateau while the member of the pair with the undisturbed hypothalamus will not eat at all, presumably because its satiety center is continuously stimulated by the excessive body fat built up by its partner's excessive food intake.

Animals and man will maintain their body weight plateaus even if they are abnormal, despite changes in temperature or activity as long as food is

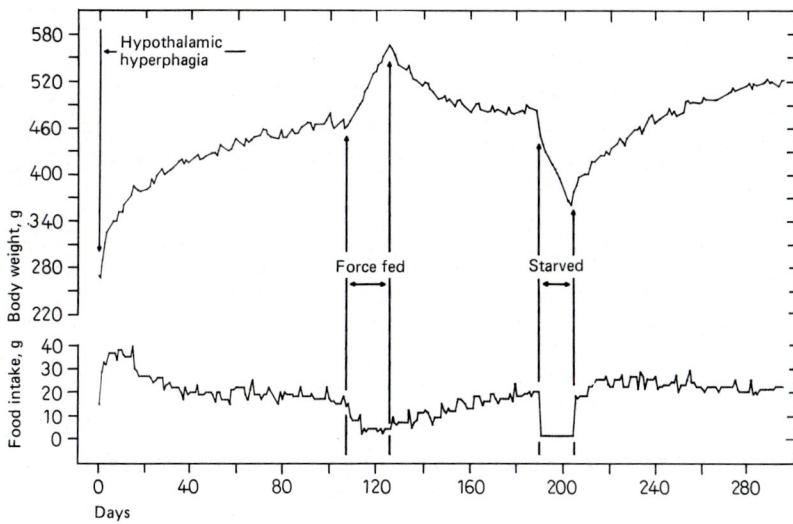

Fig. 13/5. To show that in hypothalamic hyperphagia there is a new lipostatic set point as after overfeeding or underfeeding the animal's body weight will return to its hyperphagic level. Reproduced, with permission, from *Hoebel and Teitelbaum* [7].

freely available. The only exception to this seems to be in the lactating animal. In rats at full lactation, in which hyperphagia is very evident, hypothalamic lesions cannot increase food intake [11]. In lactating animals only reduction in body temperature will increase food intake. This must mean that at full lactation the animal's food intake is maximal and that temperature will increase food intake. Regulation of temperature is independent of the regulation of body energy stores. Ablation of the satiety center does not alter the change in food intake induced by changing environmental temperature; this is a function of the temperature regulating center. Long-term exposure to changes in temperature will, of course, affect the fat depots which will in turn result in dietary adaptation.

Fat and Glucose Metabolism

There is a close connection between the metabolism of fat and that of glucose. It is reciprocal so that a glucostatic theory for short-term regulation and a lipostatic one for long-term regulation are quite compatible. Indeed, they are so interconnected that it is hardly feasible that they could act independently.

Elevation of blood glucose causes insulin release. This hormone increases glucose utilization by cells including those of the hypothalamus, fills up the carbohydrate stores and brings about fat synthesis and deposition with a consequent fall in circulating FFA. When blood glucose levels fall, as happens in starvation, the converse occurs: lipolysis begins, fat stores fall and FFA levels in blood rise. There are a few findings which make insulin a better indicator of both long- (size of the fat depots) and short-term (satiety) conditions than either FFA or glucose. One is that obese people and animals metabolise more fat, and less glucose than normals. This should cause a positive feedback which of course does not occur. In addition, progressively increasing levels of obesity are accompanied by increasing blood levels of insulin despite decreasing glucose utilization. This can explain why the force-fed fat rat will not eat until its body weight has returned to normal, only then are its blood insulin levels back to normal. Why then does any animal remain obese? In the case of the animal with the damaged satiety center, reduction in the receptor cell population requires that the signal be more intense, i.e. more insulin, which in this instance means more body fat. In the obese human perhaps a new equilibrium is reached between insulin sensitivity and insulin levels because concomitant with the increase in insulin levels there is a decrease in insulin sensitivity to the extent that the very obese often become diabetic. It has been suggested that the increase in circulating FFA in the obese is responsible for this decline in insulin sensitivity [15]. Such shifts in insulin sensitivity have been demonstrated in the rat. The rat eats at night, when, because it is depositing fat, its insulin responsiveness is high and its plasma FFA low. During the day, lipolysis rather than lipogenesis is taking place and the insulin responsiveness is low.

None of this really explains why fat people, in contrast to normal rats and most normal dogs and cats, stay fat. Many experiments have been done with lean volunteers who have forced themselves to eat until they were 20–25% above their normal weights. Their insulin levels rose and their responsiveness to it decreased, but these people all retuned to normal weight and insulin sensitivity at the end of the experiment.

The claim is often made, at least in popular literature, that carbohydrates are more likely to produce obesity than other foods, that is that carbohydrate calories are more fattening than those from fats or proteins. There is no physiological evidence to sustain this contention. The popular notion that the potato is fattening is especially fatuous. Many populations have subsisted almost enirely on potatoes in good health. Obesity was never a problem. Three thousand calories per day would require that between 5 and 6 kg of potatoes be eaten daily. Of course, the present day concoctions used for garnishing the potato (consisting as they do for the most part of butter fat) do turn it into a calorie-dense food.

Palatability

Palatability does influence food intake in man and in experimental animals, but for the most part in the old and the obese. Normal fasted rats will adapt to an unpalatable diet (flavored with quinine or mixed with kaolin) in 3–4 days. Obese rats will take 10 days and even so will not completely regain their previous weight plateaus. Given more palatable food, the obese animals will reach a plateau higher than that attained on the control diet. The same is seen in man. The obese seem to be much fussier about likes and dislikes than the lean. A lean person after a long fast will eat almost any source of calories. An obese person will only eat that for which he has an appetite; possibly because the obese person does not have nearly as intense a sensation of hunger (caloric deficiency) as the lean.

During Infancy and Growth

The regulation of energy intake poses special problems in growing and in suckling animals [11]. Most of the work on sucklings has been done in rats, but in general it may be taken as applying to man.

The very young infant is totally dependent on milk or some liquid substitute. Attempts to wean infant rats before 14 days will result in death. The young are usually poikilothermic and in addition have difficulty regulating fluids and electrolytes; this is the cause of death in early weaning. On milk they are spared the necessity of any of this, since in slaking thirst with milk they consume an isotonic fluid with the electrolytes in the correct proportions and, incidentally, adequate calories. Milk, therefore, serves as both food and drink. The infant rat does not distinguish between food and drink until weaning. It is doubtful that satiety has any meaning in the suckling animal. The only limit is the available supply. When this becomes insufficient the young animal becomes aware of the difference between food and drink and begins foraging for food. By this time the kidney has matured to the extent that water drinking is no longer lethal as it was earlier.

The food intake of the growing animal keeps pace with the growth rate [11], and the normal animal does not become obese because, according to the lipostatic theory which we have favored here, food intake is regulated to maintain the proportion of body fat. There is excellent evidence now that

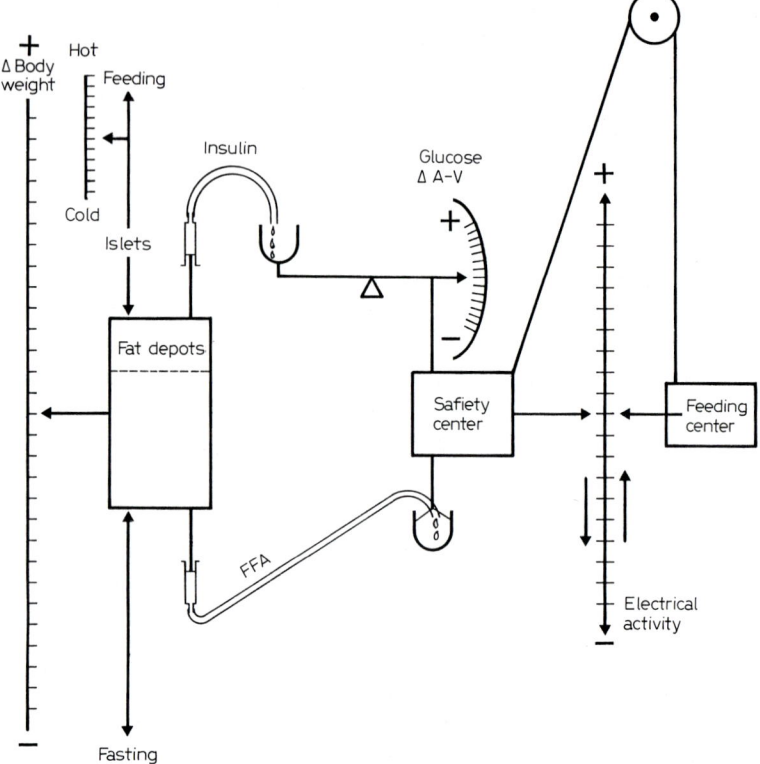

Fig. 13/6. Summary of factors in the regulation of food intake. During fasting, fat depots and body weight will fall. Plasma free fatty acids FFA will rise and insulin will fall. The A-V glucose difference will fall and so will the electrical activity of the satiety center. That of the feeding center on the contrary will rise. Returning to the upper left hand corner it will be seen that the animal will get cold. When the animal eats the converse will occur.

overfeeding the human infant can result in an increase in the fat cell population and an elevated lipostatic set point. There is a strong probability that this can be the start of a life of obesity. The new element in growth is the pituitary growth hormone. This has the effect of retaining nitrogen and causing loss of body fat; thus the growing animal is in positive nitrogen balance, but to maintain its shrinking fat depots it must eat more. When growth stops, fat stores increase and the upper limit for satiety falls. Damage to the satiety center does not interfere with growth. It will simply produce an animal which eats more than the normal growing animal and it will consequently become an obese adolescent. Destruction of the hunger center will of course prevent growth as even the adult animal loses its drive to eat and a depressed satiety center no longer has a hunger center to liberate (fig. 13/6).

Thirst

Animals become thirsty in response to water deprivation (dehydration), to an excess of solutes in the body *(hyperosmosis)* and when they are eating dry food.

Dehydration
Dehydration as measured by weight loss is corrected precisely and promptly when drinking water is made available in most animals; man is an exception. The usual means by which water is lost (sweating, urination, transpiration, respiration) result in an elevation in the osmotic pressure of body fluids. But a simple isosmotic reduction in fluid volume will also stimulate thirst. This occurs after hemorrhage and after peritoneal dialysis. It indicates the existence of both osmoreceptors and volume receptors. The threshold for thirst in man has been estimated as occurring when the water deficit reaches 0.8% of the body weight. Many species will repair such a deficit promptly but man does it only slowly, drinking and stopping again before the deficit has been completely made good only to start up again later.

Hyperosmosis
Administration of hypertonic NaCl by vein, stomach tube or intravenously will induce drinking as a result of stimulation of osmoreceptors in the supraoptic nucleus of the hypothalamus. In this case the volume of water drunk is never sufficient to restore the osmotic pressure. The ultimate adjustment is made by the kidney. In this manner hyperosmolality is corrected without as large an increase in body fluid volume as would occur if the kidney played no part.

Dry Food
Drinking with dry food is simply to facilitate swallowing. It increases if salivary secretion is suppressed and decreases if either the food or mouth are moistened.

Satiety

As we noted in the case of food, so also for drinking: satiation occurs before osmotic or volume deficits have been corrected and must presumably result from signals from the alimentary canal. Some of these arise from the mouth and pharynx. A smaller volume of water satiates if drunk than if placed directly in the stomach through a fistula. Esophagostomised dogs can achieve satiation even though none of the water drunk enters the stomach. This does not last long of course and the animal has to drink twice as much to achieve it as it would if the water entered the stomach.

In thirsty animals drinking is greatly and promptly reduced by placing water or an inflated balloon in the stomach. The effect of the balloon is of course only temporary because final satiation is only achieved when the deficits in the tissues are corrected. The gastric signal seems to be purely mechanical since the composition or osmolality of fluids placed in the stomach do not influence it. How the signals reach the hypothalamic drinking center is mysterious since denervation of the stomach does not eliminate them. Since water, as mentioned above, runs straight through the stomach to the duodenum it is likely that under physiological circumstances satiety signals from the small bowel are more important than from the stomach. Perhaps this is how the alimentary canal distinguishes between food and water. Thirst after a good meal is common and probably the consequence of copious secretion into the gut. How can this be when the stomach is distended with food? The distinction between water and food seems to be a sharp one. A well-hydrated hungry animal will not reject wet food even though it has no need for the contained water.

Thirst can be abolished by administering water intravenously.

References

1   Adolph, G.; Stevenson, J.; Towbin, E.; Epstein, N. M.: Alimentary canal, sect. 1: Food and water intake. Handbook of physiology, chap. 12–15 (Williams & Wilkins, Baltimore 1967).
2   Anand, B.; Subberwal, U.; Manchanda, S.; Singh, B.: Glucoreceptor mechanisms in the hypothalamic feeding center. Indian J. med. Res. *49:* 717–724 (1961).
3   Anand, B.; Chhina, G.; Sharma, K.; Dua, S.; Singh, B.: Activity of single neurones in the hypothalamic feeding centers. Am. J. Physiol. *207:* 1146–1154 (1964).
4   Anand, B.; Bannerjee, M.; Chhina, G.: Single neurone activity of hypothalamic feeding centres: effect of local heating. Brain Res. *1:* 269–278 (1966).
5   Brobeck, J.: Food and temperature; in Pincus, Recent progress in hormone research, pp. 439–466 (Academic Press, New York 1960).

6   Harper, A: Proteins and amino acids in the regulation of food intake; in Novin, Wyrwicka, Bray, Hunger: basic mechanisms and clinical implications (Raven Press, New York 1976).
7   Hoebel, B.; Teitelbaum, P.: Weight regulation in normal and hypothalamic hyperphagic rats. J. comp. Physiol. Psychol. *61*: 189–195 (1966).
8   Janowitz, H.; Hollander, F.: The time factor in the adjustment of food to varied caloric requirements in the dog. A study of the precision of appetite regulation. Ann. N.Y. Acad. Sci. *63*: 56–67 (1955).
9   Janowitz, H.; Grossman, M.: Some factors affecting food intake of normal dogs and dogs with esophagostomy and gastric fistula. Am. J. Physiol. *159*: 143–148 (1949).
10 Kennedy, G.: The role of depot fat in the hypothalamic control of food intake in the rat. Proc. R. Soc. Biol. *140*: 578–497 (1953).
11 Kennedy, G.: Ontogeny of mechanisms controlling food and water intake, chap. 25; in Handbook of physiology, sect. 6: alimentary canal, vol. 1; food and water intake (Williams & Wilkins, Baltimore 1967).
12 Le Magnen, J.: Interactions of glucostatic and lipostatic mechanisms in the regulatory control of feeding; in Novin, Wyrwicka, Bray, Hunger: basic mechanisms and clinical implications (Raven Press, New York 1976).
13 Mayer, J.: Regulation of energy intake and the body weight; the glucostatic theory and the lipostatic hypothesis. Ann. N.Y. Acad. Sci. *63:* 15–42 (1955).
14 Paintal, A.: A study of gastric stretch receptors: their role in the peripheral mechanisms of satiation of hunger and thirst. J. Physiol., Lond. *126*: 255–270 (1955).
15 Randle, M.; Hales, C.; Garland, P.; Newsholm, E.: The glucose fatty acid cycle, its role in insulin sensitivity and the metabolic disturbances of diabetes mellitus. Lancet *i*: 785–789 (1963).
16 Sharma, K.; Anand, B.; Dua, S.; Singh, B.: Role of stomach in regulating activities of the hypothalamic feeding centers. Am. J. Physiol. *201*: 593–598 (1961).
17 Smith, A.; Gibbs, J.: Cholecystokinin and satiety. Theoretic and therapeutic implications; in Novin, Wyrwicka, Bray, Hunger: basic mechanisms and clinical implications (Raven Press, New York 1976).
18 Strominger, J.; Brobeck, J.: A mechanism of regulation of food intake. Yale J. Biol. Med. *25*: 383–390 (1953).
19 Teitelbaum, P.: Sensory control of hypothalamic hyperphagia. J. comp. Physiol. Psychol. *48*: 156–163 (1955).
20 Teitelbaum, P.; Stevenson, J.; Towbin, E.; Epstein, N. M.: Alimentary canal, sect. 1: Food and water intake, chap. 12–15. Handbook of physiology (Williams & Wilkins, Baltimore 1967).

Subject Index

i = Illustration, t = table

Absorbability, factors contributing 231, 232
Absorption
  amino acids 240, 241, 269
  ammonia 272
  bile salts 248, 249
  calcium 250
  carbohydrates 237–239
    deficiencies 238, 239
  chloride 272
  colonic 269–271
    chloride 270
    lower fatty acids 272
    potassium 270
    sodium 269
    water 269, 270
  fat-soluble vitamins 248
  fats 245–247
  fructose 238
  galactose 238
  gallbladder, water and electrolytes 144
  glucose 238, 269
  in the infant 179, 180
  iron 250, 251
  lower fatty acids 272
  micellar 245–247
  peptides 239–242
  potassium 270
  small intestinal 231–252
    factors contributing 231, 232
    surface, digestion 236–251
    water and electrolytes 232–236
  sodium 269
  water 232–236, 269, 270
  water-soluble vitamins 249, 250
Absorptive adaptation 251, 252
Absorptive cell(s), intestine, see Cells, columnar absorptive, intestinal

Accommodation response, rectal ampulla 284, 285, 285i
Acetylcholine
  action on isolated gallbladder musculature 140
  distribution, intestinal wall 198i
  gastrin acting, release 201, 201i, 202
  intestinal motility 198, 201
  rate of manufacture, gut 198, 198i
  receptor site, parietal cells 104
  salivary flow 37–39
  stimulation
    colonic sympathetics 266
    gastrin-producing cells 117–119
Achalasia 71, 72
Achlorhydria 121
  pancreatic secretions 160
Acholuric jaundice 147
Acid
  ascorbic 250
  bile, pool, see Bile acid pool
  chenodeoxycholic, see Acids, bile
  deoxycholic, see Acids, bile
  gastric
    peptic ulceration 105
    site of secretion, stomach 80
    volume produced, and parietal cell population 105
    see also Secretion, gastric
  hydrochloric, see Acid, gastric
  lithocholic, see Acids, bile
  uric 271
  ursodeoxycholic, see Acids, bile
Acid load 159
Acids, bile 142, 149, 150i, 151
  synthesis 150i
Acinus
  compound gland 3

pancreas 153, 154i, 155, 156
salivary glands 27, 28i, 34t, 40–42
Actin
 microvilli, intestinal absorptive
  cells 225
 microvilli, parietal cells 103
 smooth muscle 13, 13i, 14
Adaptation, absorptive 251, 252
Adenomeres, see Units, secretory,
 salivary glands
Adenosine triphosphate (ATP), inhibitory
 neurohumor 191, 199
Adenosine triphosphatase (ATPase) 223
Adipsia 327
Adrenalin, relaxation reflex, internal anal
 sphincter 292
Adrenergic agents, depression,
 gastrin-stimulated secretion 119
Adrenergic blocking agents, pancreatic
 secretion 166
Adventitia, general structure,
 function 18, 19t, 20i, 23
Aerophagia 274
Aganglionosis, see Disease,
 Hirschsprung's
Aldosterone, sodium absorption/luminal
 potassium 270
Alkaline tide 108
Alkaloids, taste 33
Alveoli, see Acinus
Amine precursor uptake and
 decarboxylation (APUD) cells, see
 Cells, amine precursor uptake and
 decarboxylation
Amines, gut, see Gut peptides and amines
Amino acids
 absorption 240, 241, 269
 essential 241
 stimulation, gastric secretion 116
Aminopeptidase 224i, 240
Ammonia 271, 272
Amphiphiles, swelling insoluble, see
 Lipids, classification
Ampulla
 hepatopancreatic (Vater) 131, 136,
  137i
 rectal 256, 278i, 279i, 283–288

accommodation response 284, 285,
 285i
Amylase
 colonic 276
 pancreatic 171, 172, 172i
 site of formation, cell 7
Amylolytic activity 171, 172
 conditions 35
 end products 172i
 fermentation 36
 minor salivary glands 33
 parotid gland 29, 33
 sublingual glands 30
 submandibular gland 29, 33
 stomach 80, 80i
Anal canal, see Canal, anal
Anal sphincter(s), see Sphincter(s), anal
Analgesic agents, nonopiate, suppression
 of mucus secretion 113
Anastomoses
 arteriovenous, hemorrhoidal venous
  plexus 297
 ileo-anal 286
Anemia, pernicious 113, 249
 gastrin levels 119
Anorectal cavernous bodies 297, 297i,
 298
Anorectal junction, see Junction(s),
 anorectal
Anorectal region 277, 298
 general structure and function 277,
  278, 278i
 mucosa 19t
 muscularis externa 19t
 see also Sphincter(s), anal
Anorexia 323
Antibodies, antigens 241, 242
Anticholinergics
 pancreatic secretion 161
 suppression, salivary flow 38
Antigens, antibodies 241, 242
Antiinflammatory agents, suppression of
 mucus production 113
Antimitotic compounds, epithelial
 renewal 219
Antiperistalsis, see Peristalsis, reverse
Antral inhibition 120, 121

Subject Index

Antral pouch, *see* Pouch, antral
Antrum, pyloric 89i, 303
Aphagia 327
Apparatus, Golgi, *see* Golgi complex
Apparatus, protein-synthesizing, *see* Endoplasmic reticulum
Appendices epiploicae 257i, 258
Appendix, vermiform 256, 257i, 267
   structure of wall 19t
Appetite vs. hunger, *see* Hunger vs. appetite
Appetite, hunger, *see* Hunger, appetite
APUD cells, *see* Cell(s), amine precursor uptake and decarboxylation
Apudomas 313
Areae gastricae 88, 89i
Argentaffin cells, *see* Cells, argentaffin
Argyrophylic cells, *see* Cells, argyrophilic
Arteriovenous anastomoses, hemorrhoidal venous plexus 297
Ascorbic acid, *see* Acid, ascorbic
ATP, *see* Adenosine triphosphate
ATPase, *see* Adenosine triphosphatase
Atropine
   cholecystokinin (CCK) 162
   colonic secretion 276
   defecation reflex 294
   peristaltic reflex 191, 192
   suppression
      effects of metoclopromide 197
      gastrin levels (G cells) 116, 119, 201
      interdigestive activity 208
      intestinal secretions 228
      motilin-mediated activity 202
      mucus secretion 113
      pancreatic enzyme secretion 161, 163
      pepsin secretion 123
      salivary flow 38
      TRH-stimulated colonic motility 266
   vomiting 187
Auerbach's plexus, *see* Plexus, Auerbach's myenteric nerve

B_{12}, *see* Vitamin B_{12}
B_{12} intrinsic factor complex, *see* Intrinsic factor

Bacteria
   colonic 267, 271, 272
   fecal 271, 272
   ileal 269
      bile salts 142, 143, 145
   intestinal, Paneth cells 227
   oral 45
Basal cells, *see* Cells, basal
Basal electrical rhythm (BER)
   colon 263, 264, 264i
   small intestine 185
   stomach 82, 82i, 83, 83i, 84
Basal lamina, *see* Lamina, basal
Basement membrane, *see* Lamina, basal
Basket cells, *see* Cells, myoepithelial
BBG, *see* Gastrin, chemical structure and types
Belching 73, 74
Bicarbonate
   colonic absorption, ammonia 272
      chloride 270
   hydrogen ion (H^+) secretion 107
   pancreatic secretion 155, 165, 166
   release, cells of bile canaliculi 144
      parietal (oxyntic) cells 104, 105, 107
Bile 142–151
   bilirubin 145–147
   composition, humans 144t
   concentration 135, 136
   daily volume, secretion 136
   effect of diversion, fecal fat 272, 273
   stimulation of goblet cell (mucus) secretion 226, 227
   water and electrolytes 144, 145
   white 147
Bile acid pool 143, 143i, 144t
Bile duct, common, *see* Duct(s), bile, common
Bile pigments, *see* Pigments, bile
Bile salts, *see* Salts, bile
Biliary tract, *see* Tract, biliary
Bilirubin 131, 145, 146, 146i, 147
Biliverdin 146, 146i
Billroth gastric resection, *see* Resection, gastric
Blood enzymes, *see* Enzymes, circulating
Blood flow, salivary 44, 45

Subject Index

Blood supply, anorectal region 277
Bodies
  dense 13
  fat, ischiorectal 279, 280i, 282, 291
Body weight
  diet 328–333
  plateaus 328, 329
Bombesin 200, 201, 309t, 311, 315t
  primary action 315t
Border, brush, intestinal absorptive
  cells 214–216, 216i, 217, 218i,
  223–225
Border, striated, see Border, brush
Bradykinin 44
Brunner's glands, see Glands, Brunner's
Brush border, see Border, brush
Buccal glands, see Glands, buccal
Buds, taste 46, 47, 47i, 48, 48i, 49
Bypass, jejunoileal 251

Caecum, see Cecum
Calcium, absorption 250
Caliculi, gustatory, see Buds, taste
Calories 321
Canal, anal 277–298
  lamina propria/submucosa 19t, 296
  sphincters, see Sphincters, anal
  structure 19t, 277, 278i
Canaliculus, intracellular, parietal
  cells 99i, 100, 100i, 102i, 103, 104
Candidate hormones, see Hormone(s),
  candidate
Carbohydrates
  absorption 237–239
  colonic degradation and
    absorption 272
  malabsorption 275
Carbon dioxide, gut 275
Carbonic anhydrase
  centroacinar and ductule cells 165
  colonic mucosa 270
  hydrogen ion (H$^+$) secretion 107
Carboxypeptidase/procarboxypeptidase
  173
Cardiac glands, see Glands, cardiac
Cavernous bodies, anorectal 297, 297i,
  298

CCK, see Cholecystokinin (CCK)
Cecum 256, 257
Celiac disease, see Sprue, non-tropical
Cell(s)
  absorptive, intestinal, see Cell(s),
    columnar, absorptive, intestinal
  acid-producing, see Cell(s), parietal
    (oxyntic)
  acinar, pancreatic, see Cell(s),
    pancreatic acinar
  acinar, parotid, see Cell(s), parotid
    acinar
  amine precursor uptake and
    decarboxylation (APUD) 305, 308,
    309t
    neoplasms 313–317
    origin 306, 308, 309t
  APUD, see Cell(s), amine precursor
    uptake and decarboxylation (APUD)
  argentaffin 305
    intestinal epithelium 218i, 220
  argyrophilic 305
  basal, taste buds 47i, 49
  basal granular, intestine, see Cell(s),
    enterochromaffin
  basket, see Cell(s), myoepithelial
  centroacinar, pancreas 154, 154i, 155,
    156
  chief (zymogenic), gastric glands 89i,
    95, 95i, 96
    location, stomach 97, 98i
  chromaffin 305, 308t
  columnar absorptive, intestinal 216i,
    222–225, 257i
    peptidase 239, 239i
  columnar epithelial, gallbladder 134,
    135i, 136
  EC, see Cell(s), enterochromaffin
  enterochromaffin (EC) 96, 199, 304t,
    305, 307, 308, 309t
    motilin 202
    serotonin (5-HT) 199, 304i, 305,
      308, 309t, 315t
    substance P 200
  enteroendocrine
    colonic epithelium 257i
    gastric glands 89i, 96

Cell(s), enteroendocrine (cont.)
    intestinal epithelium 218i, 220, 267
    pyloric glands 89i, 97
  G, see Cell(s), gastrin-producing
  gastrin-producing (G) 96, 97, 116, 117, 118i, 119–125, 304i, 307, 309t, 314t, 316t
    acid inhibition 120, 120i, 121
    APUD cells 124
    chemical stimulants 117, 118
    location 304i
    mechanical stimulation 116–118
    vagal innervation 118i
  gastroenteropancreatic (GEP) 306–309, 309t, 310, 310i, 311
    basal secretion 307–310, 310i, 311
    classification 308, 309t, 311
    distribution 303, 304i
    neoplasias 313–317
    'open' and 'closed' types 307, 310i
    origin 306, 308, 309t
    secretion 311–317
    secretions, see Secretions, GEP cell
    ultrastructure 306, 307, 310i
  GEP, see Cell(s), gastroenteropancreatic goblet
    colon 257i, 267, 276
    small intestine 218i, 220, 225, 226i, 227
      function 226, 227
      structure 225, 226i
  gustatory sensory 47i, 48
  immature columnar, gastric glands 94
  intercalated duct 28i, 30
  Kulchitsky 306, 309t
  M 242, 267
  mucous neck 94, 95
    dedifferentiation, parietal cells 99
  mucous-secreting
    stomach 91, 92, 92i
    structure 9, 10, 10i, 20i, 28i
  myoepithelial 28i, 31, 32, 40
  neuroepithelial 48
  oxyntic, see Cell(s), parietal (oxyntic), gastric glands
  pancreatic acinar 154i, 155
  Paneth, crypts (Lieberkuhn), small intestine 218i, 220, 227, 228
  parietal (oxyntic), gastric glands 89i, 94, 96–108, 119
    basal border 104, 105
    formation of hydrochloric acid 106–108, 108i
    intracellular canaliculi 99i, 100, 100i, 102i, 103, 104
    location, stomach 97, 98i
    microvilli 100i, 101, 102i
    mitochondria 104
    population, volume of acid produced 105
    tubulovesicular system 100i, 101, 102i, 103, 104
    ultrastructure 99, 99i, 100, 100i, 101, 102, 102i, 103–105
  parotid acinar 28i, 29
  pepsin-producing, see Cell(s), chief (zymogenic)
  peripheral, taste buds 47i, 49
  protein-secreting, see Cell(s), serous
  S, see Cell(s), secretin-producing
  secretin-producing (S) 158–160, 304i, 309t, 314t
    distribution 304i
    stimulation 158–160, 160i
  seromucous 9, 28i
    parotid gland, see Cell(s), parotid acinar
  serous
    pancreatic, see Cell(s), pancreatic acinar
    structure 8, 8i, 9
  SIF 306
  smooth muscle 10–12, 12i, 13, 13i, 14, 15, 15i, 16, 17
    contraction 13, 14
    gallbladder 133
    organization, tissue 14, 15, 15i, 16, 17
    structure 11, 12, 12i, 13, 13i
  stellate, see Cell(s), myoepithelial
  striated duct 28i, 30, 31
  striated muscle, red and white 288–290

Subject Index

supporting ('sustentacular'), taste buds 47i, 49
surface mucous, stomach 89i, 91, 92, 92i
type A 309t
type B 309t
type D 304i, 309t, 313, 315t
  cancer 313
type D1 304i, 309t, 314t
type EC2 304i, 309t, 314t
type ECL 309t, 315t
type G, *see* Cell(s), gastrin-producing
type I 304i, 309t, 314t
type K 304i, 309t, 314t
type L 304i, 309t, 314t
type N 304i, 309t, 315t
type P 309t, 315t
type PP 304i, 309t, 315t
type S, *see* Cell(s), secretin-producing
undifferentiated columnar, crypts of small intestine 220, 222
vacuolated, colon 268
zymogenic, gastric glands, *see* Cell(s), chief (zymogenic), gastric glands
Cellular migration, intestinal epithelium 219, 220
Cellulases 26, 275
Cellulose 25, 171
Central nervous system, GI function 197
Centroacinar cells, *see* Cell(s), centroacinar
Chemotherapy, epithelial renewal 219, 220
Chenodeoxycholate, *see* Salts, bile
Chewing 25, 26
Chief cells, *see* Cell(s), chief (zymogenic)
Chloride
  colonic absorption 270
  pancreatic secretion 165
Cholate, *see* Salts, bile
Cholecystectomy 151
Cholecystokinin (CCK) 162–164, 202, 204i, 309t, 311, 314t, 316t
  amino acid sequence 316t
  biochemistry 163, 164
  chemical structure and mechanisms 141, 142, 316t

  distribution of GEP (type I) cells producing 304i
  gallbladder emptying 139–142
  gastrin 124, 202
  gastrocolic reflex 265
  gut motility 197, 202
  mechanism of release 163
  pancreatic polypeptide (PP) 202, 203
  primary actions 314t
  satiety 324, 327, 328
  sphincter of hepatopancreatic ampulla (Oddi) 138
  stimuli for release 162
Choledochal sphincter, *see* Sphincter(s), choledochal
Cholera 229
Cholera endotoxins, *see* Entotoxins, cholera
Cholesterol 149–151, 246
Cholesterol esterase 246, 248
Cholinomimetic drugs
  action on stomach 79
  mucus secretion 113
  pepsin secretion 123
Chyle 14
Chylomicrons 143i, 221, 246, 247
Chyme 106, 269
Chymotrypsin/chymotrypsinogen 173
Circulation, enterohepatic 143, 143i
  after cholecystectomy 151
Circumanal glands, *see* Glands, circumanal
Cisternae, rER 5, 6
  mucous-secreting cells 9
Coat, surface, intestinal brush border, *see* Glycocalyx
Colipase 242, 243
Colon 256, 277
  absorption, *see* Absorption, colonic
  absorptive capacity 269
  digestion 271–274
  electrical activity 263, 264
  epithelium 19t, 257i, 268
  general structure and function 256, 257i, 263
  histology of wall of 19t, 20i, 257i, 267, 268

Colon (cont.)
  innervation 263, 266, 267
    motor 266
    pain 266, 267
  lamina propria 19t, 20i, 267
  mass movements 264, 265
  motility 262–265
  mucosa 19t, 20i, 257i, 267, 268
  muscularis externa 19t, 20i, 258, 259
  muscularis mucosa 19t, 20i, 267
  secretion, see Secretion(s), colonic
  serosa 19t, 20i
  submucosa 19t, 20i, 257i, 267
  taeniae coli 257i, 258
  transmucosal potential difference (PD) 269
Colostrum 241
Common bile duct, see Duct(s), bile, common
Complex(es)
  B_{12} intrinsic factor, see Intrinsic factor
  Golgi, see Golgi complex
  junctional, see Junction(s), tight
  migrating myoelectrical, see Interdigestive activity
Condensing vacuoles, see Vacuoles, condensing
Constrictors, pharyngeal 57i, 58, 62
Continence, fecal 279–297
  mechanisms
    primary 279–293
    secondary 294–297
Contraction(s)
  gallbladder, see Gallbladder, contraction
  gluteal, anal canal closure 279, 280i
  smooth muscle, see Muscle, smooth, contraction
  striated muscle, tonic vs. phasic 288–290
  standing 184
Corpora cavernosa recti 297, 297i
Creatinine 271
Crigler-Najjar syndrome, see Syndrome Crigler-Najjar
Critical micellar concentration 244

Crypts (Lieberkuhn)
  large intestine 19t, 267, 268
  small intestine 18, 19t, 217, 218, 218i, 219–221
Cyclic AMP 229
Cystic duct, see Duct(s), cystic
Cystinuria 240
Cystosol peptidases 239, 239i, 240

Defecation, urgent 286
Deficiency
  caloric 320, 321
  immunoglobulins A 242
  lactase 237, 238
  sucrase-isomaltase 238, 239
  vitamin K 274
Dehydration
  acute 269, 271
  thirst 333
Demilunes 28i, 29, 30
Denervation
  gut 24
  hypersensitivity 228, 229
  salivary glands 38
  stomach, see Vagotomy
Dentate line, see Line, pectinate
Deoxyribonuclease (DNAase) 155, 173, 174
  site of formation, cell 7
DES, see Diffuse endocrine system
Dextrinase 237
Dextrins, limit 171
Diabetes 325, 326
Diarrhea 192, 196, 229, 270–272
  sequela of irradiation/chemotherapy 219, 220
Diet, body weight 328–333
Diffuse endocrine system (DES) 303–317
  APUD cells 305
  introduction and historical perspective 303–306
  neoplasias 313
  nervous and endocrine systems 312
Digestion
  colonic 271–274
  fats 242, 243, 243i, 244, 245
  infant 179–180

Disease
  celiac, see Sprue, non-tropical
  diverticular, colon 274
  Hartnup's 240
  Hirschsprung's 192, 266, 267, 286
  liver 275
  protein deficiency 177–179
Distension, luminal
  colonic 266
  gastric 322–324
    hunger and satiety 322, 322i, 323
    vagus (X) nerve 324
  luminal, small intestine 191
  rectal 283–286, 284i, 285i, 288
Diverticuli/diverticulum, pulsion
  colon 274
  pharynx 62
Diverticulitis, see Disease, diverticular, colon
DNase, see Deoxyribonuclease
Dubin Johnson syndrome, see Syndrome, Dubin Johnson
Duct(s)
  bile 131, 132i
    common 131, 132i
  compound glands 3, 5i
  cystic 131, 132i
  excretory, salivary glands 27, 32, 34t
  hepatic 131, 132i
  intercalated (intercalary)
    pancreas 154, 154i, 156
    salivary glands 27, 30, 34t
  lymphatic, intestinal villi, see Lacteals
  pancreatic
    accessory (Santorini) 136, 137i
    ligation 179
    main (Wirsung) 136, 137i, 156
  parotid (Stenson's) 34t
  striated, salivary glands 30–32, 34t, 40, 41
  sublingual (Bartholin's/Rivinian) 34t
  submandibular (Wharton's) 34t
Duct system
  compound glands 3, 5i
  intralobular, salivary glands 27
Ductus choledochus, see Duct(s), bile, common

Dumping syndrome, see Syndrome, dumping
Duocrinin 230
Duodeno-jejunal junction, see Junction, duodeno-jejunal
Duodenum 211
  structure of wall 18t, 20i

EC cells, see Cell(s), enterochromaffin
Elastase/proelastase 173
Electrolyte transport
  across epithelium of gallbladder 144, 145
  characteristics of cells engaged 31
  colonic 269, 270
  intestinal columnar absorptive cells 223, 234–236
  production of gastric secretions 105–108, 108i
  striated ducts of salivary glands 31
Elements, transitional, endoplasmic reticulum 6, 6i
Emaciation 320
Emetics, centrally-acting, effect on BER, intestines 187
Emotions, effect on GI function 196, 197
En passant innervation, see Innervation, en passant
Encephalins 200
End-piece, secretory, see Acinus
Endoamylases 171
Endocrine glands, see Glands, endocrine, and specific gland
Endrocrine secretion, see Secretion, endocrine
Endopeptidases 110
Endoplasmic reticulum
  mucous-secreting cells 9
  structure and function 5, 6, 6i, 7
Endorphins 200
Endotoxins, cholera 227, 229
Enkephalin(s) 304i, 309t, 311, 316t
  amino acid sequence 316t
Enterochromaffin system, see Diffuse endocrine system
Enterocrinin 229

Enterocytes, *see* Cells, columnar absorptive, intestinal
Enterogastrone 167, 169, 170
Enteroglucagon 304i, 309t, 311, 314t
  distribution of GEP (type L) cells producing 304i
  primary action 314t
Enterohepatic circulation, *see* Circulation, enterohepatic
Enterokinase 172
Enzyme(s)
  circulating ('blood enzymes') 175
  colonic 276
  microvillus membrane/glycocalyx 223, 224, 224i, 225
  mucolytic 277
  pancreatic, intraluminal digestion 170–177, 225
Epiglottis, role in swallowing 60i, 61
Epithelial renewal, intestinal 219, 220
Epithelium
  anal canal 19t, 277
  colonic, *see* Colon, epithelium
  esophageal, *see* Esophagus, epithelium
  gallbladder, *see* Gallbladder, epithelium/mucosa
  gastric, *see* Stomach, epithelium
  general structure, function, GI canal 17, 18, 18t, 19, 19t, 20, 20i, 21
  glandular 3–10
  intestinal, *see* Intestines, small, epithelium
  secretory, major salivary glands 29, 30, 34t
  secretory, *see also* Epithelium, glandular
Eructation 73, 74, 274
Escherichia coli 226
Esophageal phase, swallowing, *see* Swallowing, esophageal phase
Esophagogastric junction, *see* Junction(s), gastroesophageal
Esophagus
  epithelium 18t, 20i, 64i, 65i
  general structure and function 18t, 20i, 64i, 63–67

  glands 64, 65, 66i
  indentation, aorta and bronchi 67, 69i
  innervation 67
  junction with stomach, *see* Junction(s), gastroesophageal
  lamina propria 64, 64i
  mucosa and submucosa 20i, 63, 64, 64i, 65, 66, 71
  muscularis externa 18t, 20i, 66, 66i, 67
  spasm, diffuse 72
  sphincters, *see* Sphincter(s), esophageal
  ulceration 71
Exocrine gland(s), *see* Glands, exocrine
Exocytosis, compound 276
Exopeptidases 173

Factor
  intrinsic, *see* Intrinsic factor
  lipocaic, *see* Lipocaic factor
Fasciculi longitudinales, gallbladder, *see* Rugae, gallbladder
Fat(s)
  absorption, intestinal 245–247, 272, 273
  digestion 242, 243, 243i, 244, 245
  fecal 247, 248
  glucose, metabolism 329–331
  inhibition, gastric motility and gastrin-stimulated secretion 168, 169
  stimulation of Brunner's gland secretion 230
Fat bodies (pads), ischiorectal 279, 280i, 282, 291
Fat set point, *see* Lipostatic set point
Fatty acid hypothesis, satiety 326, 327, 332i
Fatty acids, free (FFA), plasma levels, satiety 324
Fecal bulk 273, 274
Fecal fat, *see* Fat(s), fecal
Feeding pattern
  gastrocolic reflex 265
  interdigestive activity 206, 209, 210
  intestinal motility 185
Feeding, *see* Hunger and/or Satiety
FFA, *see* Fatty acids, free
Fiber, dietary, diverticular disease 274

Fibers, muscle, *see* Cell(s), smooth muscle
Filaments
   actin, *see* Actin
   myosin, *see* Myosin
Flatus 274, 275
Flavor 46
Flexure, perineal 280, 281, 281i, 282, 289, 291, 294
Flora, *see* Bacteria
Flutter-valve mechanism, anal canal 279, 279i, 297
Flux, water and electrolytes 233, 233i
Foetor hepaticus 275
Folate/Folic acid, absorption 249
Folds, *see* Plicae
Follicles, solitary lymphatic, lamina propria 21, 89i, 93, 217, 257i, 267
Food, dry, thirst 333
Food intake
   factors regulating 332i
   growth 331, 332, 332i, 333
Forrest pouch, *see* Pouch, Forrest
Foveolae gastricae, *see* Pits, gastric
Fructose, absorption 238

G cells, *see* Cells, gastrin-producing
G-17, *see* Gastrin, chemical structure and types
G-34, *see* Gastrin, chemical structure and types
Galactose, absorption 238
Gallbladder
   concentration of bile 135, 135i, 136
   emptying 139–142
   epithelium/mucosa 134, 135, 135i
   general structure and function 131, 132, 132i, 133
   innervation 134, 140
   interdigestive activity 141, 142, 206, 209
   microstructure 134, 135, 135i, 136
   removal, *see* Cholecystectomy
   structure 132i, 134
   vagotomy 141
   volume 136
Gallstones 148–151
Gammaglobulins 21

Ganglionic blockade, relaxation reflex 294
Ganglionic blocking agents
   intestinal motility 201, 201i
   pepsin secretion 123
   prevention, gastrin release 118, 119
   suppression, CCK-mediated pancreatic stimulation 162
   suppression, interdigestive activity 208
   suppression, motilin-mediated activity 202
Gap junctions, *see* Nexuses
Gaps, endothelial, lacteals, intestinal villi 221, 247
Gastrectomy
   gastrocolic reflex 265
   subtotal 126
Gastric acid, *see* Acid, gastric
Gastric digestion and secretion 105, 114
Gastric emptying, *see* Stomach, emptying
Gastric glands, *see* Glands, gastric
Gastric inhibitory peptide (GIP) 169, 170, 176, 177, 201, 304i, 309t, 314t, 316t
   amino acid sequence 316t
   distribution, cells producing 304i
   primary actions 314t
Gastric lining epithelium, *see* Stomach, epithelium
Gastric motility, *see* Stomach, motility
Gastric pits, *see* Pits, gastric
Gastrin 96, 116–125, 201, 202, 304i, 309t, 311, 314t, 316t
   acid inhibition, release 120, 121, 121i
   amino acid sequence 316t
   antral distension and blood levels 116–118
   catabolism 124, 125
   chemical structure and types 123
   cholecystokinin (CCK) 124
   distribution, GEP (G) cells producing 304i
   gastrocolic reflex 265
   gut motility 197, 201, 201i, 202
   history 303
   pancreatic secretion 157
   pepsin secretion 123

Subject Index

Gastrin (cont.)
  plasma levels 123
    following bowel resection 124
    pernicious anemia 119
    renal failure 124
  primary actions 314t
  receptor site, parietal (oxyntic) cells 104
  satiety 328
  secretin 125
  trophic action 124
Gastrinoma 125
Gastritis, atrophic
  HCl production 97
  intrinsic factor 113
Gastrocolic reflex, see Reflex, gastrocolic
Gastroenterostomy 84, 126, 127 127i
Gastroileal reflex, see Reflex, gastroileal
Gastrojejunostomy 126, 127, 127i
Gastropancreatic reflex, see Reflex, gastropancreatic
GEP cells, see Cell(s), gastroenteropancreatic
Glands
  acinar 5i
  Brunner's 18t, 229–231
  buccal 32, 35t
  cardiac, esophageal 65
  cardiac, stomach 89i, 93
  circumanal 268
  classification 3–5, 5i
  compound 3, 5i
  esophageal 64i, 65, 66i
  exocrine 3, 4i
  gastric, structure 89i, 93–96
  glossopalatine 33
  heterocrine 29
  homocrine 29
  intermediate, stomach 97
  intestinal, see Crypts (Lieberkuhn), large/small intestine
  labial 32, 35t
  Lieberkuhn's, see Crypts (Lieberkuhn)
  lingual 32, 35t
  merocrine 3
  mixed 4, 29
  mucous 4
  oral cavity, see Glands, salivary, minor
  multicellular 3, 4i
  oxyntic, see Glands, gastric
  palatine 33, 35t
  parotid 29, 34, 35t
  pharyngeal 58
  pyloric 89i, 96, 97
  salivary
    classification 26
    development 26
    iodine 36
    major 27–32
      cells 27, 28, 28i, 29, 30, 34t
      secretory units 27, 28i
    minor 32, 33, 35t
    nerves 35t
    secretory units 27, 28i
    species variation 26
    structure 26–33
  seromucous 4, 29, 30
  serous 4, 29, 30
  stomach, see Glands, gastric
  sublingual 29, 30, 34, 35t
  submandibular 29, 34, 35t
  submaxillary, see Glands, submandibular
  tubular 5i
  tubuloacinar 5i
  unicellular 3, 5i
  Von Ebner 32, 35, 35t
Glandular epithelial tissues 3–10
Glucagon 229, 309t, 316t
  amino acid sequence 316t
  mucosal cyclic AMP/intestinal secretion 229
Glucoamylase, see Maltase
Glucose
  absorption 238, 269
  blood
    effect, gastric secretion 115, 115i
    insulin-mediated fat deposition/lipolysis 330
  fat, metabolism 329–331
  satiety 324–326, 326i

Glucose availability theory, *see*
  Glucostatic theory of satiety
Glucostatic theory of satiety 325, 325i,
  326, 326i
Glucuronyl transferase 148
Gluten enteropathy, *see* Sprue,
  non-tropical
Glycocalyx 216i, 223, 224, 224i, 225
  peptidases 224i, 239
  saccharidases 224i, 237
Glycoconjugate, neutral, parietal (oxyntic)
  cells 101
Glycogen, hydrolysis, *see* Amylolytic
  activity
Glycose aminoglycan 16
Glycosylation, production of salivary
  mucins 33
Goblet cell, *see* Cell(s), goblet
Golgi complex
  function, exocrine gland secretion 7
  parietal cells 104
  relationship, endoplasmic reticulum 6,
    7
  structure 6i, 7
Granules
  secretory 6i, 7, 8i
    major salivary glands 34t
    pancreatic acinar cells 154i, 155,
      170, 171
    parotid acinar cells 33
  storage, *see* Granules, secretory
  zymogen, *see* Granules, secretory
Growth, food intake 331, 332, 332i, 333
Growth hormone 332
Gustatory cell, *see* Cell(s), gustatory
  sensory
Gustatory pores, *see* Pores, gustatory
Gut peptides and amines 311–317
Gynemic acid 51

Hairs, gustatory 48, 48i
Hamburger chloride shift 270
Hartnup's disease, *see* Disease, Hartnup's
Haustra/haustrations 257i, 258, 264
Heidenhain pouch, *see* Pouch,
  Heidenhain
Heme iron 250

Hemochromatosis 251
Hemolytic jaundice, *see* Jaundice
Hemorrhoids 296
Hepatopancreatic junction, *see* Junction,
  hepatopancreatic
Hirschsprung's disease, *see* Disease,
  Hirschsprung's
Histamine
  primary action 315t
  receptor site, parietal (oxyntic)
    cells 104
  stimulation of gastric secretion 122,
    123
Hollander test, *see* Test, Hollander
Hormone(s)
  candidate 311, 312
  growth 332
  gut 303–305
  pancreatic, *see* Cholecystokinin,
    Secretin 158i
  thyrotropic releasing, *see* Thyrotropic
    releasing hormone
5-HT, *see* Serotonin
Hunger
  appetite 320–334
    socioeconomic factors 320–331
  center, *see* Hypothalamus, satiety and
    hunger centers
  satiety 321–328
    effects of distension/bulk 322, 322i,
      323
    oropharyngeal factor 324
    thermoregulation 327, 332i
  vs. appetite 320
Hydrochloric acid, *see* Acid gastric
Hydrogen gas, gut 275
Hydrogen ion (H^+) secretion 107, 108
Hydrolases 171
Hydroxytryptamine
  (5-Hydroxytryptamine), *see* Serotonin
Hypergastrinemia 124, 125
Hyperglycemia/hypoglycemia 325, 326
Hyperosmosis, thirst 333
Hyperphagia 323
  hypothalamic 328, 329, 329i
Hyperplasia, intestine 251
Hypertrophy, intestine 251

Subject Index

Hypoglycemia, effect on gastric
  secretion 115i
Hypothalamus, satiety and hunger
  centers 321, 323, 323i, 324
  electrical activity 325i
  insulin 325, 326

Icterus, see Jaundice
IgA, see Immunoglobulins A
Ileocolic junction, see Junction, ileocolic
Ileostomy 271
Ileum 211
  structure of wall 18t, 20i
  villi, see Villi/villus
  see also Intestines, small
Ileus 195, 196, 231
Immunoglobulins A 241, 242, 267, 268
Inclusions, intranuclear 9, 10i
Incontinence, clinical studies 292
Infant
  antigens and antibodies 241, 242
  digestion and absorption 179, 180
  pancreatic juice 170
  suckling, food intake 331
Inhibitory peptide
  gastric, see Gastric inhibitory peptide
  vasomotor, see Vasoactive intestinal
    polypeptide
Innervation
  en passant 13i
  gut, change, pectinate line 277, 278i
  intestinal muscularis externa 189, 190,
    190i, 191
  multi-unit 16
  pain
    colon 266, 267
    stomach/intestines 193–195
  purinergic 191
  sympathetic
    gastric secretion 120
    ileocolic junction 261
    internal anal sphincter 286, 292
    intestinal motility 193, 194, 194i,
      228
  unitary 16
  vagal, see Nerve, vagus (X)
Insulin 309t

hypothalamic satiety/hunger
  centers 325, 326, 332i
  levels, body weight 330
  pancreatic exocrine secretion 176, 177
Interdigestive activity 205–210
  alterations in pattern 206–208
  gallbladder 141, 142, 206, 209
  gastrocolic reflex 265
  Heidenhain pouch 207i, 208
  hypothetical mechanisms 208–210
  ileocolic junction 261, 262
  intrinsic nerve plexuses 208, 209
  motilin 202, 208, 209, 209i
  pancreas 206, 207i
  small intestine 185, 205, 207i, 208,
    209
    intestinal secretion 228
  ruminants 210
  secretory studies 210
  stomach 86, 205, 207i, 209
    nonparietal secretion 114, 206
  vagus (X) nerve 208, 209
Intestinal juice, see Succus entericus
Intestines
  glands, see Crypts (Lieberkuhn)
  large, see Colon
  small
    absorptive surface,
      augmentation 212i, 213–217
    crypts (Lieberkuhn) 18, 19, 19t, 217,
      218, 218i, 219–221
    differentiation of cells 220
    epithelium 219–228
      cellular migration and
        renewal 219, 220
    general structure and function 211,
      212, 212i, 213
    hyperplasia/hypertrophy 251
    interdigestive activity 185
    lamina propria 217, 220, 221
    length 213
    motility 183–203
      CNS effects 196–199
      directional polarity 186
      effect of pharmacological
        agents 192
      hormonal regulators 197–199

Subject Index 349

patterns, intact animals 183, 184,
  184i, 185–192
peptidergic transmitters 199–203
radiological studies 184, 184i, 185
mucosa 217–228
muscularis
  externa, innervation 189, 190,
    190i, 191
  mucosa 217
Peyer's patches 217, 242
plicae circulares 212i, 213, 214
secretion 228–231
  see also Secretions, intestinal
serosa 18, 19, 19t
submucosa 18, 19, 19t
transmucosal potential difference
  (PD) 236
valves of Kerckring, see intestines,
  small, plicae circulares
villi, see Villi/villus
  see also Duodenum, Ileum,
    Jejunum
Intrinsic factor 99, 113, 114, 224, 249
Ionization (pK), intestinal
  absorption 231, 232, 232i
Iron, absorption 250, 251
Islets of Langerhans 153
Isomaltase 237

Jaundice 146–148
Jejunoileal bypass surgery 251
Jejunum 211
  structure of wall 18t, 20i
  villi, see Villi/villus
  see also Intestine, small
Juice
  gastric 105–110
  principle constituents 105
  intestinal, see Succus entericus
  pancreatic, see Secretion, pancreatic
Junction(s)
  anorectal 268
  duodenojejunal 211
  esophagogastric, see Junctions(s),
    gastroesophageal
  gap, see Nexuses
  gastroduodenal 76i, 77

innervation 77, 78, 78i, 79
gastroesophageal 63, 65i
hepatopancreatic 136, 137, 137i, 138,
  139
ileocolic 259, 260, 260i, 261, 262
  function 261
pharyngoesophageal, see Sphincter(s),
  esophageal, upper
tight
  absorption at different levels of GI
    tract 233
  endothelium of lacteals of intestinal
    villi 221
  gallbladder columnar epithelial
    cells 135i, 145
  gastroenteropancreatic (GEP)
    cells 307–310, 310i, 311
  intestinal columnar absorptive
    cells 216i, 222, 223

Kallidin, see Bradykinin
Kallikrein 44
Kerckring, valves, see Plicae circulares
Killian's dehiscence 62

Labial glands, see Glands, labial
Lactase 237
  deficiency 237, 238
Lacteals, intestinal villus 14, 143i, 212i,
  218i, 220, 221
Lactoperoxidase system 46
Lamina propria
  anorectal region 19t
  colon 19t, 20i, 267
  esophagus 18t, 20i, 63i, 64, 64i, 65, 66
  general structure and function 21
  pharynx 58, 59
  small intestine 18t, 20i, 217, 220, 221
  stomach 18t, 20i, 93–97
Lamina, basal, epithelium 21
Langerhans, Islets, see Islets of
  Langerhans
Laryngopharynx 55, 56i, 62
Larynx, swallowing 61, 62
Latarjet, nerve, see Nerve of Latarjet
Lecithins 149, 244
Lectins 46

Lieberkuhn, crypts, *see* Crypts (Lieberkuhn)
Ligament, suspensory, duodenum (Treitz) 211
Limit dextrins, *see* Dextrins, limit
Line
 dentate, *see* Line, pectinate
 pectinate 268, 296
 Z, gastroesophageal junction 64
Lingual glands, *see* Glands, lingual
Lining, *see* Mucosa
Lip, glands, *see* Glands, labial
Lipase/lipase activity 242, 243, 243i
 colonic 276
 in the infant 179, 180
 pancreatic 155, 173, 174, 242
Lipid(s)
 absorption, *see* Absorption, fats
 bimolecular layer, cell membrane 224i
 classification 243, 244
 non-polar, *see* Lipids, classification
 polar, *see* Lipids, classification
Lipid-water partition coefficient 231
Lipocaic factor 179
Lipostatic
 hypothesis, body weight 328
 set point 328, 332
Liver
 fatty infiltration 179
 general function, as excretory organ 131
Loop, Thiry-Vella, *see* Thiry-Vella loops
Lymphatic drainage, anorectal region 277
Lymphatic follicles, *see* Follicles, lymphatic
Lymphatic vessels, intestinal villi, *see* Lacteals
Lymphatic/lymphoid tissue
 esophagus 63
 stomach 89i, 93
Lysozyme 227

M cells, *see* Cells, M
Macrophages, lamina propria 21
Magenstrasse 75, 88, 105

Malabsorption
 carbohydrate, hydrogen 275
 sequela of irradiation/chemotherapy 219–220
Maltase 237
Manellin 51
Maneuver, Valsalva, *see* Valsalva maneuver
Mastication, *see* Chewing
McBurney's point 259
Mechanism(s), transport, *see* Transport mechanism(s)
Megacolon, *see* Disease, Hirschsprung's
Membrane peptidases 239, 239i, 240
Membrane
 basement, *see* Lamina, basal
 mucous, *see* Mucosa
Meromyosin
Methane, gut 275
Metoclopromide 197, 266
Micellar concentration, critical 244
Micelles 149, 244, 245
 absorption 245–247
Microvilli
 gastric parietal cells 99i, 100, 100i, 101, 102, 102i, 103, 104
 intestinal absorptive cells 214–216, 216i, 217, 223–225
 structure of membrane/glycocalyx 216i, 223, 224, 224i, 225
 motility 225
Migrating myoelectrical complexes, *see* Interdigestive activity
Migration, intestinal epithelial cells 219, 220
Miraculin 51
Molecular weight, intestinal absorption 231
Morphine/morphine compounds, intestinal motility 192
Motilin 202, 304i, 309t, 311, 314t, 316t
 amino acid sequence 316t
 distribution of GEP (type EC2) cells producing 304i
 interdigestive activity 202, 208, 209, 209i
Motility,*see* Specific organ

Movement(s)
  mass, colonic 264, 265
  net, water and electrolytes 233, 233i
  pendular, intestine 184
Mucin(s)
  nonparietal secretion 106
  saliva 33
Mucocutaneous junction, see Junction, mucocutaneous
Mucolytic enzyme 277
Mucosa
  colon 257i
  esophagus 63, 64, 64i, 65
  general structure and function 17, 18, 18t, 19, 19t, 20, 20i, 21
  pharynx 58, 59i
  small intestine 18t, 20i, 212i, 215i, 217-228
  stomach 18t, 20i, 88, 89, 89i, 90, 90i, 91-105
Mucous membrane, see Mucosa
Mucus 110-113
  effect of blood group on secretion 112
  function
    small intestine 226, 227
    stomach, GI tract 110
  properties 110-112
  structure 110, 111, 111i
  surface tension, maintaining closure of anal canal 279, 280i
Muscle(s)/musculature
  aryepiglottic 61
  conjoined longitudinal, anorectal mechanism 294, 295, 295i, 296
  corrugator cutis ani 294-296
  cricopharyngeus 62
  digastric 59
  evertor ani 280i, 296
  geniohyoid 59
  GI system 10-17
  gluteus maximus 280i, 282, 291
  involuntary (nonstriated, unstriped), see Muscle(s), smooth
  levator ani 278i, 294-296
  longitudinal anal 294, 295, 295i, 296
  palatopharyngeus 58
  puborectalis 280, 281i
  smooth
    biliary tract 141
    cells (myocytes), see Cells, smooth muscle
    contraction 16, 192
    distribution 11
    innervation 13i, 16, 17
    intestinal villi 221
    organization 14, 15, 15i
  striated
    anal sphincteric complex 287, 288, 288i, 289
    esophagus 67, 68
    striated, pharynx 55-58
  stylopharyngeus 58
  suspensory (ligament) of duodenum (Treitz) 211
  thyropharyngeus 62
Muscularis
  externa
    arrangement of smooth muscle 14
    colon 19t, 258, 259
    esophagus 18t, 20i, 66, 66i, 67
    gallbladder 133, 134
    general structure and function 18t, 19t, 20i, 22
    pharynx 55-57, 57i, 58
    small intestine, innervation 109i, 189-191
    stomach 18t, 20i, 75-77
  mucosa
    biliary tract 133, 134
    esophagus 20i, 64i, 66i
    general structure and function 18t, 19t, 20i, 21
    small intestine 217
Myocytes, see Cell(s), smooth muscle
Myoepitheliocytes, see Cell(s), myoepithelial
Myofilaments, myofibrils 13
Myogenic contraction, smooth muscle 16
Myosin 13, 13i, 14, 225

Naloxone 200
Nasopharynx 55, 56i

Nerve(s)
  fibers, vasodilatory, salivary glands 44, 45
  chorda tympani, salivary secretion 37
  extrinsic, see Nerve(s), vagus (X), Nerve(s), pelvic splanchnic, Innervation, sympathetic
  glossopharyngeal (IX), swallowing 58
    taste 49, 51
  intrinsic, see Plexus, Auerbach's, Plexus, Meissner's
  Latarget 78, 78i
  lingual, taste 49
  pelvic splanchnic
    anorectal mechanism 278i
    defecation reflex 294
    innervation of left colon 263, 266
    stimulation of colonic secretion 276
  phrenic, swallowing 59
  pudendal 277, 278i, 289
  trigeminal (V), taste 49
  vagus (X)
    gastric
      distension 324
      motility 84
      secretion 115, 116
    GI tract 23, 24
    ileocolic junction 261
    innervation
      right colon 263, 266
      stomach 77, 78i, 79
    interdigestive activity 208, 209
    intestinal
      motility 193, 194i
      secretion 228, 229
    mucus secretion 113
    pancreatic secretion 161–163
    pyloric G cells 116, 118i, 120
    swallowing 58, 67, 69–72
    taste 49
    see also Vagotomy
Neuroepithelial cells, see Cells, neuroepithelial
Neurogenic contraction, smooth muscle cells 16
Neurotensin 200, 304i, 309i, 311, 315t, 316t
  amino acid sequence 316t
  distribution of GEP (type N) cells producing 304i
  primary action 315t
Nexuses 13i, 16, 133
Nitrogen
  balance 240, 241
  gut 275
Norepinephrine, intestinal motility 198, 199
Nucleases 173, 174
Nutrition, protein 239–242

Obesity 320, 323, 330, 331
  dehydration 330, 331
  insulin 330
  palatability 331
Obstructive jaundice, see Jaundice
Oesophagus, see Esophagus
Oropharyngeal factor, satiety 324
Oropharynx 55, 56i, 59
Oscillator, relaxation, see Pacemaker
Osmoreceptors 333

Pacemaker
  colonic 264
  intestinal 187–189
  potential (PP), see Basal electrical rhythm (BER)
Pads, threshold 279i, 296, 297, 297i, 298
Pain
  colonic causes 266
  sensitivity, see Innervation, pain
Palatability 331
Pancreas
  innervation 157
  protein deficiency disease 177–179
  secretion from, see Secretion, pancreatic
  structure 153–157
Pancreatectomy 179
Pancreatic polypeptide (PP) 202, 203, 304i, 309i, 315t, 316t
  amino acid sequence 316t
  distribution of type PP cells producing 304i
  primary action 315t

Subject Index

Pancreatitis, acute 139, 174
Pancreozymin
  see also Cholecystokinin 140, 304i
Papillae
  circumvallate, foliate 46
  esophageal 63, 64i
  fungiform 46, 47, 51
Paracasein 110
Paracellular pathway 223
Paracrine secretion, see Secretion, paracrine
Paralysis, intestinal, see Ileus
Paraneurons 311
Passavant's ridge 60i
Pattern, feeding, see Feeding pattern
Pavlov pouch, see Pouch, Pavlov
Pectinate line, see Line, pectinate
Pegs, esophageal, see Papillae, esophageal
Pendular movement, intestine 184
Pepsin 108–110
  digestion of protein, conditions 80, 80i
  secretion, see Secretion, pepsin
  site of secretion, chief cells 80
Pepsinogen
  activation 109
  chemical structure and types (I and II) 109, 110
  intracellular 95
Peptidase/peptidase activity 239, 239i, 240
Peptide absorption 239–242
Peptidergic transmitters, see Transmitters, peptidergic
Perineal flexure, see Flexure, perineal
Periodic activity, see Interdigestive activity
Peristalsis
  directional polarity 186
  reverse 187
Peristaltic
  reflex, see Reflex, peristaltic
  rushes 184, 185
  waves, propagated 185–187
Pernicious anemia, see Anemia, pernicious
Peyer's patches 217, 242

Pharyngeal
  constrictors, see Constrictors, pharyngeal
  phase of swallowing, see Swallowing, pharyngeal phase
Pharynx, structure 55, 56, 56i, 57, 57i, 58, 59, 59i
Phenylaline (L- and d-), cholecystokinin (CCK) 327, 328
Phlorizin 238
Phospholipase(s) 174, 248
Pigments, bile 145, 146i
Piles, see Hemorrhoids
Pilocarpine, stimulation of colonic secretion 247
Pits, gastric 89i, 90, 90i, 93, 94
pK, see Ionization
Plaques, attachment 13
Plexus(es)
  Auerbach's myenteric nerve 20i, 22, 23
  colon 266
  esophagus 67, 70, 71
  gallbladder 134
  internal hemorrhoidal venous 278i, 296, 297
  intrinsic, interdigestive activity 208, 209
  Meissner's submucosal nerve 20i, 22, 23
    gallbladder 134
  substance P 200
Plicae
  circulares/semicirculares/semilunares 22, 212i, 213, 214, 257i, 258–268
  transversales, rectum 259
Point, McBurney's 259
Poisoning, heavy metal, see Stomatitis
Pool
  bile acid 143, 143i, 144t
  cholesterol 143i
Pores, gustatory 47i, 48, 48i
Position, effect
  continence/defecation 281, 282
  tonic contraction of anal sphincteric complex 288
Potassium
  colonic absorption 270

Potassium (cont.)
  transient 43
Pouch
  antral 114i
  Forrest 114i
  Heidenhain 114i, 119, 120
  Pavlov 114i
  pyloric 114i
PP, see Pacemaker potential, Pancreatic polypeptide
Pressures, intraluminal
  colonic, effect of dietary fiber 274
  diverticular disease 274
Proenzymes 155
Propagated peristaltic waves 185–187
Prostaglandins
  mucus secretion 113
  possible role, intestinal motility 198
Protease(s) 172–175, 276
  in the infant 180
Protein
  deficiency disease, see Disease, protein deficiency
  nutrition 239–242
Pteroylpolyglutamate hydrolase 249
Pudendal block, external anal sphincter (inflation reflex) 289
Pyloric
  antrum, see Antrum, pyloric
  exclusion operation 127, 127i
  pouch, see Pouch, pyloric
Pyloroplasty 127, 127i
Pyridoxine 250

R protein 114, 249
Radiation therapy, epithelial renewal 219, 220
Raphe, pterygomandibular 62
Receptive relaxation
  rectal ampulla, see Accommodation response
  stomach 79, 80, 80i
Receptors, volume 333
Rectal ampulla, see Ampulla, rectal
Reflex
  anal 289
  defecation 283, 293, 294
  gag 58
  gastrocolic 185, 265, 293
    inhibitory 195
  gastroileal 261, 262, 293
  gastropancreatic 164
  inflation 289
  intestino-intestinal 195, 266
  orthocolic 293
  peristaltic 186, 189–191, 267
  relaxation, internal anal sphincter 283–286
  secretion, colonic 276
  swallowing 58, 59
Regulators, GI activity 197–199
Regurgitation, see Vomiting
Relaxation oscillator, see Pacemaker
Renin 110
rER, see Endoplasmic reticulum
Resection, gastric 126, 127, 127i
Response
  accommodation, rectal ampulla 284, 285, 285i
  sampling 285i, 291
Riboflavin 250
Ribonuclease (RNase) 155, 173
  site of formation, cell 7
Ribosomes, endoplasmic reticulum 5
RNase, see Ribonuclease
Rugae
  gallbladder 132i, 134
  stomach 88
Ruminant
  digestion 210
  pancreatic secretions 170
Rushes, peristaltic 185

S cells, see Cells, secretin-producing
Saccharidases 224i, 237–239
Sacculations, colonic, see Haustra haustrations
Sacral outflow, see Nerves, pelvic splanchnic
Saliva
  amylase 33
  diet 36
  digestion 33–36
  functions 33

Subject Index

lipase 35, 36
oral hygiene 45, 46
secretion 36–44
  daily volume 37
  hormonal regulation 40
  nervous regulation 37–40
  stimulation 36, 37
  substance P 200
swallowing 33, 58
taste 33
water and electrolytes 40–44
see also Glands, salivary
Salivary blood flow, see Blood flow, salivary
Salivons, see Units, secretory, salivary glands
Salts
  bile 131, 142, 143, 143i, 144, 145, 148, 149, 150i
    absorption 248, 249
    function 242, 272
    lipase activity 242–244
  chenodeoxycholic 142
  cholic 142
  Kreb's cycle acid, absorption 272
Sampling response 285i, 291
Sarcolemma 11
Sarcoplasm 11
Satiety
  center, see Hypothalamus, satiety and hunger
  glucostatic theory 325, 325i, 326, 326i
  hunger, see Hunger, satiety
  theories 324–328
  thirst 334
Secretin 158–163, 202, 309t
  amino acid sequence 316t
  bicarbonate secretion 144
  biochemistry 163, 164
  distribution of GEP (S) cells producing 304i
  gastrin 125, 202
  gut motility 197, 202
  history 303
  mechanism of release 160–162
  mucosal cyclic AMP/intestinal secretion 229

mucus secretion 113
satiety 328
stimuli for release 158–160
suppression of release 160
Secretion(s)
  apocrine 33
  colonic 275–277
  endocrine 310i, 311, 312
  exocrine 3
  gastric 88–127
    gastrin 122–125
    glucose level 115, 115i
    histamine 122, 123
    regulation 114–125
      cephalic phase 114–116
      gastric phase 116–120
      intestinal phase 121
    sympathetic influences 120
    see also Acid, gastric, Secretion(s), parietal
  GEP cell 304i, 309t, 311–314, 314t, 315, 315t, 316, 316t, 317
    bicarbonate in bile canaliculi 144
    biochemistry 313, 316t
    cells of origin/actions 313, 314t, 315t
    mode of action 311, 312
  Brunner's glands 229–231
    electrolyte composition 230, 230i
    inhibition/stimulation 229, 230
  holocrine 219
  hydrogen ion (H$^+$) stomach 107, 108
  hypotonic 40, 41
  intestinal
    electrolyte composition 230, 230i
    from Brunner's glands 229–231
    inhibition/stimulation 228, 229
    water and electrolytes 232–236
  isotonic 105
  merocrine 33
  mucous 9, 10, 112, 113
  nonparietal 105, 106, 106t
  pancreatic 157–170
    anticholinergic agents 161
    cephalic phase 164
    composition 164–166
    daily volume 170

Secretion(s), pancreatic (cont.)
    dietary adaption 175, 176, 176i, 177
    duodenal phase 157, 158
    effect of impairment, fecal fat 272, 273, 273i
    fat 169
    gastric phase 157
    inhibition 166–170
    interdigestive activity 161
    pH 158, 159, 159i, 160, 166, 167
    ruminants 170
    suckling infants 170
    sympathetic influences 166
    vagus (X) nerve stimulation 161, 162
  paracrine 310i, 312
  parietal 106–108
    see also Secretion, gastric
  pepsin, inhibition/stimulation 123
  protease, in the infant 180
  protein, synthesis, storage, release 5–9
  salivary, see Saliva, secretion
Secretory
  granules, see Granules, secretory
  sheet, see Sheet, secretory
  units, see Units, secretory
Segmentation/rhythmic segmentation 184, 184i
Sensitivity of gut, various stimuli 24
Septum (Septae)
  fibrous, pyloric sphincter 76i, 77
  interfascicular, smooth muscle 16
Serosa, general structure and function 18t, 19t, 20i, 23
Serotonin (5-HT) 96, 304i, 305, 309t
  distribution, intestinal wall 198i
    GEP (type EC2) cells producing 304i
  intestinal motility 198, 199
  primary action 315t
  suppression of TRH-stimulated colonic motility 266
Sheet, secretory 5i
Sialomucin 9
SIF cells, see Cells, SIF
Sling, pudendal, see Muscle, puborectalis
Sodium
  colonic absorption 269

  pump 235, 236
  carbohydrate absorption 238
  peptide absorption 240
Solubility, intestinal absorption 231
Somatostatin 200, 304i, 309t, 311, 315t, 316t
  amino acid sequence 316t
  distribution of GEP (type D) cells producing 304i
  primary action 315t
Spasm, esophageal 72
Sphincter(s)
  ampullae, see Sphincter(s), hepatopancreatic ampulla (Oddi)
  anal 278i, 279i, 282–293
    external 287–293
      electrical activity 287–290
      function 289, 290
      organization 287, 288i
    internal 282–287, 290–293
      function 286, 287
      innervation 278i, 283
    synchrony 290–293
  choledochal 137i, 138, 139, 142
  choledochoduodenal, see Sphincter(s), hepatopancreatic ampulla (Oddi)
  esophageal
    lower 67i, 70–72
    upper 61i, 62, 63, 68, 70
      pressure changes 61i
    vomiting 187
  general structure and function 22
  hepatopancreatic ampulla (Oddi) 131, 137i, 138, 139
    cholecystokinin (CCK) 138
    interdigestive activity 206
  ileocolic 259, 260i
  pancreatic 137i, 138
  pyloric
    function, emptying of stomach 84, 85
    innervation 77, 78, 78i, 79
    structure 75, 76, 76i, 77
Sphincterotomy
  choledochal 151
  internal anal 292
Sprue, non-tropical 217, 242

Subject Index

Squatting, continence/defecation 281, 282, 293
Standing contractions 184
Starch 171
   digestion, see Amylolytic activity
Stasis, gastric 79
Steatorrhea 178, 179, 248, 273
Stercobilin/stercobilinogen, see Pigments, bile
Stomach
   basal electrical rhythm (BER) 82, 82i, 83, 84
   denervation, see Vagotomy
   effect of cholinomimetic drugs 79
   emptying 84–86
      caloric density 86, 86i
      nature of gastric contents 85, 86
      particle size 84, 85
      pyloric sphincter 84, 85
   epithelium 91, 92
   innervation 77–79
   interdigestive activity 86
   lamina propria 93–97
   luminal volume 79
   mixing (mechanical digestive) activity 80–84
      distal portion 82
      proximal portion 81
   motility 75–87
      depression, fat 169
      fasting 86
   mucosa 88, 89, 89i, 90, 90i, 91–105
      amino acid stimulation 116
      high oxidative activity 104
      mechanical stimulation 116, 117, 117i
   muscularis externa 75, 76, 76i, 77
   pacemaker region 82, 83i
   receptive relaxation 79, 80, 80i
   regions 89i, 90, 91, 91i
   secretion 88–127
      see also Secretion, gastric
   hydrogen ion (H$^+$) 107, 108
   storage function 79, 80, 80i
   structure of wall 20i, 75–77, 88–105
   transmucosal potential difference (PD) 106, 107
Stomatitis 36
Submucosa
   esophagus 63, 64, 64i, 65, 66
   general structure and function 18t, 19t, 20i, 21, 22
Submucosal nerve plexus, see Plexus, Meissner's submucosal nerve
Substance P 200
   possible role, intestinal motility 198
Succus entericus 219, 229
Sucrase 237, 276
Sucrase-isomaltase deficiency 238, 239
Sugars, see Carbohydrates
Sulfomucin/sulphomucin 9, 28i
Surface absorption, digestion 236–251
Surface coat, intestinal brush border, see Glycocalyx
Suspensory muscle (ligament) (Treitz) 211
Swallowing
   disorders 72–74
   esophageal phase 67–74
      Auerbach's plexus 67, 70, 71
      lower esophagus during 70, 71
      pressures 68i
      primary peristalsis 68
      secondary peristalsis 70
      vagi 67–72
   oral phase 58, 60i
   pharyngeal phase 58–60, 60i, 61–63
      musculature 61
      nerves 58, 59
      pressures during 58–61, 61i, 62, 63
      respiration 55, 56i
   role of saliva 33
Sympathectomy
   colonic secretion 276
   effect on intestinal motility 193, 195, 228
Sympathetic innervation, see Innervation, sympathetic
Syndrome(s)
   Crigler-Najjar 148
   Dubin Johnson 148
   dumping 126
   psychosomatic 24
   Zollinger-Ellison 313

Subject Index

System, diffuse endocrine, *see* Diffuse endocrine system

Taeniae coli 257i, 258
Taste 46–52
 blood-borne 51
 buds, *see* Buds, taste
 nerves 49, 51
 physiology 50–52
 salt depletion 52
Tastes, four basic 50
Teeth, closing force between, chewing 25
Test, Hollander 115
Tetrodotoxin
 peristalsis 188, 192, 267
 suppression of gastrin-mediated effects 201
Thaumatin 51
Thermoregulation, hunger/satiety 327, 332i
Thiamine 249, 250
Thirst 331, 333, 334
Thiry-Vella loops 187, 195, 196, 196i, 228, 229
 colonic 275
Thymidine, tritiated 219
Thyrotropic releasing hormone (TRH)/factor (TRF) 197, 266, 308t
Tide, alkaline 108
Tongue
 function, swallowing 58, 60i
 glands, *see* Glands, lingual
TRH, *see* Thyrotropic releasing hormone
Tract, biliary 131, 132i, 133–151
 continuous digesters and intermittent eaters 131
 gross structure and general function 131–133
 interdigestive activity 141, 142
 microstructure of wall, vs. GI canal 132i, 133, 134
Transcobalamin I, *see* R protein
Transcobalamin II 249
Transection, spinal cord
 defecation reflex 294
 gastrocolic reflex 265
 relaxation reflex 286

Transfer vesicles, *see* Vesicles, transfer
Transferrin 251
Transit time
 colon 262
 factors affecting 183
 mouth to anus 183
 effect of stool weight 274
 small intestine 183
Transitional elements, *see* Elements, transitional
Transmitters, peptidergic 199–203
Transmucosal potential difference (PD)
 colon 269
 small intestine 236
 stomach 106, 107
Transport mechanism(s)
 intestinal electrolytes 232–236
 iron and calcium 250, 251
 microvillus membrane 223, 224, 224i, 225
 water-soluble vitamins 249, 250
Trypsin/trypsinogen 172, 172i, 173
 activation peptide 172
 inhibitors, pancreatic secretion 166, 167, 174, 175
 intrapancreatic (intraglandular) activation 174
Tubules, terminal, *see* Acinus
Tubulovesicular system, parietal (oxyntic) cells 100i, 101, 102, 102i, 103, 104
Tumors, multiple endocrine, *see* Apudomas
Tunica
 adventitia, *see* Adventitia
 mucosa, *see* Mucosa
 muscularis, *see* Muscularis externa
 serosa, *see* Serosa
 submucosa, *see* Submucosa

Ulcers, peptic 125–127
 anti-inflammatory/analgesic agents 113
 gastric acid 105
 operative therapy 126, 127, 127i
 relationship blood group/mucus production 112
Units, secretory, salivary glands 27
Urea 271

Uremia 271
Urgent defecation 286
Uric acid 271
Urobilinogen, see Pigments, bile
Uropepsinogen 110

Vacuoles, condensing 7
Vagal block, suppression of pancreatic enzyme secretion 7, 161, 163
Vagotomy 78i, 79, 81, 127
   cephalic phase of gastric secretory regulation 116
   gallbladder 141
   gastrin 123, 126
   gastrocolic reflex 265
   Hollander test of success 115
   intestinal motility 193
   parietal cell 78i, 119
   suppression, effects of metoclopromide 197
   vomiting 187
Vagus, see Nerve, vagus (X)
Valsalva maneuver 289, 291, 293
Valve(s)
   Houston, see Plicae transversales
   ileocolic junction 259, 260i
   Kerckring, see Plicae circulares
Vasculature
   crypts of intestine (Lieberkuhn) 220, 221
   GI tract 23, 277
   intestinal villi 220, 221
Vasoactive intestinal polypeptide (VIP)
   amino acid sequence 316t
   cyclic AMP/intestinal secretion 229
   distribution of GEP (D) cells producing 304i
   primary action 314t
   vagus stimulation 162
Vasodilation, see Nerve fibers, vasodilatory
Veins, hemorrhoidal 278i, 296, 297
Venous drainage, anorectal region 277

Vesicles, transfer 6, 6i
Villi/villus
   motility 221
   structure 215i, 216i, 220, 221
   vasculature 220, 221
Villikinin 221
Vitamin(s)
   B_1, see Thiamine
   B_2, see Riboflavin
   B_6, see Pyrodoxine
   B_{12}
      absorption 113, 249
      intrinsic factor, see Intrinsic factor
   C, see Acid, ascorbic
   colonic absorption 274
   D 250
   fat-soluble, small-intestinal absorption 248
   K, colonic absorption 274
   water-soluble, absorption 249, 250
Volume receptors 333
Vomiting 72, 73, 187

Water
   colonic absorption 269, 270
   intestinal absorption 232–236
Water mobilization, satiety 327
Waves
   primary, swallowing 68
   propagated peristaltic, intestine 185-187
   secondary, swallowing 70
   slow, colonic 263i, 264, 264i
Weaning 331
White bile, see Bile, white

Z-line, gastroesophageal junction 64
Zollinger-Ellison syndrome, see Syndrome, Zollinger-Ellison
Zone, intermediate, stomach 89i, 91i
Zymogen granules, see Granules, secretory